Industrial Tribology

Covering energy-saving technologies and how these are incorporated into component design, this book is relevant to many industries, including automotive engineering, and discusses the topical issue of sustainability in industry. This book details recent fundamental developments in the field of tribology in industrial systems.

Tribology has advanced significantly in recent years. Tribological performance depends on external parameters such as contact pressure at the interface, system temperature, relative speed between bodies and contact behaviour. Through ensuring that mechanisms work in an energy-efficient manner and minimizing wear, engineers should seek to implement the study of tribology to improve the life of machinery within industry. Essential to the study of component design and condition monitoring, the book touches upon topics such as gears, bearings and clutches. Additionally, it discusses tribology's relation to Industry 4.0 and incorporates the results from cutting-edge research.

Industrial Tribology: Sustainable Machinery and Industry 4.0 will be of interest to all engineers working in industry and involved in mechanical engineering, material engineering, mechanisms and component design and automotive engineering.

Manufacturing Design and Technology
Series Editor
J. Paulo Davim

Industrial Tribology: Sustainable Machinery and Industry 4.0
Edited by Jitendra Kumar Katiyar, Alessandro Ruggiero, T.V.V.L.N. Rao and J. Paulo Davim

Industrial Tribology
Sustainable Machinery and Industry 4.0

Edited by
Jitendra Kumar Katiyar
Alessandro Ruggiero
T.V.V.L.N. Rao
J. Paulo Davim

CRC Press is an imprint of the
Taylor & Francis Group, an **informa** business

First edition published 2023
by CRC Press
6000 Broken Sound Parkway NW, Suite 300, Boca Raton, FL 33487-2742

and by CRC Press
4 Park Square, Milton Park, Abingdon, Oxon, OX14 4RN

CRC Press is an imprint of Taylor & Francis Group, LLC

Library of Congress Cataloging-in-Publication Data
Names: Katiyar, Jitendra Kumar, editor. | Ruggiero, Alessandro G., editor. | Rao, T.V.V.L.N., editor. | Davim, J. Paulo, editor.
Title: Industrial tribology : sustainable machinery and industry 4.0 / edited by Jitendra Kumar Katiyar, Alessandro Ruggiero, T.V.V.L.N. Rao and J. Paulo Davim.
Other titles: Industrial tribology (CRC Press)
Description: First edition. | Boca Raton : CRC Press, [2023] |
Series: Manufacturing design and technology | Includes bibliographical references and index. |
Identifiers: LCCN 2022022582 (print) | LCCN 2022022583 (ebook) | ISBN 9781032152349 (hbk) | ISBN 9781032152387 (pbk) | ISBN 9781003243205 (ebk)
Subjects: LCSH: Tribology. | Industry 4.0.
Classification: LCC TJ1075 .I474 2023 (print) | LCC TJ1075 (ebook) |
DDC 621.8/9--dc23/eng/20220801
LC record available at https://lccn.loc.gov/2022022582
LC ebook record available at https://lccn.loc.gov/2022022583

ISBN: 978-1-032-15234-9 (hbk)
ISBN: 978-1-032-15238-7 (pbk)
ISBN: 978-1-003-24320-5 (ebk)

DOI: 10.1201/9781003243205

Typeset in Times
by SPi Technologies India Pvt Ltd (Straive)

Contents

Preface

The friction and wear is not very old for industries but the word tribology was first used in the Jost report in 1960. It derives from two Greek words: tribo (rubbing) and logy (science). Hence, tribology has analyzed the science and technology behind 'rubbing'. This involves the study of interaction of solid-solid, interaction of fluid-fluid and interaction of solid-fluid when bodies are in relative motion. It covers the study of friction and wear and the application of lubrication. It has been observed that any tribological performance depends upon a huge number of external parameters, including contact pressure at the interface, the temperature of the system, the relative speed between bodies and contact behavior. Over the past few decades in all industrial applications, friction and wear came to be very important aspects due to the increased number of moving parts to be found in any piece of machinery. It is very important to know in machinery the kind of mechanism involved in the improvement of energy and life of the machine. Therefore, the correct application of friction and wear can save and improve the life of machinery and industry. In industry, there are numerous examples where tribology plays a very important role, including gears, bearings, clutches, lubrication and so on. It is unfortunate, however, that the term tribology is little known in industrial sectors despite its huge involvement in daily life as well as in industrial systems. Hence, it is required to communicate properly to all industrialists due to whom they can gain the knowledge of tribology in industrial systems.

Therefore, the proposed book is very important to the entire industrial world. It covers the fundamentals and advancement occurring in the field of tribology in industrial systems. Furthermore, the proposed book explores the importance of component design and condition monitoring which can improve the performance and the life of individual components. The present book also explains the role of tribology in the so-called Industry 4.0; that is, the new era of industries. Moreover, the proposed book integrates the results obtained from the various researchers as the latest development in the reduction of wear and energy savings and presents a series of conclusions to promote better understanding of the sector across the industrial world.

Preface

[illegible]

Editors

Jitendra Kumar Katiyar, PhD, presently works as a Research Assistant Professor in the Department of Mechanical Engineering, SRM Institute of Science and Technology Kattankulathur Chennai, India. His research interests include modern manufacturing techniques, tribology of carbon materials, polymer composites, self-lubricating polymers, lubrication tribology and coatings for advanced technologies. He earned a bachelor's degree at UPTU Lucknow with honours in 2007. He earned a master's at the Indian Institute of Technology Kanpur in 2010 and a PhD from the same institution in 2017. He is a life member of the Tribology Society of India, the Malaysian Society of Tribology, the Institute of Engineers, India and the Indian Society for Technical Education (ISTE) etc. He has authored, co-authored or published more than 40+ articles in reputed journals, 35+ articles in international/national conferences, 15+ book chapters and 12 books with Springer and CRC Press, with at least 4 more books to come with CRC Press and Springer Nature. He serves as a guest editor for a special issue in *Tribology Materials, Surfaces and Interfaces, Journal of Engineering Tribology Part J, Journal of Process Mechanical Engineering Part E, Arabian Journal for Science and Engineering, Industrial Lubrication and Tribology, Material Science and Engineering Technology* and *Frontier in Mechanical Engineering: Tribology*. Further, he is a member of the Editorial Board of *Tribology Materials, Surfaces and Interfaces* and the Review Editor of *Frontier in Mechanical Engineering*. He is also an active reviewer in 20+ reputed journals related to materials, manufacturing and tribology. He has also delivered more than 35+ invited talks on various research fields related to tribology, composite materials, surface engineering and machining. He has organized 7+ FDP/Short Term Courses and the International Tribology Research Symposium in tribology.

Alessandro Ruggiero, PhD, from 1999 to 2005 was Assistant Professor in the Department of Mechanical Engineering, University of Salerno, where he has been an Associate Professor with the Department of Industrial Engineering from 2005 until the present. In 2017 he received the National Qualification to Full Professor in Applied Mechanics. He has authored more than 220 scientific papers on prestigious indexed international journals and national and international proceedings. He serves as the editor, a member of the Editorial Board and a reviewer for many WoS/SCOPUS indexed international scientific journals and as an evaluator of national and international research projects. He cooperates with numerous international universities and prestigious research centers. He is a member of the IFToMM Technical Committee for Tribology and the National Coordinator of Italian AIMETA Tribology Group. His research interests focus on (bio)tribology, biomechanics, and in-silico wear calculation of artificial human synovial joints, and they also include the dynamics of mechanical systems, noise and vibration measurements, mechanical measurement and diagnostic on mechanical systems.

T.V.V.L.N. Rao, PhD, earned a PhD in the tribology of fluid film bearings at the Indian Institute of Technology Delhi in 2000 and an MTech in mechanical manufacturing technology at the National Institute of Technology (formerly known as Regional Engineering College) Calicut in 1994. Dr. Rao's research interests include bearings, lubrication and tribology. He has authored and co-authored over 120 publications to date. He has secured (as PI and Co-PI) several research grants from the Ministry of Higher Education, Malaysia and a research grant from The Sumitomo Foundation, Japan. Dr. Rao is a member of the Editorial Board (2022) of *Tribology & Lubrication Technology* (TLT). He is a guest associate editor of *Journal of Engineering Tribology, Industrial Lubrication and Tribology, Tribology – Materials, Surfaces & Interfaces* and *Arabian Journal for Science and Engineering*. Dr. Rao is an executive member (2015–2021) of the Malaysian Tribology Society. He is a member of the Society of Tribologists and Lubrication Engineers, the Malaysian Tribology Society and the Tribology Society of India. Dr. Rao is currently Professor and Dean of the Department of Mechanical Engineering, School of Engineering and Technology at the Assam Kaziranga University. Prior to joining this university, he served as Professor and HOD at MITS (2021–2022), Research Associate Professor at SRMIST (2017–2020), Visiting Faculty at LNMIIT (2016–2017), Associate Professor at Universiti Teknologi Petronas (2010–2016), and Assistant Professor in BITS Pilani at the Pilani (2000–2004, 2007–2010) and Dubai (2004–2007) campuses.

J. Paulo Davim, PhD, earned a PhD in mechanical engineering in 1997, an MSc in mechanical engineering (materials and manufacturing processes) in 1991, a mechanical engineering degree (5 years) in 1986 at the University of Porto (FEUP), the aggregate title (Full Habilitation) at the University of Coimbra in 2005 and a DSc from London Metropolitan University in 2013. He is a Senior Chartered Engineer by the Portuguese Institution of Engineers, with an MBA and specialist title in engineering and industrial management. He is also Eur Ing by FEANI-Brussels and Fellow (FIET) by IET-London. Currently, he is a Professor in the Department of Mechanical Engineering at the University of Aveiro, Portugal. He has more than 30 years of teaching and research experience in manufacturing, materials, mechanical and industrial engineering, with a special emphasis in machining and tribology. He also has an interest in management, engineering education and higher education for sustainability. He has guided large numbers of postdoc, PhD and master's students as well as coordinated and participated in several financed research projects. He has received several scientific awards. He has worked as an evaluator of projects for ERC-European Research Council and other international research agencies as well as an examiner of PhD theses for many universities in different countries. He is the Editor in Chief of several international journals, Guest Editor of journals, and Books Editor, Book Series Editor and Scientific Advisor for many international journals and conferences. Presently, he is an Editorial Board member of 30 international journals and acts as a reviewer for more than 100 prestigious Web of Science journals. In addition, he has also published, as editor and co-editor, more than 120 books and as author and co-author of more than 10 books, 80 book chapters and 400 articles in journals and conferences (more than 250 articles in journals indexed in Web of Science core collection/h-index 51+/8500+ citations, SCOPUS/h-index 56+/10500+ citations, Google Scholar/h-index 72+/17000+).

Contributors

Mohammed Shabbir Ahmed
Department of Mechanical Engineering
Biju Patnaik University of Technology
Rourkela, Odisha, India

Ali Algahtani
Department of Mechanical Engineering
King Khalid University
Abha, Kingdom of Saudi Arabia

S. Arulvel
School of Mechanical Engineering
Vellore Institute of Technology University
Vellore, Tamil Nadu, India

Pradyumn Kumar Arya
Department of Mechanical Engineering
Indian Institute of Technology
Indore, Madhya Pradesh, India

Sudheer Reddy Beyanagari
School of Mechanical Engineering
Vellore Institute of Technology University
Vellore, Tamil Nadu, India

Skylab P. Bhore
Department of Mechanical Engineering
Motilal Nehru National Institute of Technology Allahabad
Prayagraj, India

Corina Birleanu
Department of Mechanical Systems Engineering
Micro-Nano Systems Laboratory
Technical University from Cluj-Napoca
Cluj-Napoca, Romania

Mircea Cioaza
Department of Mechanical Systems Engineering
Micro-Nano Systems Laboratory
Technical University from Cluj-Napoca
Cluj-Napoca, Romania

Cosmin Cosma
Department of Manufacturing Engineering
Technical University from Cluj-Napoca
Cluj-Napoca, Romania

Marco De Stefano
Department of Industrial Engineering
University of Salerno
Salerno, Italy

Dhiraj Kumar Reddy Gongati
School of Mechanical Engineering
Vellore Institute of Technology University
Vellore, Tamil Nadu, India

Vipin Goyal
Department of Mechanical Engineering
Indian Institute of Technology
Indore, Madhya Pradesh, India

M. Hanief
Department of Mechanical Engineering
National Institute of Technology
Srinagar, Jammu and Kashmir, India

Nilesh D. Hingawe
Department of Mechanical Engineering
Motilal Nehru National Institute of Technology Allahabad
Prayagraj, India

Jayakrishna Kandasamy
School of Mechanical Engineering
Vellore Institute of Technology University
Vellore, Tamil Nadu, India

Saravanan Karuppanan
Department of Mechanical Engineering
Universiti Teknologi Petronas
Perak, Malaysia

Jitendra Kumar Katiyar
Department of Mechanical Engineering
SRM Institute of Science and Technology, Kattankulathur
Chennai, Tamil Nadu, India

Mithun V. Kulkarni
Department of Mechanical Engineering
Vijaya Vittal Institute of Technology
Bangalore, India

M. Phani Kumar
Surface Engineering and Tribology Group
CSIR-Central Mechanical Engineering Research Institute
Durgapur, West Bengal, India

Pankaj Kumar
Department of Mechanical Engineering
Indian Institute of Technology
Indore, Madhya Pradesh, India

S. Anand Kumar
Department of Mechanical Engineering
Indian Institute of Technology
Jammu, India

P. Kumaravelu
School of Mechanical Engineering
Vellore Institute of Technology University
Vellore, Tamil Nadu, India

Yashwanth Maddini
School of Mechanical Engineering
Vellore Institute of Technology University
Vellore, Tamil Nadu, India

Naresh Chandra Murmu
Surface Engineering and Tribology Group
CSIR-Central Mechanical Engineering Research Institute
Durgapur, West Bengal, India

Zahid Mushtaq
Department of Mechanical Engineering
National Institute of Technology
Srinagar, Jammu and Kashmir, India

Ponnekanti Nagendramma
Bio Fuels Division
CSIR-Indian Institute of Petroleum
Dehradun, Uttarakhand, India

K. Prabhakaran Nair
Department of Mechanical Engineering
National Institute of Technology
Calicut, India

Kaleem Ahmad Najar
Department of Mechanical Engineering
National Institute of Technology
Srinagar, Jammu and Kashmir, India

Santosh S. Patil
Department of Mechanical Engineering
Manipal University Jaipur
Jaipur, India

Florin Popa
Department of Materials Science and Engineering
Technical University from Cluj-Napoca
Cluj-Napoca, Romania

Marius Pustan
Department of Mechanical Systems Engineering
Micro-Nano Systems Laboratory
Technical University from Cluj-Napoca
Cluj-Napoca, Romania

R. Rajesh
Department of Mechanical Engineering
Global Academy of Technology
Bangalore, India

T.V.V.L.N. Rao
Department of Mechanical Engineering
School of Engineering and Technology
The Assam Kaziranga University
Jorhat, Assam, India

Pravesh Ravi
Department of Mechanical Engineering
SRM Institute of Science and Technology, Kattankulathur
Chennai, Tamil Nadu, India

Alessandro Ruggiero
Department of Industrial Engineering
University of Salerno
Salerno, Italy

Pranab Samanta
Surface Engineering and Tribology Group
CSIR-Central Mechanical Engineering Research Institute
Durgapur, West Bengal, India

P. Sampathkumaran
Department of Mechanical Engineering
Sambhram Institute of Technology
Bangalore, India

Prabakaran Saravanan
Department of Mechanical Engineering
BITS-Pilani
Hyderabad Campus
Hyderabad, India

Dan Sathiaraj
Department of Mechanical Engineering
Indian Institute of Technology
Indore, India

S. Seetharamu
Department of Mechanical Engineering
RV College of Engineering
Bangalore, India

Alessandro Sicilia
Department of Industrial Engineering
University of Salerno
Salerno, Italy

Girish Verma
Department of Mechanical Engineering
Indian Institute of Technology
Indore, India

1 Real Contact Area and Friction
An Overview of Different Approaches

Marco De Stefano and Alessandro Ruggiero
University of Salerno, Italy

CONTENTS

1.1 INTRODUCTION

Today, contact mechanics is one of the most interesting topics in the engineering field given its various applications [1]. Phenomena such as friction, wear, and lubrication are at the core of this study. The first formulation of the friction force [2] was provided by Amontons and Coulomb:

$$F_t = \mu F_n \tag{1.1}$$

where μ represents friction coefficient, F_n normal load and F_t friction force. All the terms are referred to in Figure 1.1, which represents a block of mass m subject to a normal load.

More precisely, Equation (1.1) is linked to these three results [3]:

1. The friction force is proportional to the applied normal load.

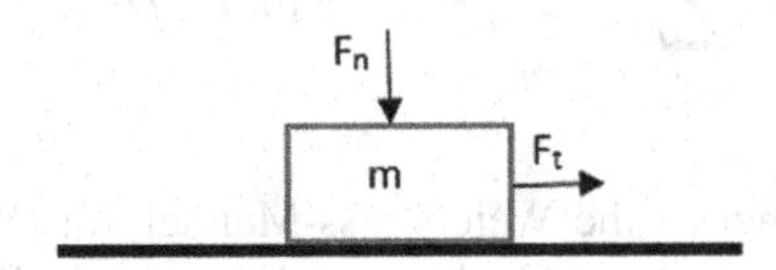

FIGURE 1.1 Example of friction force.

DOI: 10.1201/9781003243205-1

2. The friction force is independent of the apparent contact area.
3. The friction force is independent of the sliding velocity.

Point 1 is almost obvious, whereas point 2 is crucial because it underlines the difference between nominal and real contact area. Indeed, for the effect of surface roughness, the real responsible of the contact between couplings are the asperities [4] who have a great influence on the friction coefficient. The final result does not find confirm in experiments since that μ is a function not only of velocity but also of other variables, such as temperature, hardness, contact pressure.

In reference to point 2, friction force can be expressed as the sum of two contributions that consider the effect of both adhesion and hysteresis [5]:

$$F_t = F_a + F_h \tag{1.2}$$

Where:

$$F_a = \tau A_r \tag{1.3}$$

Equation (1.3) highlights the dependence of the RCA (real contact area) by the shear strength τ and it originates from intermolecular force between bodies, whereas the hysteresis term is connected to the energy dissipation due to the deformation of the bodies. Solving these formulations is not a trivial task, firstly, because the determination of RCA is a very complex issue. Statistical approaches such as Greenwood-Williamson [6] tried to investigate the multi-asperity contact assuming a Gaussian distribution of the peaks, FEM techniques [7] that simulate the coupling between real surfaces or experimental techniques by in-situ machines realized ad hoc [8]. Secondly, many variables should be considered, but it depends on the case of study: adhesion [9], wear [10] are all phenomena associated with friction, involving particular relations. Finally, the modelling of the contact, fractals [11], experimental [12] and numerical [13] are all satisfactory choices. In particular, fractals tried to overcome the main limitation of statistical approaches: the scale-measurement dependence. Each real object, unfortunately, cannot be considered completely smooth since it presents some degree of surface irregularities that vary according to the scale of measurement. In order to avoid this problem, fractals are objects whose shapes do not change with the scale. Indeed, the fractal function, described by Equation (1.4), possesses the important properties of continuity, non-differentiability and self-affinity and so it is suitable for profile modelling and simulations [14]:

$$z(x) = A^{D-1} \sum_{n=n_1}^{\infty} \frac{\cos\left(2\pi\gamma^n x\right)}{\gamma^{(2-D)n}} \quad \text{with } 1 < D < 2 \quad \gamma > 1 \tag{1.4}$$

This relation is known as the Weierstrass-Mandelbrot (W-M) function and the profile $z(x)$ is related to a constant A, the fractal dimension D and frequencies γ. On the other hand, there are some important disadvantages, such as the complexity of the approach as well as the construction of the profile and the calculation of the parameter D. Regarding that, the power spectrum method or innovative techniques [15] are good choices, whereas for the construction of the profile algorithms such as Random Midpoint Displacement [16] could help.

In any case, in this work, the idea is to use statistical approaches presented in literature and compare them with a FEM model.

For all statistical models, the distribution of the asperities is considered Gaussian:

$$\phi(z)=\frac{1}{\sqrt{2\pi\sigma_s}}\exp\left[-0.5\left(\frac{z}{\sigma_s}\right)^2\right] \tag{1.5}$$

With $\phi(z)$ being the normal probability density function of standard deviation σ_s and asperity heights z. This statement, obviously, is not the only one hypothesis but:

1. Surfaces are homogenous and isotropic
2. Peaks are modelled as spheres with the same radius of curvature
3. No interactions between asperities
4. Range of deformation elastic and plastic

Moreover, according to Johnson [17], two elastic rough surfaces can be simplified as a rigid smooth flat and a deformable rough surface characterized by a series of spherical asperities. This coupling is subjected to a normal and tangential load (P and Q), as shown in Figure 1.2.

The forces applied are usually solved by dimensionless method, i.e., by dividing the variables with σ (standard deviation of surface heights). Equation (1.5) becomes:

$$\phi^*(z^*)=\frac{1}{\sqrt{2\pi}}\frac{\sigma}{\sigma_s}\exp\left[-0.5\left(\frac{\sigma}{\sigma_s}\right)^2*(z^*)^2\right] \tag{1.6}$$

Where P, A and Q are:

$$P(d^*)=\eta A_n\int_{d^*}^{\infty}\overline{P}(z^*-d^*)\phi^*(z^*)dz^* \tag{1.7}$$

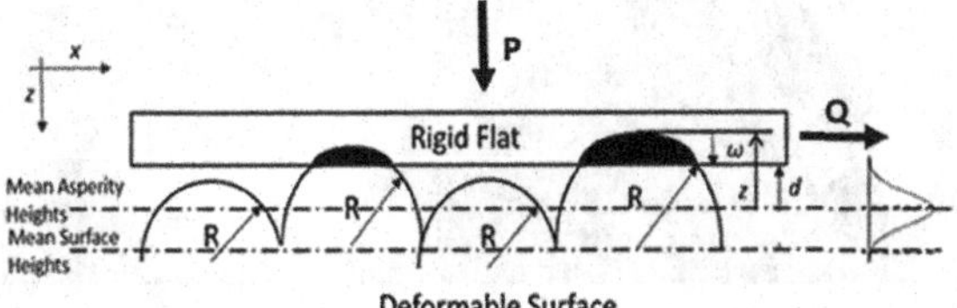

FIGURE 1.2 Schematization of the contact.

$$A\left(d^*\right) = \eta A_n \int_{d^*}^{\infty} \bar{A}\left(z^* - d^*\right)\phi^*\left(z^*\right)dz^* \quad (1.8)$$

$$Q\left(d^*\right) = \eta A_n \int_{d^*}^{\infty} \bar{Q}\left(z^* - d^*\right)\phi^*\left(z^*\right)dz^* \quad (1.9)$$

With η density of the peaks, A_n nominal area, d separation between surfaces. $Q(d^*)$ represents the tangential force necessary to destroy all the junctions between the contacting surface asperities.

They represent the forces exchanged between asperities, but only when they are in contact. This happens when:

$$\omega^* > 0 \quad (1.10)$$

Where ω^* is defined as (always in terms of dimensionless parameters):

$$\omega^* = z^* - d^* \quad (1.11)$$

And it is essentially the deformation of asperities, related to the radius of the sphere (R) and the radius of the contact (a) as shown in Figure 1.3.

With the relation:

$$a = \sqrt[2]{R\omega} \quad (1.12)$$

The three different models that will be discussed (in order of time) immediately, differ from each other for constitutive relations, but with the same starting point.

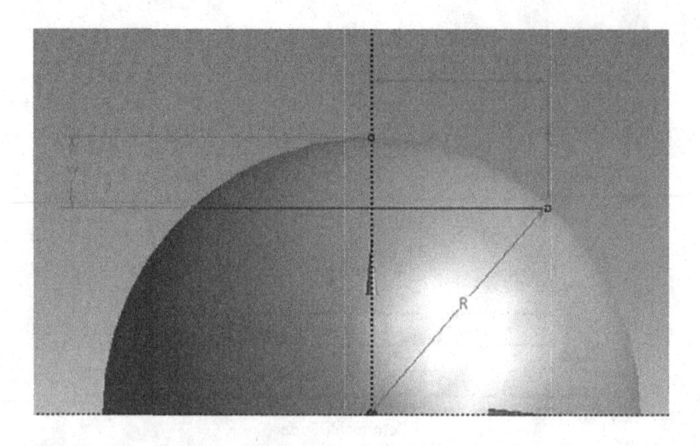

FIGURE 1.3 Asperity modelled as sphere in Ansys environment.

1.2 MATERIALS AND METHOD

In 2004, two authors, Kogut and Etsion, proposed an innovative model (KE) [18], trying to overcome the limit considered in another method (realized by Chang et al. [19]) about the underestimation of friction coefficient value related to the possibility of asperities to deform elasto-plastically, such as resisting to an additional load before failure. By study of a single asperity deformation [20] under a rigid flat, they provided different relations according to precise deformations regime:

$$P^* = \frac{P}{A_n H} = \frac{2}{3}\pi\beta K\omega_c^* \left(\int_{d^*}^{d^*+\omega_c^*} I^{1.5} + 1.03 \int_{d^*+\omega_c^*}^{d^*+6\omega_c^*} I^{1.425} + 1.4 \int_{d^*+6\omega_c^*}^{d^*+110\omega_c^*} I^{1.263} + \frac{3}{K} \int_{d^*+110\omega_c^*}^{\infty} I^1 \right) \quad (1.13)$$

- With H hardness of the softer material and where:

$$I^b = \left(\frac{z^* - d^*}{\omega_c^*} \right)^b \phi^*\left(z^*\right) dz^* \quad (1.14)$$

- K hardness coefficient is equal to [19]:

$$K = 0.454 + 0.41\nu \quad (1.15)$$

With ν Poisson coefficient

- β surface roughness parameter:

$$\beta = \eta R \sigma \quad (1.16)$$

- ω_c^* dimensionless critical interference [21]:

$$\omega_c = \left(\frac{\pi K H}{2E^*} \right)^2 R \quad (1.17)$$

- E^* Hertz elastic modulus of the two bodies:

$$\frac{1}{E^*} = \frac{1-\nu_1^2}{E_1} + \frac{1-\nu_2^2}{E_2} \quad (1.18)$$

The four integrals are related respectively to elastic, elasto-plastic (second and third) and plastic region.

For the friction force, instead:

$$Q^* = \frac{Q}{A_n H} = \frac{2}{3}\pi\beta K\omega_c^*(0.52 \int_{d^*}^{d^*+\omega_c^*} I^{0.982} + \int_{d^*+\omega_c^*}^{d^*+6\omega_c^*} -0.01I^{4.425} \\ +0.09I^{3.425} - 0.4I^{2.425} + 0.85I^{1.425}) \tag{1.19}$$

In this case the limit expressed in the integral is not infinite but finite and equals to the maximum tangential load acceptable.

Equations (1.13) and (1.19) are expressed by friction coefficient μ:

$$\mu = \frac{Q^*}{F^*} = \frac{Q^*}{P^* - F_a^*} = \frac{Q^*}{P^*\left(1 - \frac{F_a^*}{P^*}\right)} \tag{1.20}$$

Actually, this relation should also consider adhesion force, but if the plasticity index (Equation (1.21)) is >1.4, adhesion effects (F_a*) are negligible compared to contact load $\left(\frac{F_a^*}{P^*} \rightarrow 0\right)$.

$$\Psi = \frac{2E^*}{\pi KH}\sqrt[2]{\frac{\sigma_s}{R}} \tag{1.21}$$

Instead in 2008, Cohen, Kligerman and Etsion [22] provided a model (CKE) considering two areas: one elastic and another elasto-plastic, governed by following equations:

$$P^* = \frac{P}{A_n Y_o} = \frac{2}{3}\pi\beta C_v \Psi' \left(\int_{d^*}^{d^*+\delta_c^*} \left(z^* - d^*\right)^{\frac{3}{2}} \phi^*\left(z^*\right) dz^* \\ + \int_{d^*+\delta_c^*}^{\infty} \left(z^* - d^*\right)^{\frac{3}{2}} \left(1 - I^{\alpha}\right) \phi^*\left(z^*\right) dz^* \right) \tag{1.22}$$

Y_o is the virgin yield strength assumed as $H = 3Y_o$.

$$A^* = \frac{A}{A_n} = \pi\beta \left(\int_{d^*}^{d^*+\delta_c^*} \left(z^* - d^*\right) \phi^*\left(z^*\right) dz^* + \int_{d^*+\delta_c^*}^{\infty} \left(z^* - d^*\right)\left(1 + I^{\gamma}\right) \phi^*\left(z^*\right) dz^* \right) \tag{1.23}$$

TABLE 1.1
List of Variables

Variables	Function
C_v	$1.234 + 1.256\,\nu$
$\overline{\delta_c}$	$6.82\nu - 7.83(\nu^2 + 0.0586)$
α	$0.174 + 0.08\,\nu$
γ	$0.25 + 0.125\,\nu$

$$Q^* = \frac{Q}{A_n Y_o} = \frac{2}{3}\pi\beta C_v \Psi' \left(\int_{d^*}^{d^*+\delta_c^*} 0.26\left(z^* - d^*\right)^{\frac{3}{2}} \right.$$
$$* \coth\left(0.27\left(\left(z^* - d^*\right)\frac{\sigma}{\sigma_s}\Psi'^2\frac{1}{\overline{\delta_c}}\right)^{0.46}\right)\phi^*\left(z^*\right)dz^*$$
$$+ \int_{d^*+\delta_c^*}^{\infty} 0.26\left(z^* - d^*\right)^{\frac{3}{2}}\left(1 - I^{\alpha}\right)$$
$$\left. * \coth\left(0.27\left(\left(z^* - d^*\right)\frac{\sigma}{\sigma_s}\Psi'^2\frac{1}{\overline{\delta_c}}\right)^{0.46}\right)\phi^*\left(z^*\right)dz^* \right) \tag{1.24}$$

Where the variables and their function in Poisson's ratio are expressed in Table 1.1. And:

$$\delta_c = \overline{\delta_c}\,\omega_c' \tag{1.25}$$

$$I^K = \exp\left[\left(1 - \left(\left(z^* - d^*\right)\frac{\sigma}{\sigma_s}\Psi'^2\frac{1}{\overline{\delta_c}}\right)^K\right)^{-1}\right] \tag{1.26}$$

With the reformulated plasticity index:

$$\Psi' = \frac{2E^*}{\pi C_v \left(1 - \nu\right)^2 Y_o}\sqrt[2]{\frac{\sigma_s}{R}} \tag{1.27}$$

With respect to the other method, the critical interference, which indicates the transition from elastic to elasto-plastic area, now is [23]:

$$\omega_c' = \left(C_v \frac{\pi\left(1 - \nu^2\right)Y_o}{2E^*}\right)^2 R \tag{1.28}$$

Finally, this approach gives important definite relations (valid for $\nu > 0.25$) between the variables involved:

$$A^* = \left(0.36 + \frac{0.41}{\Psi'}\right)\left(P^*\right)^{0.97-0.1\exp\left(-\Psi'^{1.5}\right)} \tag{1.29}$$

$$Q^* = \left(0.26 + \frac{0.43}{\Psi'}\right)\left(P^*\right)^{0.0095\Psi'+0.91} \tag{1.30}$$

$$\mu = \left(0.26 + \frac{0.43}{\Psi'}\right)\left(P^*\right)^{0.0095\Psi'-0.09} \tag{1.31}$$

It is clear that all the factors are function of the plasticity index (in addition to the load) in the sense that if Ψ' is high (contact essentially plastic between asperities) the tangential load is small (because asperities can tolerate less stress) and the force required to break the junctions is lower, as well as for the real contact area. Besides, Equation (1.29) confirms the proportional relationship between area and load: for a given plasticity index, the greater the load, the greater the number of contact peaks, and the greater the tangential force, but the less the coefficient of friction. This is true only for the assumptions made, i.e., for the negligible adhesion effect (valid for not smooth surfaces).

The final model considered was realized in 2010 by [24] trio Li, Etsion and Talke (LET) with the aim to consider again the literature results for very high values of the plasticity index (the models discussed before have the limit of 8), so when the contact is essentially fully plastic. Therefore, the new equations become, in reference to three precise regimes of deformation (elastic, elasto-plastic and plastic):

$$\begin{aligned} P^* = \frac{P}{A_n Y_o} = \frac{2}{3}\pi\beta C_\nu \Psi' \Bigg(& \int_{d^*}^{d^*+\delta_c^*} \left(z^* - d^*\right)^{\frac{3}{2}} \phi^*\left(z^*\right) dz^* \\ & + \int_{d^*+\delta_c^*}^{d^*+110\delta_c^*} \left(z^* - d^*\right)^{\frac{3}{2}} \left(1 - I^\alpha\right) \phi^*\left(z^*\right) dz^* \\ & + 4.6 \frac{\overline{\delta_c}^{-1/2}}{\Psi'} \int_{d^*+110\delta_c^*}^{\infty} \left(z^* - d^*\right) \phi^*\left(z^*\right) dz^* \Bigg) \end{aligned} \tag{1.32}$$

$$\begin{aligned} A^* = \frac{A}{A_n} = \pi\beta \Bigg(& \int_{d^*}^{d^*+\delta_c^*} \left(z^* - d^*\right) \phi^*\left(z^*\right) dz^* \\ & + \int_{d^*+\delta_c^*}^{d^*+110\delta_c^*} \left(z^* - d^*\right)\left(1 + I^\gamma\right) \phi^*\left(z^*\right) dz^* + 2.2 \int_{d^*+110\delta_c^*}^{\infty} \left(z^* - d^*\right) \phi^*\left(z^*\right) dz^* \Bigg) \end{aligned} \tag{1.33}$$

$$Q^* = \frac{Q}{A_n Y_o} = \frac{2}{3}\pi\beta C_v \Psi' \Bigg(\int_{d^*}^{d^*+\delta_c^*} 0.26\left(z^*-d^*\right)^{\frac{3}{2}} * \coth\left(0.27\left(\left(z^*-d^*\right)\frac{\sigma}{\sigma_s}\Psi'^2\frac{1}{\overline{\delta_c}}\right)^{0.46}\right)\phi^*\left(z^*\right)dz^* + \int_{d^*+\delta_c^*}^{\infty} 0.26\left(z^*-d^*\right)^{\frac{3}{2}}\left(1-I^{\alpha}\right) * \coth\left(0.27\left(\left(z^*-d^*\right)\frac{\sigma}{\sigma_s}\Psi'^2\frac{1}{\overline{\delta_c}}\right)^{0.46}\right)\phi^*\left(z^*\right)dz^* \Bigg) + 4.6\frac{\overline{\delta_c}^{-1/2}}{\Psi'}\int_{d^*+110\delta_c^*}^{\infty} 0.26\left(z^*-d^*\right)^{\frac{3}{2}}\left(1-I^{\alpha}\right) * \coth\left(0.27\left(\left(z^*-d^*\right)\frac{\sigma}{\sigma_s}\Psi'^2\frac{1}{\overline{\delta_c}}\right)^{0.46}\right)\phi^*\left(z^*\right)dz^* \tag{1.34}$$

These are quite similar to Equations (1.22)–(1.24) but except for the plastic zone and the integral extremes. Similarly, relations (1.29)–(1.31) cannot be accepted in this case. Indeed, when the plasticity index is greater than 9.5, a contradiction with friction laws happens: increasing the load, the friction coefficient increases. For these reasons, the new relationships become:

$$A^* = \left(0.47 + 0.53\exp(-0.87\Psi'^{1.12}\right)\left(P^*\right)^{1-\exp\left(-2.7\Psi'^{0.35}\right)} \tag{1.35}$$

$$Q^* = \left(0.26 + 0.32\exp(-0.34\Psi'^{1.19}\right)\left(P^*\right)^{1-\exp\left(-1.9\Psi'^{0.4}\right)} \tag{1.36}$$

$$\mu = \left(0.26 + 0.32\exp(-0.34\Psi'^{1.19}\right)\left(P^*\right)^{-\exp\left(-1.9\Psi'^{0.4}\right)} \tag{1.37}$$

That describes the behavior of μ for Ψ'>8 as a constant trend, whereas for lower values as indicates by Equation (1.31).

Having discussed the different models and considering the others explained in the work [4], the next figures try to summarize and to drive researchers to choose the best approach. The input is the mechanical and topographical properties of the samples. These properties are combined and considered in Tabor coefficient [25], that evaluates the impact of adhesion:

$$\vartheta_a = \frac{E^*\sigma_s}{\Delta y}\left(\frac{\sigma_s}{r}\right)^{1/2} \tag{1.38}$$

Where σ_s is the Gaussian standard deviation of the peaks, r radius of the asperities and Δy is the specific work per unit area required to separate two surfaces when they are in contact. When ϑ_a is >10, the adhesion is negligible; otherwise it must be evaluated (first diamond). Going on the left (the part of the diagram where adhesion is not involved), Greenwood–Williamson (GW), Chang–Etsion–Bogy (CEB), and Zhao–Maietta–Chang (ZMC) are the choices (Figure 1.4). The differences are related to the GW plasticity index:

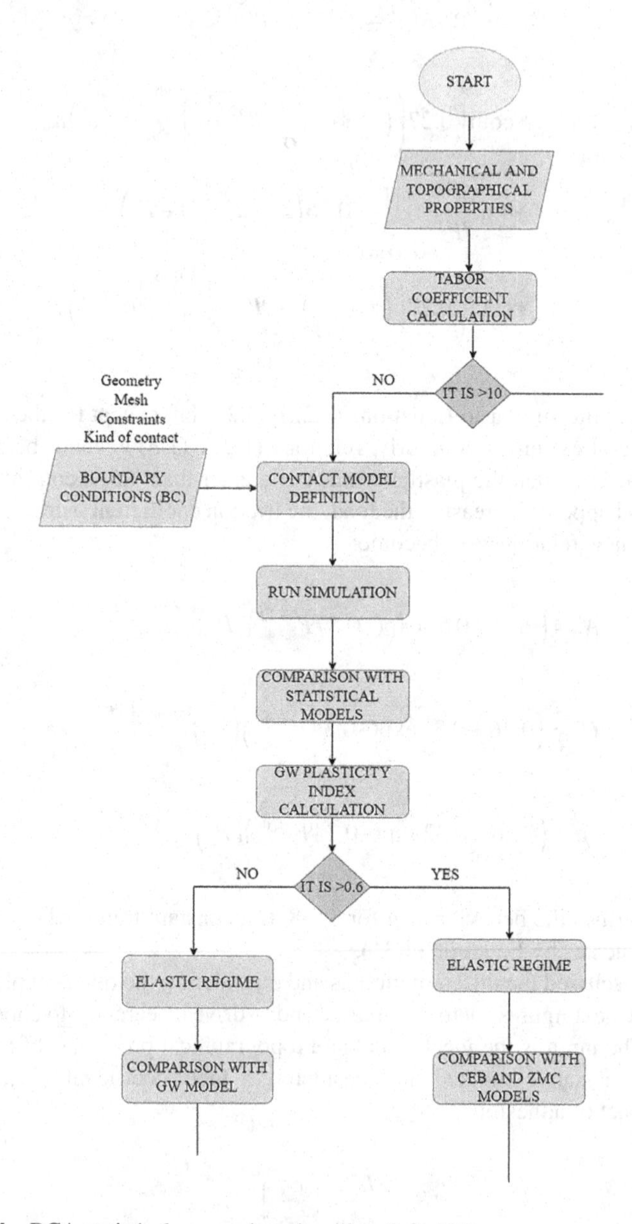

FIGURE 1.4 RCA statistical approaches algorithm (left side).

$$\Psi = \frac{E^*}{H} \sqrt[2]{\frac{\sigma_s}{R}} \tag{1.39}$$

When Ψ is >0.6 plastic regime is added to the elastic one, otherwise GW should be used.

On the right-hand side, instead, the adhesion and friction variables are involved (Figure 1.5). Firstly, the adhesion force must be calculated. About that, considering again to the sample properties, and examining the adhesion map, the right model is chosen. The next step is the calculation of the friction coefficient and, in particular, the potential impact of adhesion force (as clarified before in Equation (1.21)). Finally, the comparison between the three models discussed in this work according to index explained in relation 27.

The last investigation could be driven to compare the impact of adhesion in RCA, asperities deformation, pressure, etc. (Figure 1.6).

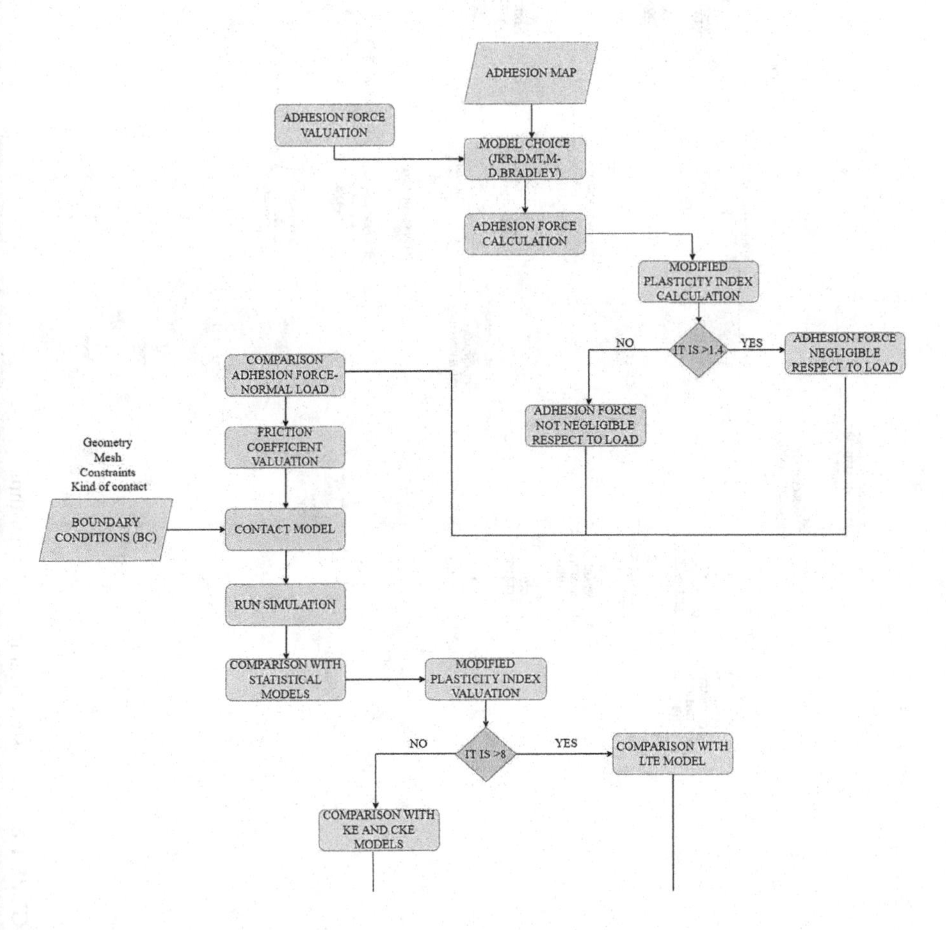

FIGURE 1.5 RCA statistical approaches algorithm (right side).

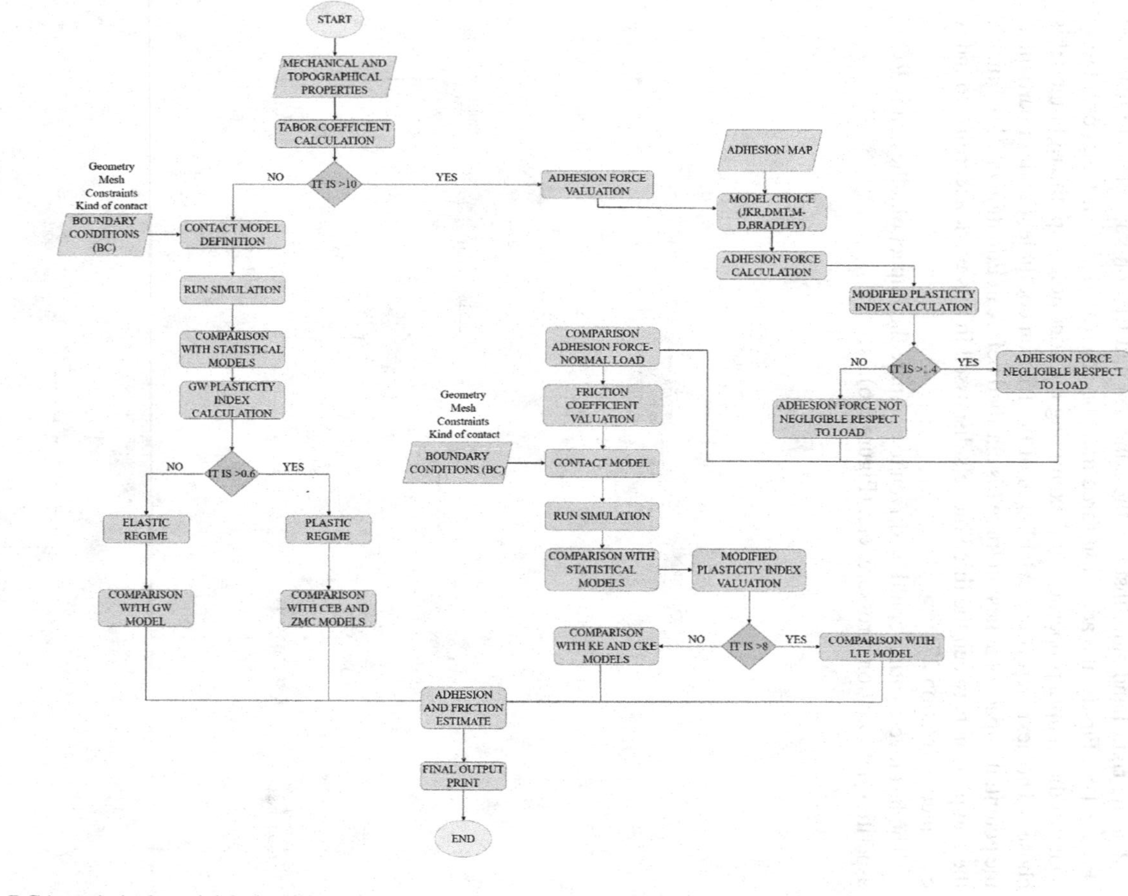

FIGURE 1.6 RCA statistical models' algorithm.

TABLE 1.2
Mechanical Properties of the Samples

	Young's Modulus (*E*) [GPa]	Poisson's Ratio (ν) [/]	Hardness (*H*) [Vickers]	Yield Strength (γ) [MPa]
Alloy EN-2030	72.5	0.33	100	250
Steel 1.2343	210	0.28	245	1240
Steel 1.3343	217	0.28	400	1850
Steel 1.2379	210	0.28	600	1650
Bronze $CuSn_{12}$	100	0.34	170	150

TABLE 1.3
Surface Parameters

R_a (μm)	0.1	0.6	1
m (μm)	0.2942	0.83	0.63
σ_s (μm)	0.0926	0.445	0.249
η (1\m²) * 10^9	2.8	2.02	1.22

MatLab codes were built for these statistical models [4] and they were applied on different specimens whose profile is a circular base with a conical tip of height 10 mm and diameter of the top contact area 200 μm (nominal area 30,000 ≅ μm²). It was cut after a milling operation. Regarding that, 5 materials with different mechanical properties (Table 1.2), 3 types of roughness and surface parameters (Table 1.3), were considered, in contact with a sapphire smooth flat (Table 1.4), for a total of 15 conformal couplings (Figure 1.7).

TABLE 1.4
Mechanical Properties of the Sapphire Flat

	Young's Modulus (*E*) [GPa]	Poisson's Ratio (ν) [/]	Hardness (*H*) [Vickers]	Yield Strength (γ) [MPa]
Sapphire	335	0.25	2710	350

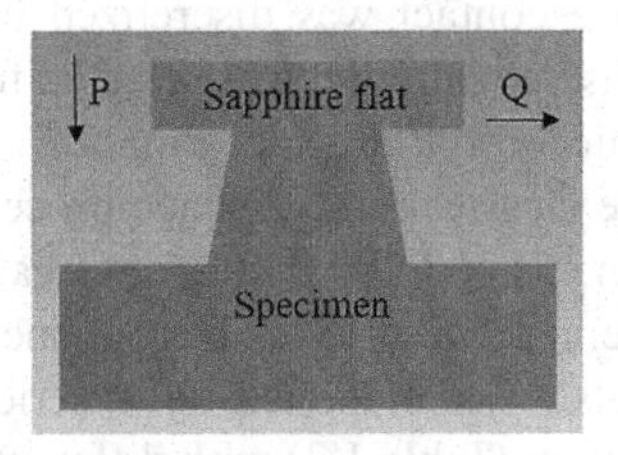

FIGURE 1.7 Scheme of contact in Ansys environment.

All the surface peaks follow a Gaussian distribution with mean m and standard deviation σ_s indicated above.

Successively, the topographical and mechanical properties were imported into the Ansys environment. Modelling by FEM is not a trivial task because of the complexity of the contact. Indeed, it can be seen as a clear example of a boundary non-linearity [26]. Nevertheless, thanks to the great improvement in scientific software, the results are accurate enough. In any case, a rigorous modelling is required. Because of the presence of friction interpreted as plastic failure mechanism, the conditions of the contacts are:

1. Non-penetration ($g_{N+}^{L} \leq 0$)
2. Stick-condition ($g_{T}^{L} = 0$)
3. Plastic slip parameter ($\lambda > 0$)
4. Friction force ($f_t \leq 0$)

Where the letter g indicates the gap function of the volume L in the normal direction N, defined as the difference between the displacements of the coupling samples (usually referred as master and slave surfaces). In our -case study, because the specimen is fixed at distance d from the sapphire flat, the gap function equates to:

$$g_{N+}^{L} = x(t) - d \tag{1.40}$$

Which must be always ≤0. Obviously, the same happens with the derivates of the vectors (impenetrability must guaranteed also for velocities [27]).

The second ones indicate that there is no movement in the tangential direction until the tangential force is lower than the static friction force ($f_t = t - \mu N \leq 0$). The last is $\lambda > 0$, which represents the amplitude of the plastic slip. Actually, conditions 3 and 4 can be combined in one relation:

$$\lambda f_t = 0 \tag{1.41}$$

These are the basis of the contact between the bodies implemented in Ansys. To ensure the non-penetration state, the Augmented Lagrange formulation was chosen as a reinforcement of the coupling. The reason was that this method is less sensitive (compared with Pure Penalty method) thanks to the extra term added and because it is usually recommended for frictional contact in large-deformation problems (that is our analysis). Moreover, the contact was discretized by Node (slave surface)-to-segment (Master surface) discretization: it was selected for its simplicity, mesh independence and range of applicability (large deformation and presence of friction).

The mesh used, after the classical convergence process, was medium-sized with an average of 100,000 elements and 200,000 nodes, realized by the tridimensional, tetrahedral second-order element Solid 187. The roughness of the specimen, for both statistical and FEM models, was modelled as a series of spheres (Figure 1.8) following Gaussian distribution (Table 1.3) with different radius and distance equals to $1/\eta$ [4].

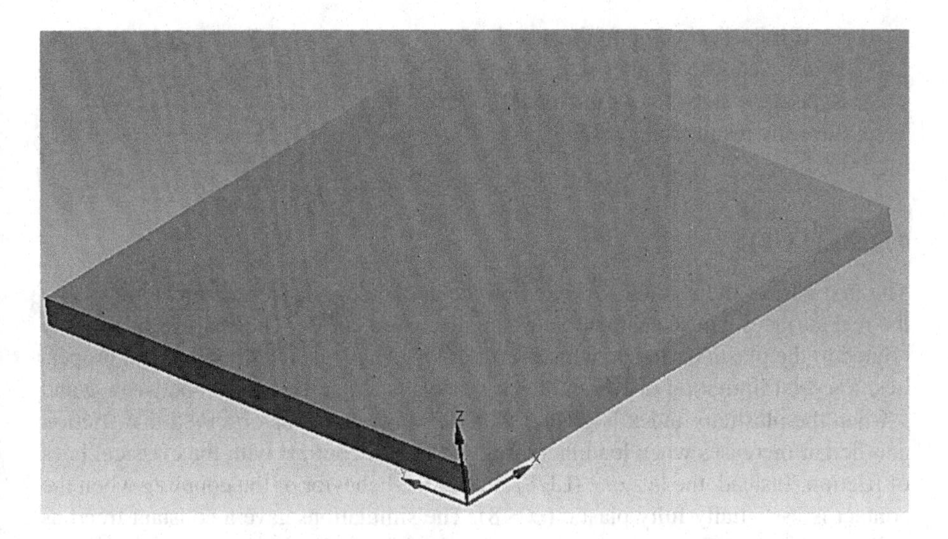

FIGURE 1.8 Model of sample roughness.

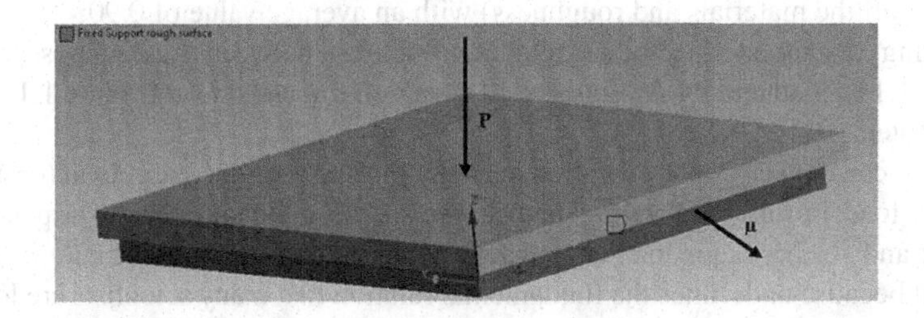

FIGURE 1.9 Ansys BC.

The BC (boundary conditions) are represented, instead, in Figure 1.9.

From this figure, it is evident that surfaces cannot move tangentially because of the presence of friction (all of the models regard static friction and stick condition), whereas the rigid surface can move along the *z* coordinate by a frictional contact.

Hence, the contact algorithm, which considers all the features described before, can be summarized in this way:

1. Start with the load applied (*P*)
2. Check for contact if the conditions of no-penetration and no-sliding are respected
   a. If nodes/segments violate contact constraint
      Apply contact force (Augmented Lagrange) and constraint (Frictional contact)
   b. Else Go to simulation
3. Solve the problem by a Newton–Raphson method (criterion chosen by program)
4. Check convergence

   a. If it is checked go to the next load
   b. Else choose another size
5. Repeat the steps for all the loads
6. Save the results in Excel Files
7. End

1.3 RESULTS

The first part of the results is focused on the application of the statistical models on the real surfaces whose characteristics are discussed before. The analysis is strongly related to the plasticity index, which sets both mechanical and topographical properties. The first limitation of KE and CKE models is about the relation between φ and μ when the plasticity index is greater than 9.5. Indeed, it was observed that friction coefficient increases when load increases and it is in contrast with the classical laws of friction. Instead, the last one (LET) studies the behavior of the coupling when the contact is essentially fully plastic ($\varphi > 8$). The simulations gave a constant trend as indicated by the LET model (our samples plasticity index is far greater than 8), as expressed in Figure 1.10 for steel 1.3343 and R_a 1 µm (this statement is, obviously, valid for all the materials and roughness) with an average value of 0.90:

Having discussed this, other relations such as between load and separation (Figure 1.11), load and RCA (Figure 1.12), friction force and load (Figure 1.13) were investigated.

As we can expect when separation between the surfaces decrease, load (or dimensionless load in this case) is higher because of the major number of asperities in contact and so the major load exchanged between them. The next figure is almost relevant because underlines the fundamental relation that many scientists are looking for: real contact area and load. Precisely both CKE and LET models are reported in the same graph for steel 1.2343 to highlight potential differences.

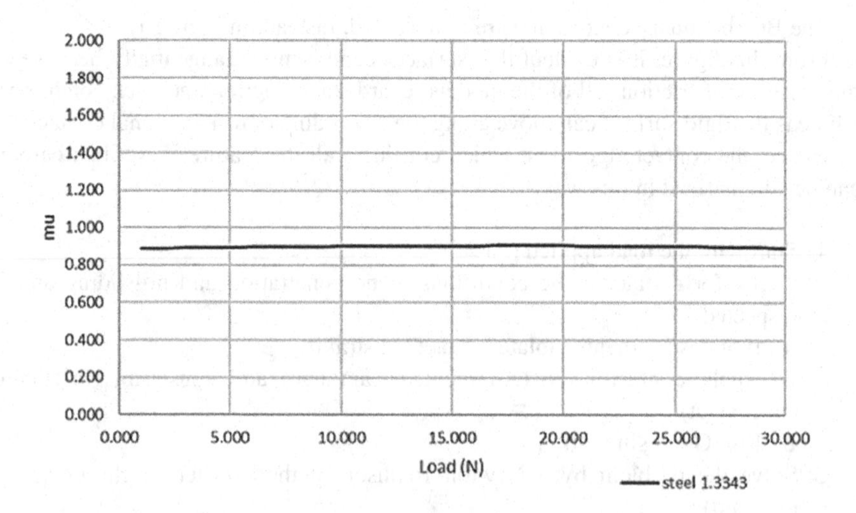

FIGURE 1.10 Friction coefficient as function of load R_a 1 µm.

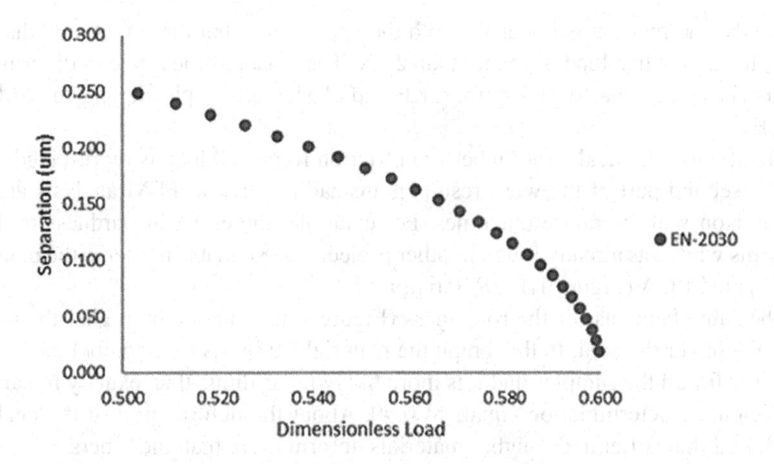

FIGURE 1.11 Separation-dimensionless load curve R_a 0.1 μm.

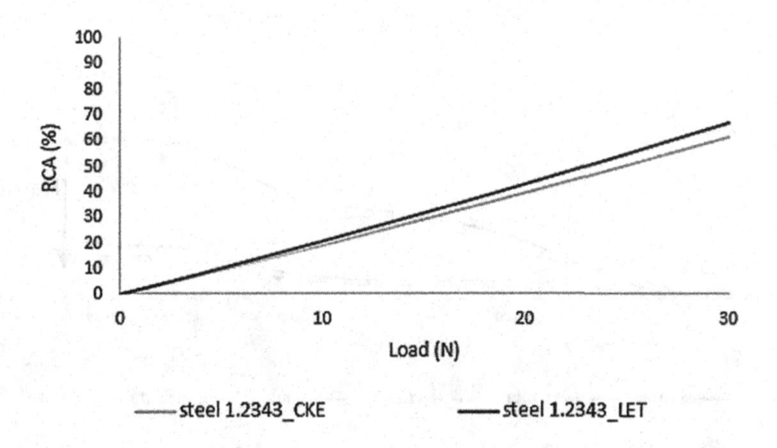

FIGURE 1.12 RCA-load curve for CKE and LET models R_a 0.1 μm.

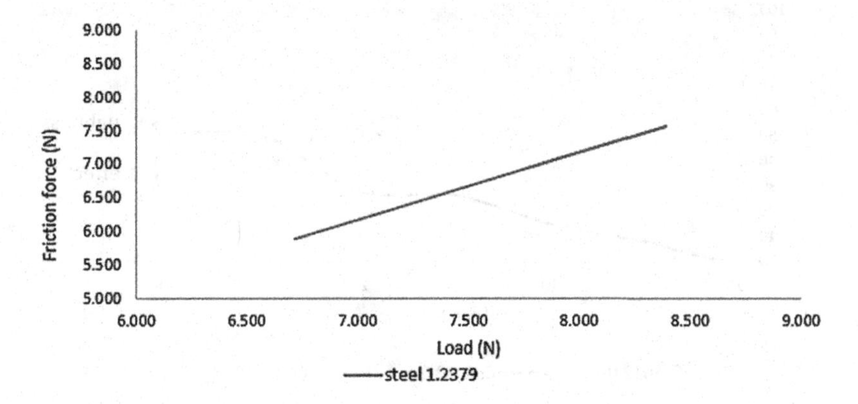

FIGURE 1.13 Friction force-load curve R_a 0.6 μm.

Firstly, the relation is linear for both the approaches, but there is a little distance about the area when load is greater than 25 N. The reason is the range of deformation that is elastic and elasto-plastic for CKE and elastic, elasto-plastic and fully plastic for LET.

Finally, the classical relation between friction force and load is represented.

The second part of the work results is instead referred to FEM analysis and the comparison with literature outcomes. For example, the effect of hardness on RCA confirms what was already found in other projects [4, 8], that is harder is the material, lower is the RCA (Figure 1.14, R_a 0.6 μm).

The same happens for the roughness (Figure 1.15), that is the higher the roughness, the lower the area. In this graph the material $CuSn_{12}$ is reported but the trend is the same for all the samples and it is more likely logarithmic than exactly linear with coefficient of determination equals to 0.90. About the deformation of thc pcaks, it was found that softer and rougher materials deform more than the others.

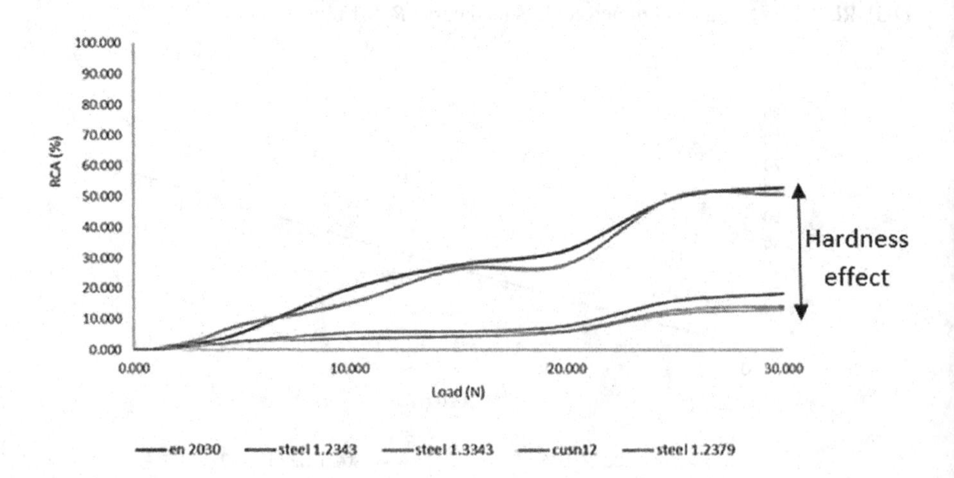

FIGURE 1.14 RCA-load curve for R_a 0.6 μm.

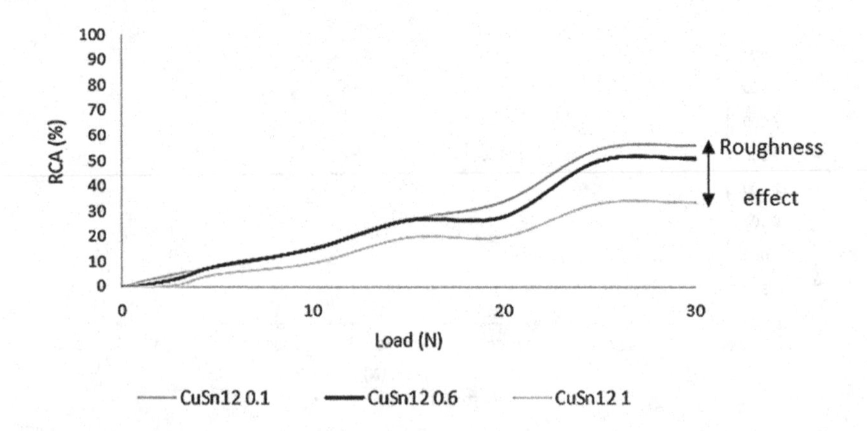

FIGURE 1.15 RCA-load curve for $CuSn_{12}$.

All these considerations, in the first instance, could determine the absence of effects of the friction on the coupling. Instead, relevant differences were found for RCA, pressure and deformation of the asperities as explained in Figures 1.16–1.18.

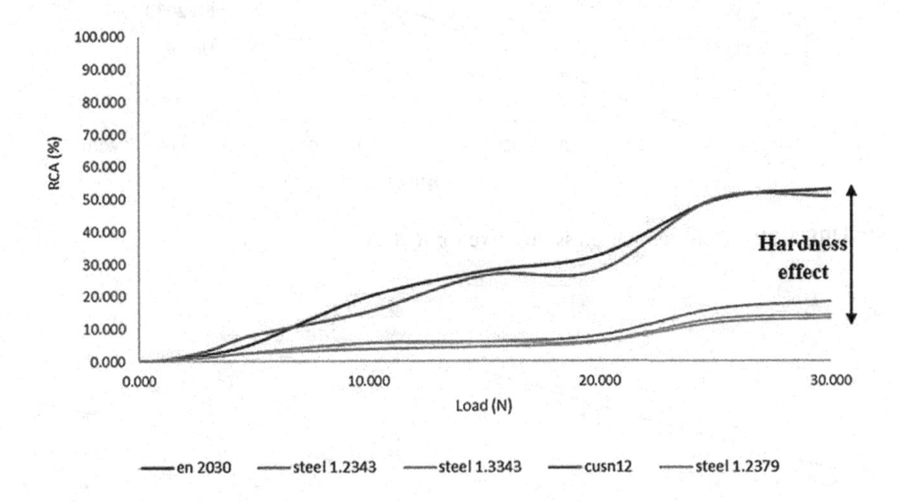

FIGURE 1.16 Comparison between RCA in the case with friction and with no friction for R_a 0.1 μm.

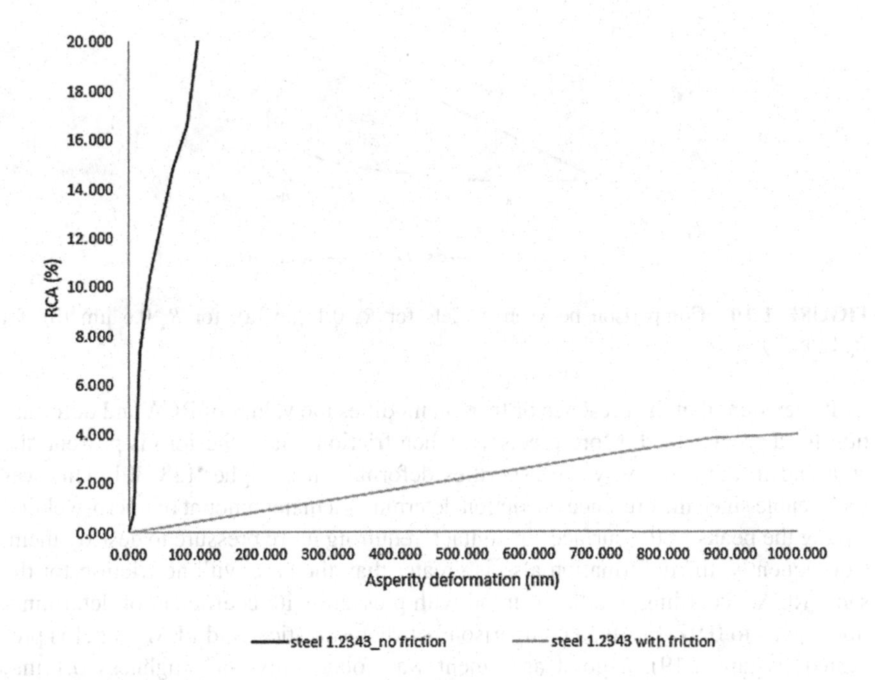

FIGURE 1.17 Comparison between asperities deformation in the case with friction and with no friction R_a 0.1 μm.

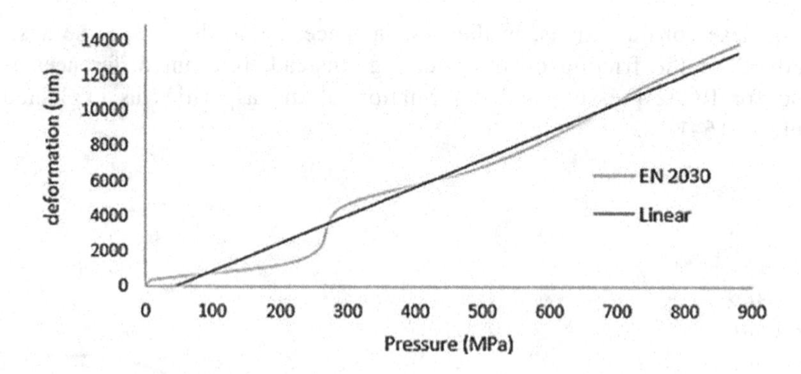

FIGURE 1.18 Deformation-pressure curve for R_a 1 μm.

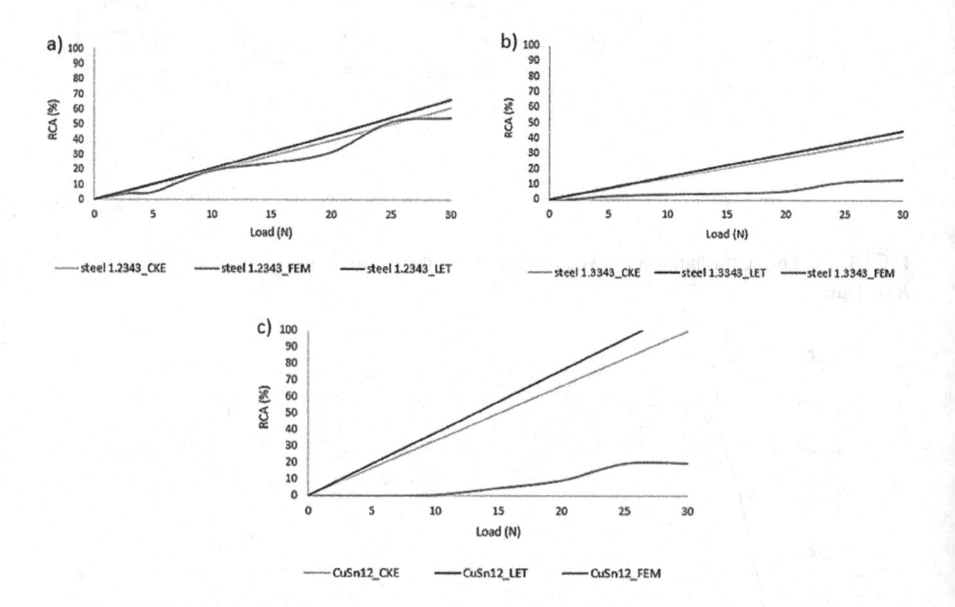

FIGURE 1.19 Comparison between models for R_a 0.1 μm (a), for R_a 0.6 μm (b), for R_a 1 μm (c).

It is evident that the presence of friction modifies the values of RCA and deformation for a precise load. More precisely, when friction (and adhesion) is present, the area and in the same way the asperities deformation is higher [28–30]. This was foreseeable since the presence of friction determines a major amount of micro-welding among the peaks of the surfaces in contact, requiring more pressure to destroy them. Consequently, the deformation also is greater than the case with no friction for the same RCA, according to a linear trend with pressure with coefficient of determination equals to 0.97. Lastly, a comparison between statistical and FEM model is presented (Figure 1.19): a good agreement was found only for roughness 0.1 μm, whereas for the other two the discrepancies are notable. In addition, this difference become higher when the roughness increases (especially from 0.6 to 1 μm).

This happens considering the random behavior of asperities against the statistical approximation of equal radius for all the peaks, which has a great influence on the output, especially when the roughness is far away from low values (surfaces not smooth).

1.4 CONCLUSIONS AND FUTURE DEVELOPMENT

In this work the influence of friction on RCA was analyzed. Literature approaches, together with a FEM model, were used to get more answers about this field. After a brief state of art, the materials and the techniques adopted were explained. The various results can be summarized in this way:

1. The relationship between RCA and mechanical and topographical properties was kept valid
2. The friction coefficient showed a constant trend with normal load
3. RCA-load and asperity deformation-pressure curves followed, respectively, a logarithmic and a linear trend
4. RCA, pressure and asperities deformation reached higher values than the case with no friction
5. The comparison between statistical models and the FEM model exhibited a good agreement only for low roughness
6. The asperities with smaller radius underwent more stress and deformation than bigger radius ones

Finally, more investigations are necessary to establish a precise and unique theory about this field. For instance, studies could validate the model experimentally on the specimens and consider other techniques such as multi-scale analysis [13] or they include in the model other variables, such as temperature or the asperity interaction by using other criterion to identify peaks [31] and representing the surfaces at a microscopical level in different ways (for example, by changing the distribution of asperities from Gaussian to Weibull [32]). Finally, analyzing the coupling for different load conditions and materials could also be a good alternative.

ACKNOWLEDGMENTS

This research was funded by MIUR, PRIN 2017 BIONIC.

REFERENCES

1. R. L. Jackson, H. Ghaednia, H. Lee, A. Rostami and X. Wang, "Contact Mechanics," in P.L. Menezes et al. (Eds.), *Tribology for Scientists and Engineers*, Springer, New York, 2013, pp. 93–140.
2. A. Ruggiero et al., "Accurate Measurement of Reciprocating Kinetic Friction Coefficient through Automatic Detection of the Running-In," *IEEE Transactions on Instrumentation and Measurement*, 69(5): 2398–2407, 2020.
3. J. Mergel, R. Sahli, J. Scheibert and R. Sauer, "Continuum Contact Models for Coupled Adhesion and Friction," *Journal of Adhesion*, 95(12):, 1101–1133, 2019.

4. A. Ruggiero and M. De Stefano, "Evaluation of the Real Contact Area of Rough Surfaces by Using a Finite Element Model," in V. Niola and A. Gasparetto (Eds.), *Advances in Italian Mechanism Science. IFToMM ITALY 2020. Mechanisms and Machine Science*, vol. 91, Springer, Cham, 2020.
5. S. Ozaki, K. Mieda, T. Matsuura and S. Maegawa, "Simple Prediction Method for Rubber Adhesive Friction by the Combining Friction Test and FE Analysis," *Lubricants*, 6(38): 38, 2018.
6. J. A. Greenwood and J. B. P. Williamson, "Contact of Nominally Flat Surfaces," *Proceedings of the Royal Society of London. Series A, Mathematical and Physical Sciences*, 295: 300–319, 1966.
7. R. L. Jackson and I. Green, "A Finite Element Study of Elasto-Plastic Hemispherical Contact Against a Rigid Flat," *Journal of Tribology*, 127: 34354, 2005.
8. B. B. Žugelj and M. Kalin, "In-Situ Observations of a Multi-Asperity Real Contact Area on a Submicron Scale," *Journal of Mechanical Engineering*, 63(6): 351–362, 2017.
9. M. Ciavarella, J. Joe, A. Papangelo and J. R. Barber, "The Role of Adhesion in Contact Mechanics," *Journal of the Royal Society Interface*, 16: 20180738, 2019.
10. G. Straffelini, *Attrito e usura: Metodologie di progettazione e controllo*, Tecniche Nuove, Milano, 2005.
11. B. N. J. Persson, I. M. Sivebaek, V. N. Samoilov, K. Zhao, A. I. Volokitin and Z. Zhang, "On the Origin of Amonton's Friction Law," *Journal of Physics: Condensed Matter*, 20: 395006, 2008.
12. N. Tas, C. Gui and M. Elwenspoek, "Static Friction in Elastic Adhesive MEMS Contacts, Models and Experiment," in *Proceedings IEEE Thirteenth Annual International Conference on Micro Electro Mechanical Systems*, 2000.
13. X. Wang, Y. Xu and R. L. Jackson, "Theoretical and Finite Element Analysis of Static Friction between Multi-Scale Rough Surfaces," *Tribology Letters*, 66: 146, 2018.
14. A. Majumdar and C. Tien, "Fractal Characterization and Simulation of Rough Surfaces," *Wear*, 136: 313–327, 1990.
15. M. Hasegawa, J. Liu, K. Okuda and M. Nunobiki, "Calculation of the Fractal Dimensions of Machined Surface Profiles," *Wear*, vol. 192:, pp. 40–45, 1996.
16. H. Zahouani, R. Vargiolu and J.-L. Loubet, "Fractal Models of Surface Topography and Contact Mechanichs," *Mathematical and Computer Modelling*, 28(4–8): 517–534, 1998.
17. K. Johnson, *Contact Mechanics*, Cambridge University Press, Cambridge, 1987.
18. L. Kogut and I. Etsion, "A Static Friction Model for Elastic-Plastic Contacting Rough Surfaces," *Journal of Tribology*, 126(1): 34–40, 2004.
19. W. R. Chang, I. Etsion and D. B. Bogy, "Static Friction Coefficient Model for Metallic Rough Surfaces," *Journal of Tribology*, 110(1): 57–63, 1988.
20. L. Kogut and I. Etsion, "Elastic-Plastic Contact Analysis of a Sphere and a Rigid Flat," *Journal of Applied Mechanics*, 69(5): 657–662, 2002.
21. W. Chang, I. Etsion and D. Bogy, "Elastic Plastic Model for the Contact of Rough Surfaces," *ASME Journal of Tribology*, 109: 257–262, 1987.
22. D. Cohen, Y. Kligerman and I. Etsion, "A Model for Contact and Static Friction of Nominally Flat Rough Surfaces Under Full Stick Contact Condition," *Journal of Tribology*, 130(3): 031401, 2008.
23. V. Brizmer, Y. Kligerman and I. Etsion, "The Effect of Contact Conditions and Material Properties on the Elasticity Terminus of a Spherical Contact," *International Journal of Solids and Structures*, 43(18): 5736–5749, 2006.
24. L. Li, I. Etsion and F.E. Talke, "Contact Area and Static Friction of Rough Surfaces with High Plasticity Index," *Journal of Tribology*, 132(3): 031401, 2010.

25. K. Fuller and D. Tabor, "The Effect of Surface Roughness on the Adhesion of Elastic Solids," *Proceedings of the Royal Society of London. A. Mathematical and Physical Sciences*, 345: 327–342, 1975.
26. P. Wriggers, "Finite Element Algorithms for Contact Problems," *Archives of Computational Methods in Engineering*, 2(4): 1–49, 1995.
27. F.-J. Wang, L.-P. Wang, J.-G. Cheng and Z.-H. Yao, "Contact Force Algorithm in Explicit Transient Analysis Using Finite-Element Method," *Finite Elements in Analysis and Design*, 43: 580–587, 2007.
28. S. Baojiang and Y. Shaoze, "Relationship between the Real Contact Area and Contact Force in Pre-Sliding Regime," *Physics B*, 26(7): 074601, 2017.
29. G. Straffellini, "A Simplified Approach to the Adhesive Theory of Friction," *Wear*, 249: 79–85, 1995.
30. G. Fortunato, V. Ciaravola and A. Furno, "On the Dependency of Rubber Friction on the Normal Force or Load: Theory and Experiment," *Tire Science and Technology*, 45(1): 25–54, 2015.
31. A. K. Waghmare and P. Sahoo, "Adhesive Friction Based on Accurate Elastic-Plastic Finite Element Analysis and n-Point Asperity Concept," *International Frontier Science Letters*, 11: 1–28, 2017.
32. N. Yu and A. A. Polycarpou, "Combining and Contacting of Two Rough Surfaces with Asymmetric Distribution of Asperity Heights," *Journal of Tribology*, 126(2): 225–232, 2004.

2 Non-Edible Biodegradable Plant Oils

The Future of the Lubricant Industry

Zahid Mushtaq and M. Hanief
National Institute of Technology, Srinagar, India

Kaleem Ahmad Najar
Kashmir University, Srinagar, India

CONTENTS

DOI: 10.1201/9781003243205-2

2.1 INTRODUCTION

Any mechanical system or machine has moving parts that come into contact after every cycle. The rubbing action induces frictional force between the sliding surfaces resulting in the wear of one or both of them and decreasing their life. The simplest and easiest way to reduce friction is lubrication. Lubrication is the process of applying solid, liquid, or gaseous lubricant between the sliding surfaces so that it generates pressure and increases their load carrying-capacity. The primary functions of a lubricant are to reduce friction and wear, reduce energy losses and extend the life of the mating materials, while other functions arc to insulate from heat, to prevent corrosion and oxidation, and to keep the parts clean(like.) [1, 2]. The demand for an effective, long-lasting, and inexpensive lubricant is escalating with time on a global level with the dire need for energy conservation. In this situation, the researchers have to carefully select the perfect lubricant which behooves a particular application. Generally, all the lubricants have almost similar desirable qualities, including high viscosity, high viscosity index, low pour point, high flash and fire points, good oxidation stability, good corrosion resistance, non-toxicity, etc. [3]. Lubricants can be applicable in either open systems or closed systems. In open systems, the lubricant is clearly visible and is directly exposed to the environment. Hence it is gradually lost to the environment, directly contributing to the pollution. Examples of open system lubricants include chain saw oils, metalworking fluids, drilling fluids, etc. In closed systems, the lubricants are enclosed and are not directly exposed to the environment. Examples include engines, compressors, and so on [4]. Mineral oils have been used as lubricants for several decades in industrial and automotive applications. They possess excellent lubricating properties, are widely available, durable and cheap [5]. However, there are certain issues related to mineral oils as they are not biodegradable, and are toxic and hazardous to the environment. They release harmful emissions to the atmosphere, which result in pollution and environmental degradation. They also face disposal problems and are harmful to terrestrial and aquatic life [6]. The mineral oil-based lubricants are derived from crude oil through the fractional distillation process and are their by-products. The energy demands of the world are rising very rapidly and the resources of crude oil are very limited which leads to price rise. Hence it is highly needed to decelerate the rapid exhaustion of resources to conserve energy. This approach does not allow the use of crude oil for lubricant production forever and some alternative should be available which should be renewable, non-toxic, and environmentally friendly [7]. Plant oil-based biolubricants are being investigated by researchers and their lubricating abilities have been compared to those of mineral oils. They have better biodegradability, low toxicity, higher viscosity index, and a higher flash point than mineral oils. The plant oils contain triglycerides from the molecules of glycerol with fatty acids attached through the linkages of esters. They are effective in both hydrodynamic and boundary lubrication due to polar groups and long-chained fatty acids in their structure. The plant oils, which have a high quantity of monounsaturated fatty acids, are considered to be possible future alternatives for lubrication [8, 9]. However, the bio-oils have certain disadvantages such as a high pour point, low oxidative stability, etc. which can be improved by certain modifications and introduction of additives [10–13]. The biolubricants can be produced from either edible or non-edible plant oils. However, due to the current policy of some countries, there are limitations on the utilization of edible feedstocks because they cause imbalances in the food cycle. Overuse of edible resources would eventually result in food scarcity, environmental imbalances, and deforestation [14]. Hence, the non-edible sources are preferable for biolubricant production as they are present in abundance. There are around 350 different types of crops around the world that can bear non-edible oils, most of which are cultivated on the wastelands and are cheaply and easily available [15, 16]. Only a few of them have been under examination for their lubricating abilities, and hence further interest on the part of researchers is needed to utilize non-edible crops for biolubricant production.

The current chapter aims to highlight the lubricating potential of bio-oils derived from non-edible resources which can mitigate the heavy dependence of the lubricant industry on petroleum oils and edible crops. It provides an overview of the benefits of using non-edible bio-oils as lubricants in place of mineral oils. Various non-edible oils are enlisted and their structural details are provided. Moreover, their extraction and conversion techniques are also discussed. This chapter will cater the idea of using non-edible oils as lubricants to the researchers which, in turn, will help in accelerating the research in the field.

2.1.1 History and Importance of Lubrication

Lubrication is of paramount importance in almost all industries which involve machinery. Lubricants are applied between the two or more surfaces in mutual contact in order to allow their smooth functioning for a longer period. The primary purpose of a lubricant is to restrict the friction and wear between the contacting surfaces in motion, to dissipate the extra heat and prevent corrosion and oxidation, to protect the contact area from dirt, dust, and other foreign particles, etc. Once a lubricant has been applied, the contacting surfaces are separated by a thin protective film. This film restricts the direct contact of the surfaces and therefore restricts the friction and wear between them. The characteristics of a good lubricant are high boiling point, low freezing point, high viscosity index, high thermal and oxidative stability, corrosion resistance, high flash and fire points, low pour point, etc. [17]. It is a challenge for researchers to allocate the most appropriate lubricant for a particular system that would increase its lifespan, efficiency, and reliability. It is widely believed that the concept of lubrication started in 1883, when oil was used to lubricate a journal bearing and its lubrication mechanism was studied [18]. In earlier history, before the discovery of crude oil, there was extensive use of vegetable oils and animal fats as lubricants in vehicles and machines. Olive oil-based calcium soaps were used by Egyptians to lubricate the wheel axes of carts [19]. Then the crude oil was discovered, and due to the excellent lubricating properties and low cost at that time, the petroleum-based oils assumed complete control of the lubrication sector. Then the rapid consumption of crude oil was started and industrial bloom was at its best. The exploitation of crude oil continued and after a few years demand increased and prices rose steeply. Moreover, there was a dire need to conserve the fossils and also protect the environment from pollution, which gave rise to the need for vegetable oils as lubricants and enhanced the scope of research into them. Because of these developments, several companies around the globe have initiated the development of biolubricants [20–22].

2.1.2 Definition of Bio-oil

Bio-oils are derived from plants, vegetables, or animal fats. Vegetable oils are primarily comprised of triacylglycerols and fatty acid molecules linked to single glycerol. The fatty acids can be divided into different groups: saturated, mono-unsaturated, di-unsaturated, and tri-unsaturated. The triglyceride structure is the reason for its highly viscous nature and the high viscosity index. The most important advantages of bio-oils are that they are renewable, biodegradable, and non-hazardous to the environment when compared with synthetic oils and mineral oils [23–25]. Vegetable oils can be differentiated into two sections – edible oils and non-edible oils. The oils which are safe for human consumption and can become a part of our food are called edible oils while those which are unsafe or unfit to be eaten are called non-edible oils. This chapter focuses on non-edible oils that can or have been converted into biolubricants. The advantages of selecting non-edible oils over edible oils are that these are generally extracted from waste plants, and thus mitigates the effect of depletion of food crops to produce lubricants. These can grow on lands receiving either low rainfall or high rainfall and their cultivation can be extended with the help of cuttings and seeds. Where wastelands that are not suitable for food crops are available, the farmers should expand the cultivation of non-edible crops which would provide enough base material for biolubricant production. The non-edible oils are renewable, biodegradable, readily available, pest resistant, disease resistant, having higher heat content with a low sulfur percentage. A large number of non-edible crops are available from which oil can be extracted. Some common non-edible oils along with their fatty acid combinations are enlisted in Table 2.1 [14, 26, 27].

2.1.3 Why Biolubricants

With the diminishing crude oil reserves and environmental pollution concerns, it has become vital to look for a substitute, something which has at least comparable properties or better. After the industrial revolution and modernization, the energy demands have spiked enormously, resulting in overdependence on petroleum oils. This has been followed by an escalation in fuel prices and increased pollution due to more consumption and emissions. This rapid consumption of petroleum oils needs to be de-escalated to conserve energy and save the environment. Moreover, due to the direct pollution in the open lubrication systems, environmentally friendly and biodegradable lubricants are in great demand. Various researchers have concluded that bio-oils have staggering lubrication capabilities. This is primarily due to their impeccable properties of high viscosity, high viscosity index, biodegradability, renewability, high flash and fire points, non-toxicity, low pour point, high metal adherence, and so on. These properties have made bio-oils strong contenders to partially replace the use of

TABLE 2.1
Fatty Acid Combinations of Various Non-Edible Bio Oils

	C14:0	C16:0	C16:1	C18:0	C18:1	C18:2	C18:3	C20:0	C20:1	C22:0	C22:1	C24:0
Jatropha oil	1.4	12.7–15.6	0.7	5.5–9.7	39–40.8	32–41.5	0.2	0.2–0.4	—	—	—	—
Neem oil	0.2–0.26	14.9	0.1	20.6	43.9	17.9	0.4	1.6	—	0.3	—	0.3
Jojoba oil	—	0.3	0.3	0.2	9.3	0.2	0.1	—	77	0.1	12	0.1
Castor oil	—	1.1	—	3.1	4.9	1.3	0.6	0.7	—	—	—	—
Karanja oil	—	3.7–7.9	—	2.4–8.9	44.5–71.3	10.8–18.3	—	4.1	2.4	5.3	—	1.1–3.5
Kusum oil	15.54	10.35	—	11.11	27.08	6.14	—	15.79	6.17	0.01	—	—
Soapnut oil	—	4.6	0.5	1.5	53	5	1.9	7	23	1.5	1	0.45
Mahua oil	0–1.0	16–28.2	—	14–25.1	41–51	8.9–17.9	—	0–3.3	—	—	—	—
Calophyllum Inophyllum	0–0.09	14.5–18	2.5	18.5–20	37.5–42.7	13.7–26.3	0.27–2.1	0–0.94	0–0.72	—	—	2.6
Yellow oleander	—	15.6	—	10.5	60.9	5.2	7.4	0.3	—	0.1	—	—
Deccan hemp oil	—	5.2	2.4	13.1	57.1	20	0.7	—	—	—	0.3	—
Rubber seed oil	2.2	10.2	—	8.7	24.6	39. 6	16.3	—	—	—	—	—

fossils for the production of lubricants [28, 29]. Over recent years, bio-oils have been used in lubricating a number of different mechanical systems. E.g, in automotives, hydraulics, chain saws, gears, tractor transmissions, compressors, gas engines, steam engines, motors, chain bars, steel industries, as metalworking fluids, as metal casting fluids, as greases, etc. However, the bio-oils are generally associated with poor low-temperature properties, low oxidation, and thermal stability. These shortcomings in bio-oils can be recuperated up to a large extent by the introduction of suitable additives [1, 30, 31].

2.2 EXTRACTION METHODS

Vegetable oil can be extracted from its seeds and kernels by any of the following three methods.

2.2.1 Mechanical Cold Pressing

This is the most commonly used method because of its simplicity. In this method, seeds are first dried at high temperatures in an oven or under the Sun and then fed to the screw press which is mechanically driven to crush them. A manual or engine-powered ram is used for this purpose and both seeds and kernels can be crushed to extract oil from them. However, this method has low efficiency and can extract only about 60 to 80% of the oil available in the seeds and kernels. Moreover, after the oil is received, filtration and degumming need to be done. Hence this method is not very feasible and other methods have been developed [14, 32, 33].

2.2.2 Solvent Extraction

This process is also called leaching. It is essentially a chemical process in which the oil is extracted from the seeds by dissolving it in a liquid solvent of low viscosity. Later the solvent is removed from the extracted oil. The speed of extraction depends upon the size of particles, type of liquid, and temperature of the solvent. This process results in the highest yield in terms of the percentage of oil extracted, but is suitable and economical at mass-scale production. Three methods can be used in the solvent extraction process: (i) Hot water extraction; (ii) Soxhlet extraction; (iii) Ultrasonication process [14, 32, 33].

2.2.3 Enzymatic Oil Extraction

In this process, the extraction of oil from the crushed seeds is done by utilizing some desirable enzymes. The biggest disadvantage of this process is that it is time-consuming. However, it doesn't generate toxic compounds and is hence environmentally friendly [14, 34, 35].

2.3 THE CONVERSION PROCESS TO BIOLUBRICANTS

Generally, the methods below are employed to convert raw vegetable oils into biolubricants.

2.3.1 Transesterification

It is a chemical process of converting triglycerides into glycerol and fatty acid methyl esters by reaction with alcohol or methanol. Acids or bases are used to catalyze the reaction. The final products have superior lubricating characteristics such as low pour point, high oxidation stability, improved low-temperature qualities, etc. Their quantity is controlled by varying the quantity of added alcohol [36, 37].

2.3.2 Hydrogenation

In this process, hydrogen is added to the oil molecule via a chemical reaction in presence of a catalyst like Ni or platinum. The properties of hydrogenated vegetable oils are very much dependent on the presence of hydrogen and the number of double bonds. It consists of three chemical reactions which take place simultaneously: (i) Saturation of double bonds; (ii) Geometric isomerization; and (iii) Positional isomerization. One disadvantage of this process is that the Ni catalyst increases the toxicity of the oil. Hydrogenation can be either partial or selective [38, 39].

2.3.3 Epoxidation

Epoxidation is a very useful process because of its ability to perform a long range of reactions in ordinary conditions. Epoxides are formed by the reaction of alkenes and peroxy acid. The epoxidized vegetable oils can be manufactured by the reaction of peroxy acids with double bonds. They have better anti-friction characteristics but have poor high-temperature stability. After the epoxidation, the epoxidized oil is subjected to the ring-opening and esterification process [40, 41].

2.3.4 Direct Mixing of Additives

Some suitable additives can directly be added to vegetable oils in definite percentages in order to ameliorate their properties. Different additives have different functions and hence should be chosen carefully and suitably for a particular application. Generally, the additives are solids or liquids in nature and are added in small proportions by volume or weight, depending on the requirement. To ensure the proper and uniform mixing of additives in the base oil, magnetic stirring or ultrasonication is employed for a few hours. The blend prepared has better rheological and tribological properties than the base oil.

2.4 TYPES OF ADDITIVES USED IN LUBRICANTS

Despite being highly sought after, pure bio-oils lag behind in certain specific properties. These properties can be improved by the addition of some proper additives. Additives can comprise from 0.1% to 30% of the base oil, depending upon the particular application. However, while selecting an additive for a bio-oil, its biodegradability and non-toxicity need to be taken into consideration. We cannot kill the primary motive of using biolubricant by adding toxic products to it. Some common types of additives used are as under:-

2.4.1 Antioxidants

Antioxidants are added to hamper the oxidation of the lubricant, ensuring that it stays for a longer time and fulfills its purpose in the industry. With increasing velocity and temperature, the lubricating oils are more susceptible to oxidation, which results in more friction and wear [42]. The oxidation rate is dependent on various factors like the presence of oxygen and water in the oil, temperature, velocity, wear debris presence, etc. Hence antioxidants are added to the oil to defer or hinder the oxidation process by enhancing the oxidation resistance of the oil. Examples of best antioxidant additives are zinc diamyl dithiocarbamate (ZDDC), zinc dialkyl dithiophosphate (ZDDP), amines, etc. [43].

2.4.2 Detergents and Dispersants

These additives are generally used in engine oils to expand the life of the lubricant, increase the efficiency of the engine and maintain its cleanliness. Engine oils are continuously subjected to high temperatures and combustion products like sludge. These additives prevent the agglomeration of sludge particles and disperse them. They also neutralize any gases formed as a result of incomplete combustion. Some additives are multifunctional and act as both detergent and dispersant. Some common examples are phosphates, sulfonates, phenates, calcium, barium, etc. Sulfonates and phenates are toxic and not environmentally friendly, and hence should be avoided [44].

2.4.3 Viscosity Modifiers

Viscosity is a very important property of a lubricant and is directly related to its performance. A highly viscous lubricant will result in temperature rise and the accumulation of foreign particles while a low viscous lubricant may be unable to lubricate the sliding surfaces properly as it will increase the contact between them. Hence we have to strike a balance between the two. Additives called viscosity modifiers are added to improve the viscosity of a lubricant. They are generally fluids of very high/low viscosities. Ethylene-vinyl acetate copolymer is reported to increase the viscosity of the vegetable oils which are having low viscosity [45].

2.4.4 Viscosity Index Improvers

Viscosity Index (VI) of a lubricant is a measure of change in its viscosity with respect to temperature. If a lubricant is having high VI, this means it will not exhibit much change in its viscosity with the changing temperatures. Generally, the VI of bio-oils is high but it can still be improved by adding VI improvers like ZnDDP, polymethacrylate, which ameliorate their low-temperature as well as high-temperature viscosity, and hence overall lubrication performance of the lubricant [46].

2.4.5 Rust and Corrosion Inhibitors

The presence of some products like sulfur, phosphorus, chlorine, iodine, oxygen, water, etc. in the oil can cause damage to the machine parts by corroding them, and hence need to be eradicated. Rust inhibitors, such as barium, calcium, etc., are used to protect ferrous metals from corrosion while corrosion inhibitors like benzotriazole, zinc diethyl dithiophosphate, zinc diethyl dithiocarbamate, etc. are used to restrain the corrosion of non-ferrous metals. They adhere to the metal surface and form a durable protective film [43].

2.4.6 Nanoparticles

Nanoparticles have staggering anti-friction and anti-wear properties and can be added directly to bio-based lubricants in small percentages in order to improve their performance. When nanoparticles are added as additives to a lubricant, it is then called a nanolubricant. The performance of a nanolubricant depends upon the size, concentration, and shape of the nanoparticles present in it. The nanoparticles should be carefully mixed and their uniform dispensation in the oil should be ensured to avoid agglomeration. If the nanoparticles form agglomerates in the oil, it results in a higher 3-body abrasive wear. Nanoparticles like TiO_2, ZnO, MoS_2, CuO, etc. are very effective as they establish a ball-bearing-type effect between the sliding surfaces and hence reduce friction between them. The nanoparticles perform their function by reacting with the oil and the environment, and thereby forming a protective tribo-layer on the surfaces. This layer prevents the direct contact of materials and reduces friction and wear [47, 48].

2.4.7 Pour Point Depressants (PPDs)

The pour point of a liquid is the minimum temperature below which it stops flowing. Generally, bio-oils have poor low-temperature properties as they cease to flow when the temperatures start dipping below zero degrees Celsius. PPDs can be added to base oils in very small quantities to enhance their low-temperature properties. They work by hindering the development of wax crystals and allow the smooth flow of oil at low temperatures. E.g, polymethyl acrylate, poly-alpha-olefin, alkyl naphthalene, etc. [44, 49].

2.4.8 Friction Modifiers

These are the most significant type of additives added to bio-oils. These additives reduce the friction between the contacting surfaces of a tribo-pair. The additive particles become adsorbed on the surface, act as ball bearings and convert the sliding friction into rolling friction. The friction modifiers are efficiently effective at low temperatures and loads.

2.4.9 Anti-Wear Additives

These additives are generally composed of phosphorus, sulfur, and chlorine [14]. They utilize the process of chemisorption instead of adsorption and react with the surface, forming a very enduring layer that can also withstand high temperatures.

2.4.10 Extreme Pressure (EP) Additives

These are specially formulated to withstand high pressures and velocities. They reduce friction and wear by forming a film of low shear strength on the surface. The concentration of EP additives should be well optimized as too a high concentration results in corrosion and too low a concentration may not be able to reduce friction and wear to a desirable level. Examples are molybdenum disulphide, lead napthalenes, etc. [43, 50].

2.5 CHALLENGES TO ENCOUNTER

The prime issues with using biolubricants are their inferior low-temperature endurance and poor oxidative stability. It ceases to flow as the temperature dips into minus degrees due to the swift and early evolution of wax crystals. Generally, raw vegetable oils cannot withstand high temperatures and they oxidize quickly. These problems can be mitigated by a selection of suitable additives according to the application. The pour point of the oil needs to be reduced in order to make it suitable for low-temperature lubrication while increasing the flash and fire points would help them withstand higher temperatures. Moreover, some other hindrances in exploiting the non-edible resources are collecting from different and far-flung locations, bad quality of plants or seeds, available only for a short period of time, inappropriate and weak marketing, poor technology in harvesting and processing, etc. [14].

2.6 CONCLUSIONS

Bio-oils have enormous capabilities to partially replace petroleum-based lubricants and be a substantial part of the lubrication industry. Bio-oils can be extracted in a number of easy ways and are easily available in large quantities. Non-edible sources are more preferred to avoid food crop depletion as most of them are useless otherwise. Even after wide-ranging research about these oils, they have been little used to date. This is because of their lagging in certain areas to match with synthetic and mineral oil-based lubricants. These areas can be fortified by the wise selection of additives for the specific application. It is inevitable to find non-edible bio-oils becoming a prime source of lubricants in the coming years.

REFERENCES

1. T.M. Panchal, A. Patel, D.D. Chauhan, M. Thomas, J.V. Patel, A methodological review on bio-lubricants from vegetable oil based resources, *Renewable and Sustainable Energy Reviews*. 70 (2017) 65–70. https://doi.org/10.1016/j.rser.2016.11.105

2. T. Naveed, R. Zahid, R.A. Mufti, M. Waqas, M.T. Hanif, A review on tribological performance of ionic liquids as additives to bio lubricants, *Proceedings of the Institution of Mechanical Engineers, Part J: Journal of Engineering Tribology*. 235 (2021) 1782–1806.
3. N.A. Zainal, N.W.M. Zulkifli, M. Gulzar, H.H. Masjuki, A review on the chemistry, production, and technological potential of bio-based lubricants, *Renewable and Sustainable Energy Reviews*. 82 (2018) 80–102. https://doi.org/10.1016/j.rser.2017.09.004
4. A.Z. Syahir, N.W.M. Zulkifli, H.H. Masjuki, M.A. Kalam, A. Alabdulkarem, M. Gulzar, L.S. Khuong, M.H. Harith, A review on bio-based lubricants and their applications, *Journal of Cleaner Production*. 168 (2017) 997–1016. https://doi.org/10.1016/j.jclepro.2017.09.106
5. A. Bahari, R. Lewis, T. Slatter, Friction and wear phenomena of vegetable oil-based lubricants with additives at severe sliding wear conditions, *Tribology Transactions*. 61 (2018) 207–219.
6. J.C. Ssempebwa, D.O. Carpenter, The generation, use and disposal of waste crankcase oil in developing countries: A case for Kampala district, Uganda, *Journal of Hazardous Materials*. 161 (2009) 835–841.
7. A. Ruggiero, R. D'Amato, M. Merola, P. Valašek, M. Müller, Tribological characterization of vegetal lubricants: Comparative experimental investigation on Jatropha curcas L. oil, Rapeseed Methyl Ester oil, Hydrotreated Rapeseed oil, *Tribology International*. 109 (2017) 529–540.
8. N.H. Jayadas, K.P. Nair, G. Ajithkumar, Tribological evaluation of coconut oil as an environment-friendly lubricant, *Tribology International*. 40 (2007) 350–354.
9. S.M. Alves, B.S. Barros, M.F. Trajano, K.S.B. Ribeiro, E. Moura, Tribological behavior of vegetable oil-based lubricants with nanoparticles of oxides in boundary lubrication conditions, *Tribology International*. 65 (2013) 28–36. https://doi.org/10.1016/j.triboint.2013.03.027
10. A. Adhvaryu, S.Z. Erhan, J.M. Perez, Tribological studies of thermally and chemically modified vegetable oils for use as environmentally friendly lubricants, *Wear*. 257 (2004) 359–367.
11. B.S. Kumar, G. Padmanabhan, P.V. Krishna, Performance assessment of vegetable oil based cutting fluids with extreme pressure additive in machining, *Journal of Advanced Research in Materials Science* 19 (2016) 1–13.
12. Z. Mushtaq, M. Hanief, Evaluation of tribological performance of jatropha oil modified with molybdenum disulphide micro-particles for steel-steel contacts, *Journal of Tribology*. 143 (2021) 1–13. https://doi.org/10.1115/1.4047752
13. Z. Mushtaq, M. Hanief, S.A. Manroo, Prediction of friction and wear during ball-on-flat sliding using multiple regression and ANN: Modeling and experimental validation, *Jurnal Tribology*. 28 (2021) 117–128.
14. A.E. Atabani, A.S. Silitonga, H.C. Ong, T.M.I. Mahlia, H.H. Masjuki, I.A. Badruddin, H. Fayaz, Non-edible vegetable oils: A critical evaluation of oil extraction, fatty acid compositions, biodiesel production, characteristics, engine performance and emissions production, *Renewable and Sustainable Energy Reviews*. 18 (2013) 211–245.
15. S.-Y. No, Inedible vegetable oils and their derivatives for alternative diesel fuels in CI engines: A review, *Renewable and Sustainable Energy Reviews*. 15 (2011) 131–149.
16. A.K. Agarwal, Biofuels (alcohols and biodiesel) applications as fuels for internal combustion engines, *Progress in Energy and Combustion Science*. 33 (2007) 233–271.
17. Y. Singh, A. Farooq, A. Raza, M.A. Mahmood, S. Jain, Sustainability of a non-edible vegetable oil based bio-lubricant for automotive applications: A review, *Process Safety and Environmental Protection*. 111 (2017) 701–713. https://doi.org/10.1016/j.psep.2017.08.041

18. F. Archibald, History of lubrication, *Tribology & Lubrication Technology*. 55 (1999) 9.
19. T. Norrby, Environmentally adapted lubricants-where are the opportunities?, *Industrial Lubrication and Tribology*. 55 (2003) 268–274.
20. A. Srivastava, P. Sahai, Vegetable oils as lube basestocks: A review, *African Journal of Biotechnology*. 12 (2013) 880–891.
21. G. Biresaw, Biolubricant production catalyzed by enzymes, in *Environmentally friendly and biobased lubricants*, CRC Press. (2016) 185–202.
22. H.M. Mobarak, E.N. Mohamad, H.H. Masjuki, M.A. Kalam, K.A.H. Al Mahmud, M. Habibullah, A.M. Ashraful, The prospects of biolubricants as alternatives in automotive applications, *Renewable and Sustainable Energy Reviews*. 33 (2014) 34–43.
23. S.Z. Erhan, S. Asadauskas, Lubricant basestocks from vegetable oils, *Industrial Crops and Products*. 11 (2000) 277–282.
24. S.C.A. De Almeida, C.R. Belchior, M.V.G. Nascimento, L. dos Vieira, G. Fleury, Performance of a diesel generator fuelled with palm oil, *Fuel*. 81 (2002) 2097–2102.
25. N. Salih, J. Salimon, B.M. Abdullah, E. Yousif, Thermo-oxidation, friction-reducing and physicochemical properties of ricinoleic acid based-diester biolubricants, *Arabian Journal of Chemistry*. 10 (2017) S2273–S2280.
26. A.L. Ahmad, N.H.M. Yasin, C.J.C. Derek, J.K. Lim, Microalgae as a sustainable energy source for biodiesel production: A review, *Renewable and Sustainable Energy Reviews*. 15 (2011) 584–593.
27. M.M. Gui, K.T. Lee, S. Bhatia, Feasibility of edible oil vs. non-edible oil vs. waste edible oil as biodiesel feedstock, *Energy*. 33 (2008) 1646–1653.
28. I.S. Tamada, P.R.M. Lopes, R.N. Montagnolli, E.D. Bidoia, Biodegradation and toxicological evaluation of lubricant oils, *Brazilian Archives of Biology and Technology*. 55 (2012) 951–956.
29. S. Rani, M.L. Joy, K.P. Nair, Evaluation of physiochemical and tribological properties of rice bran oil--biodegradable and potential base stoke for industrial lubricants, *Industrial Crops and Products*. 65 (2015) 328–333.
30. F.M.T. Luna, J.B. Cavalcante, F.O.N. Silva, C.L. Cavalcante Jr, Studies on biodegradability of bio-based lubricants, *Tribology International*. 92 (2015) 301–306.
31. S.P. Darminesh, N.A.C. Sidik, G. Najafi, R. Mamat, T.L. Ken, Y. Asako, Recent development on biodegradable nanolubricant: A review, *International Communications in Heat and Mass Transfer*. 86 (2017) 159–165. https://doi.org/10.1016/j.icheatmasstransfer.2017.05.022
32. W.M.J. Achten, L. Verchot, Y.J. Franken, E. Mathijs, V.P. Singh, R. Aerts, B. Muys, Jatropha bio-diesel production and use, *Biomass and Bioenergy*. 32 (2008) 1063–1084.
33. A.E. Atabani, A.S. Silitonga, I.A. Badruddin, T.M.I. Mahlia, H.H. Masjuki, S. Mekhilef, A comprehensive review on biodiesel as an alternative energy resource and its characteristics, *Renewable and Sustainable Energy Reviews*. 16 (2012) 2070–2093.
34. P. Mahanta, A. Shrivastava, *Technology development of bio-diesel as an energy alternative*, Department of Mechanical Engineering Indian Institute of Technology. (2004).
35. S. Shah, A. Sharma, M.N. Gupta, Extraction of oil from Jatropha curcas L. seed kernels by combination of ultrasonication and aqueous enzymatic oil extraction, *Bioresource Technology*. 96 (2005) 121–123.
36. D.Y.C. Leung, X. Wu, M.K.H. Leung, A review on biodiesel production using catalyzed transesterification, *Applied Energy*. 87 (2010) 1083–1095.
37. C.S. Madankar, S. Pradhan, S.N. Naik, Parametric study of reactive extraction of castor seed (Ricinus communis L.) for methyl ester production and its potential use as bio lubricant, *Industrial Crops and Products*. 43 (2013) 283–290.

38. M.B. Fernández, G.M. Tonetto, D.E. Damiani, Hydrogenation of sunflower oil over different palladium supported catalysts: Activity and selectivity, *Chemical Engineering Journal*. 155 (2009) 941–949.
39. B. Shomchoam, B. Yoosuk, Eco-friendly lubricant by partial hydrogenation of palm oil over Pd/γ-Al_2O_3 catalyst, *Industrial Crops and Products*. 62 (2014) 395–399.
40. S.G. Tan, W.S. Chow, Biobased epoxidized vegetable oils and its greener epoxy blends: A review, *Polymer-Plastics Technology and Engineering*. 49 (2010) 1581–1590.
41. B. Mudhaffar, J. Salimon, Epoxidation of vegetable oils and fatty acids: Catalysts, methods and advantages, *Journal of Applied Sciences*. 10 (2010) 1545–1553.
42. P. Wanasundara, F. Shahidi, *Antioxidants: Science, technology, and applications*, Bailey's Industrial Oil and Fat Products. (2005).
43. G. Stachowiak, A.W. Batchelor, *Engineering tribology*, Butterworth-Heinemann. (2013).
44. S.Q.A. Rizvi, *A comprehensive review of lubricant chemistry, technology, selection, and design*. (2009) 100–112.
45. L.A. Quinchia, M.A. Delgado, C. Valencia, J.M. Franco, C. Gallegos, Viscosity modification of different vegetable oils with EVA copolymer for lubricant applications, *Industrial Crops and Products*. 32 (2010) 607–612.
46. L.R. Rudnick, *Lubricant additives: Chemistry and applications*, CRC Press. (2009).
47. Z. Mushtaq, M. Hanief, Enhancing the tribological characteristics of Jatropha oil using graphene nanoflakes, *Journal of Tribology*. 28 (2021) 129–143.
48. Z. Mushtaq, M. Hanief, *Friction and wear performance of jatropha oil added with molybdenum disulphide nanoparticles*. (2021). https://doi.org/10.1007/978-981-33-4443-3_69
49. H.-S. Hwang, S.Z. Erhan, Modification of epoxidized soybean oil for lubricant formulations with improved oxidative stability and low pour point, *Journal of the American Oil Chemists' Society*. 78 (2001) 1179–1184.
50. N. Canter, Special report: Trends in extreme pressure additives, *Tribology and Lubrication Technology*. 63 (2007) 10.

3 Static Analysis of Journal Bearing with Bionanolubricants Featuring Three-Layered Nanoadditive Couple Stress Fluids

Mohammed Shabbir Ahmed
BijuPatnaik University of Technology, Rourkela, India

K. Prabhakaran Nair
National Institute of Technology, Calicut, India

T.V.V.L.N. Rao
The Assam Kaziranga University, Jorhat India

Ali Algahtani
King Khalid University, Abha, Kingdom of Saudi Arabia

CONTENTS

DOI: 10.1201/9781003243205-3

3.1 INTRODUCTION

Rotary machines use the hydrodynamic journal bearing to support heavy loads in the turbines of power plants. The bearings of such machines are failing due to their inability to sustain the heavy loads in power plants. The sustainability of rotating machinery running at higher speeds is attributed to the design of journal bearings for the smooth functioning of power plants. The use of nanoparticles in biolubricants in journal bearings provide enhanced load-carrying capacity. The addition of nanoparticles enhances the viscosity which thickens the lubricant and affects performance.

In the conventional bearing design, the lubricant was assumed to be Newtonian and the bearing shell to be rigid. Soni et al. [1] and Tayal et al. [2] reported the performance characteristics with these assumptions. When heavy loads are considered in the bearings the deformations affect the clearance space geometry of the bearing to such an extent that the performance characteristics may differ for those computed with rigid bearings. The elastohydrodynamic analysis of the circular and non-circular bearings for Newtonian lubricants were reported by Nair et al. [3, 4]. To improve certain characteristics, the small particles of solids/liquids were added to the lubricants and these behaved as couple stress fluids. The suspended particles thicken the lubricant and, in turn, enhance the performance characteristics of the journal bearings. This also increases the viscosity of the lubricant film. Prakash and Sinha [5] explored the static characteristics of a journal bearing with micro polar fluids. Nair et al. [6] studied micro polar lubricant affects in the static performance characteristics of elastohydrodynamic journal bearings.

The performance (static and dynamic) characteristics of multi-lobe (journal) bearings are significantly influenced by limited (partial) texture/slip patterns on the lobed configurations. Therefore, the analysis of lobed configuration with limited (partial) texture for the improvement of the bearing (multi-lobe) performance is presented. The analysis of texturing to improve the static, dynamic and stability characteristics of multi-lobe journal bearings encompasses tribology for sustainability. Based on the literature available, it can be concluded that the chemical modification of vegetable oils along with proper nanoadditives have the potential to replace depleting and harmful mineral oil lubricants. The addition of nanoparticles enhances the tribological properties and, in turn, the performance characteristics of the journal bearings. The viscosity of the biolubricants increases with nanoadditives and thus there is also an improvement in the static performance of bearings. There is a shortage of existing studies which show the analyses of hydrodynamic journal bearing with nanoadditive biolubricants. Therefore it is useful if the performance characteristics of circular journal bearings with biolubricants' nanoadditives are computed.

3.1.1 Biolubricants

The depletion of fossil fuel resources for energy purposes has led to a demand for alternative biodegradable oil feedstock and sustainable additives. The biolubricants selected should be environmentally friendly. Nagendramma and Kaul [7] reported that vegetable oils have higher flashpoints, have a viscosity index, are less toxic, and are characterized by lubricity. They are also biodegradable. Coconut oil [8], sesame oil [9], rice bran oil [10], date seed oil [11], castor oil [12], jathropa oil [13] and rubber seed oil [14] have all been cited by previous studies as potentially environment-friendly feedstocks for biolubricants. These vegetable oil lubricants are biodegradable, nontoxic, environment-friendly, adherent to metals, and have improved tribological properties. The major disadvantages with biolubricants being poor oxidation stability and high viscosity at lower temperatures [7]. The nanoparticle additives to vegetable oils can overcome these disadvantages. Several researchers have reported the use of nanomaterial additives in biolubricants. Thottackkad et al. [15] investigated the addition of CuO to the coconut oil-enhanced tribological properties of biolubricants. Gemsprim et al. [16] discussed the enhanced tribological evaluation of blends of vegetable oils. Nair et al. [17] studied the effect of tribological properties with TiO_2 and ZnO in sesame oil. Ahmed et al. [18] reported improved tribological properties of date seed and castor oil blends with halloysite clay nanotubes as additives.

3.1.2 Couple Stress Fluids

The bearing performance characteristics in hydrodynamic lubrication are enhanced with the use of lubricants with additives, such as couple stress fluids which take into account the additive properties in lubricants. The journal-bearing performance characteristics are widely obtained using micro-continuum theory based on a simplified couple stress fluid model [19]. The couple stress fluid-lubricated journal bearing were widely investigated in literature: analysis of couple stresses on finite journal bearing characteristics [20], couple stress fluid lubrication with the elastic deformation of finite journal bearing liner [21], porous media and couple stress models on thin film lubrication of journal bearing [22]. The performance characteristics of journal bearing are significantly improved by taking into consideration of surface-adsorbent layers and couple stress fluids.

3.1.3 Nanoparticle Additive Lubricants

The viscosity plays an important role in the load-carrying capacity of the journal bearings. The addition of nanoparticles to biolubricants enhances the viscosity, which also alters the performance characteristics of journal bearings. Based on Einstein's work applicable to relatively low particle volume fraction, the viscosity of spherical particle suspensions in fluids has been derived. Einstein's equation is extended by Brinkman [23] to a more generalized form considering moderate particle volume fraction. The Brownian motion effects are considered by Batchelor [24] to predict the viscosity of spherically-shaped nanoparticle additives. The performance characteristics of fluid film journal bearings with nanoparticle additive lubricants are enhanced due to the higher viscosity of nanofluids compared to base fluids. The enhancement in the performance characteristics of hydrodynamic journal bearing with nanoparticle additive lubricants are investigated [25].

3.1.4 Three-Layered Journal Bearing

The structure and properties of fluid film have a significant impact on the load capacity and the coefficient of friction in a three-layered journal-bearing lubrication. A composite-film journal bearing that consists of immiscible high/low viscosity layers adjacent to bearing/journal surfaces respectively offers an approach to the friction reduction in hydrodynamic bearings [26]. A rheological model applicable for thin films includes the analysis of adsorbent layer at solid surfaces of higher viscosity than a base core film [27]. Based on the generalized Reynolds equation for thin films of adsorbent layer at solid surfaces with a base core film [27, 28], higher load capacity and lower friction coefficient are obtained in a journal bearing. The nanoadditive couple stress fluids with the surface-adsorbent layer increases nondimensional load capacity and reduces the coefficient of friction [29]. The influence of nanoadditive couple stress fluids with the surface-adsorbent layer is of considerable interest in the analysis of hydrodynamic lubrication.

This study presents an analysis of load capacity and the coefficient of friction in a three-layered journal bearing with nanoparticle additive couple stress fluids. The authors derive a modified form of the Reynolds equation using different bearing/journal surface-adsorbent fluid film layer thickness ratios. The nondimensional pressure and shear stress expressions are obtained from the integration of modified Reynolds equation based on the Reynolds boundary conditions. The nondimensional load capacity and coefficient of friction are analyzed for: (i) journal eccentricity ratio (ε); (ii) couple stress parameter (λ); (iii) dynamic viscosity ratio of bearing/journal surface to base (core) layer (β_{s1}/β_{s3}); (iv) bearing/journal surface-adsorbent fluid film layer thickness ratio (Δ_1/Δ_2); and (v) nanoparticle volume fraction (ψ).

3.2 METHODOLOGY

The journal bearing schematic with three-layered nanoadditive couple stress fluids is shown in Figure 3.1. The couple stress fluid film region I: $0 \le y \le \delta_1$ (adsorbent layers) and the couple stress fluid film region III: $h - \delta_2 \le y \le h$ (adsorbent layers) are of higher viscosity than couple stress fluid film region II: $\delta_1 \le y \le h - \delta_2$ (core layer) $\left(\beta_{s1} = \frac{\mu_{s1}}{\mu_c}, \beta_{s3} = \frac{\mu_{s3}}{\mu_c}, \mu_c = \mu_n \text{ and } \beta_n = \frac{\mu_n}{\mu_f}\right)$. The assumptions in the one-dimensional

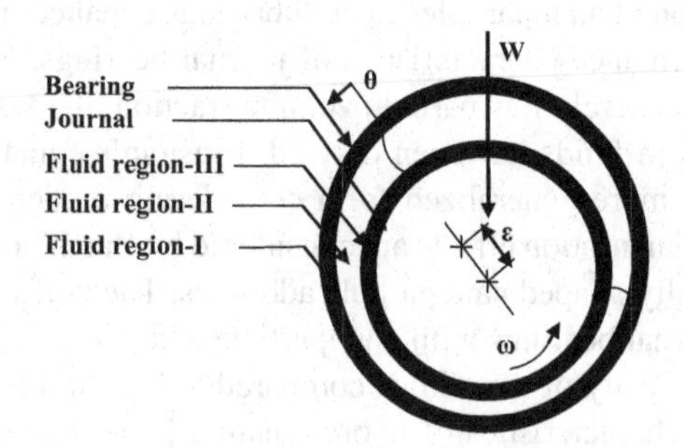

FIGURE 3.1 Journal bearing with three-layered nanoadditive couple stress fluids.

analysis are as follows: the pressure in the journal bearing is a function of sliding direction and the variation of pressure across the fluid film is assumed to be negligible.

3.2.1 Modified Reynolds Equation

Based on the couple stress fluid theory, the simplified momentum equation for $0 \le y \le \delta_1$ for $j = 1$, $\delta_1 \le y \le h - \delta_2$ for $j = 2$ and $h - \delta_2 \le y \le h$ for $j = 3$ is

$$\frac{1}{\eta}\frac{dp}{dx} = \frac{\mu_j}{\eta}\frac{d^2u_j}{dy^2} - \frac{d^4u_j}{dy^4} \tag{3.1}$$

The boundary conditions for velocity used in the analysis are

$$y = 0 \text{ at the at the bearing surface: } u_1 = 0 \text{ and } \frac{d^2u_1}{dy^2} = 0 \tag{3.2}$$

$$y = \delta_1 \text{ at the interface of bearing adsorbent and core layer: } u_1 = u_2 = u_{12},$$
$$\mu_1\frac{du_1}{dy} = \mu_2\frac{du_2}{dy}, \frac{d^2u_1}{dy^2} = 0 \text{ and } \frac{d^2u_2}{dy^2} = 0 \tag{3.3}$$

$$y = h - \delta_2 \text{ at the interface of the core and the journal adsorbent layer:}$$
$$u_2 = u_3 = u_{23} \text{ and } \mu_2\frac{du_2}{dy} = \mu\frac{du_3}{dy}, \frac{d^2u_2}{dy^2} = 0 \text{ and } \frac{d^2u_3}{dy^2} = 0 \tag{3.4}$$

$$y = h \text{ at the journal surface: } u_3 = u_j \text{ and } \frac{d^2u_3}{dy^2} = 0 \tag{3.5}$$

Based on Batchelor's [24] model, the dynamic viscosity ratio of nanoparticle additive fluid to base fluid is

$$\beta_n = \frac{\mu_n}{\mu_f} = \left(1 + 2.5\psi + 6.5\psi^2\right) \tag{3.6}$$

Integrating Equation (3.1) using the boundary conditions in Equations (3.2)–(3.5), the at the non-dimensional velocity distribution are expressed as

$$0 \le Y \le \Delta_1 \text{ in bearing adsorbent layer:}$$
$$U_1 = U_{12}\frac{Y}{\Delta_1} + \frac{1}{2\beta_{s1}\beta_n}\frac{dP}{d\theta}Y\left(Y - \Delta_1\right) + \lambda_{s1}^2\frac{1}{\beta_{s1}\beta_n} \tag{3.7}$$

$\Delta_1 \le Y \le H-\Delta_2$ in core layer:

$$U_2 = U_{12} + \left(U_{23}-U_{12}\right)\left(\frac{Y-\Delta_1}{H-\Delta_1-\Delta_2}\right) + \lambda_n^2 \frac{1}{\beta_n}\frac{dP}{d\theta}C_2 + \frac{1}{2\beta_n}\frac{dP}{d\theta}\left(Y-\Delta_1\right)\left(Y-H+\Delta_2\right) \tag{3.8}$$

$H-\Delta_2 \le Y \le H$ in journal adsorbent layer:

$$U_3 = U_{23} + \left(1-U_{23}\right)\left(\frac{Y-H+\Delta_2}{\Delta_2}\right) + \frac{1}{2\beta_{s3}\beta_n}\frac{dP}{d\theta}\left(Y-\mathrm{H}\right)\left(Y-H+\Delta_2\right) + \lambda_{s3}^2 \frac{1}{\beta_{s3}\beta_n}\frac{dP}{d\theta}C_3 \tag{3.9}$$

where

$$C_1 = \left[1 + \frac{\sinh\left(\frac{Y-\Delta_1}{\lambda_{s1}}\right) - \sinh\left(\frac{Y}{\lambda_{s1}}\right)}{\sinh\left(\frac{\Delta_1}{\lambda_{s1}}\right)}\right],$$

$$C_2 = \left[1 + \frac{\sinh\left(\frac{Y-H+\Delta_2}{\lambda_n}\right) - \sinh\left(\frac{Y-\Delta_1}{\lambda_n}\right)}{\sinh\left(\frac{H-\Delta_1-\Delta_2}{\lambda_n}\right)}\right], \tag{3.10}$$

$$C_3 = \left[1 + \frac{\sinh\left(\frac{Y-H}{\lambda_{s3}}\right) - \sinh\left(\frac{Y-H+\Delta_2}{\lambda_{s3}}\right)}{\sinh\left(\frac{\Delta_2}{\lambda_{s3}}\right)}\right]$$

$$U_{12} = F_1 - \frac{1}{\beta_n}\frac{dP}{d\theta}F_2, U_{23} = F_3 - \frac{1}{\beta_n}\frac{dP}{d\theta}F_4 \tag{3.11}$$

$$F_1 = \frac{-E_{12}E_{231}}{E_{11}E_{22}-E_{12}E_{21}}, F_2 = \frac{E_{22}E_{13}-E_{12}E_{232}}{E_{11}E_{22}-E_{12}E_{21}},$$
$$F_3 = \frac{E_{11}E_{231}}{E_{11}E_{22}-E_{12}E_{21}}, F_4 = \frac{-E_{21}E_{13}+E_{11}E_{232}}{E_{11}E_{22}-E_{12}E_{21}} \tag{3.12}$$

$$E_{11} = \frac{\beta_{s1}}{\Delta_1} + \frac{1}{(H-\Delta_1-\Delta_2)}, E_{12} = E_{21} = -\frac{1}{(H-\Delta_1-\Delta_2)},$$
$$E_{22} = \frac{\beta_{s3}}{\Delta_2} + \frac{1}{(H-\Delta_1-\Delta_2)}, E_{13} = -\lambda_{s1}H_1^* - \lambda_n H_2^* + \frac{1}{2}(H-\Delta_2), \tag{3.13}$$
$$E_{231} = \frac{\beta_s}{\Delta}, E_{232} = -\lambda_{s3}H_3^* - \lambda_n H_2^* + \frac{1}{2}(H-\Delta_1)$$

$$H_1^* = \left[\coth\left(\frac{\Delta_1}{\lambda_{s1}}\right) - \operatorname{csch}\left(\frac{\Delta_1}{\lambda_{s1}}\right)\right],$$
$$H_2^* = \left[\coth\left(\frac{H-\Delta_1-\Delta_2}{\lambda_n}\right) - \operatorname{csch}\left(\frac{H-\Delta_1-\Delta_2}{\lambda_n}\right)\right], \tag{3.14}$$
$$H_3^* = \left[\coth\left(\frac{\Delta_2}{\lambda_{s3}}\right) - \operatorname{csch}\left(\frac{\Delta_2}{\lambda_{s3}}\right)\right]$$

$$\lambda_{s1} = \frac{\lambda}{\sqrt{\beta_{s1}\beta_n}}, \lambda_{s3} = \frac{\lambda}{\sqrt{\beta_{s3}\beta_n}}, \lambda_n = \frac{\lambda}{\sqrt{\beta_n}} \tag{3.15}$$

$$H = 1 + \varepsilon \cos\theta \tag{3.16}$$

The continuity equation across the film is

$$Q = \int_0^{\Delta} U_1 dY + \int_{\Delta}^{H-\Delta} U_2 dY + \int_{H-\Delta}^{H} U_3 dY \tag{3.17}$$

The continuity equation across the film in Equation (3.17) is simplified as

$$\frac{dP}{d\theta} = \beta_n \left(\frac{G_1 - Q}{G_2}\right) \tag{3.18}$$

where

$$G_1 = \frac{1}{2}\Delta_2 + \frac{1}{2}F_1(H-\Delta_2) + \frac{1}{2}F_3(H-\Delta_1) \tag{3.19}$$

$$G_2 = \frac{1}{2}F_2(H-\Delta_2) + \frac{1}{2}F_4(H-\Delta_1) + \frac{\Delta_1^3}{12\beta_{s1}} + \frac{(H-\Delta_1-\Delta_2)^3}{12} + \frac{\Delta_2^3}{12\beta_{s3}} - \frac{\lambda_{s1}^2\Delta_1}{\beta_{s1}}$$
$$-\lambda_n^2(H-\Delta_1-\Delta_2) - \frac{\lambda_{s3}^2\Delta_2}{\beta_{s3}} + \frac{2\lambda_{s1}^3 H_1^*}{\beta_{s1}} + 2\lambda_n^3 H_2^* + \frac{2\lambda_{s3}^3 H_3^*}{\beta_{s3}} \tag{3.20}$$

3.2.2 Load Capacity

The Reynolds boundary conditions are used in the analysis. Integrating Equation (3.18) and substituting the Reynolds boundary conditions for nondimensional pressure at film reformation ($P|_{\theta=0} = 0$) yields the nondimensional pressure profile as

$$P = \beta_n \left(\int_0^\theta \frac{G_1}{G_2} d\theta - Q \int_0^\theta \frac{1}{G_2} d\theta \right) \tag{3.21}$$

Substitution of the Reynolds boundary conditions for nondimensional pressure at film rupture ($P|_{\theta=\theta r} = 0$) in Equation (3.21) and simplifying results in Q as

$$Q = \frac{\int_0^{\theta_r} \frac{G_1}{G_2} d\theta}{\int_{\theta_g}^{\theta_r} \frac{1}{G_2} d\theta} \tag{3.22}$$

Substituting the Reynolds boundary conditions for nondimensional pressure at film rupture $\left(\frac{dP}{d\theta} \Big|_{\theta=\theta_r} = 0 \right)$ in the expression for nondimensional pressure gradient in Equation (3.18), results in

$$Q = G_1|_{\theta=\theta_r} \tag{3.23}$$

The Newton-Raphson iterative procedure is used to solve simultaneously both θ_r and Q using Equations (3.22) and (3.23).

The nondimensional load capacity is

$$W = \sqrt{W_\varepsilon^2 + W_\phi^2} \tag{3.24}$$

where $W_\varepsilon = -\int_0^{\theta_r} P \cos\theta \, d\theta$ and $W_\phi = \int_0^{\theta_r} P \sin\theta \, d\theta$

3.2.3 Coefficient of Friction

The nondimensional shear stress in the journal bearing at $Y = H$ is obtained as

$$\begin{aligned} \Pi\big|_{Y=H} &= \beta_s \beta_n \frac{dU_3}{dY}\bigg|_{Y=H} \\ &= \beta_n \left[(1 - F_3) \left(\frac{\beta_{s3}}{\Delta_2} \right) + \left(\frac{G_1 - Q}{G_2} \right) \left(F_4 \frac{\beta_{s3}}{\Delta_2} + \frac{\Delta_2}{2} - \lambda_{s3} H_3^* \right) \right] \end{aligned} \tag{3.25}$$

The nondimensional friction force is

$$F = \int_{0}^{\theta_r} \Pi \, d\theta \tag{3.26}$$

The nondimensional friction coefficient is $C_f = \left(\frac{R}{C}\right)\frac{f}{w} = \frac{F}{W}$.

3.3 RESULTS AND DISCUSSION

An analysis of journal bearing with three-layered nanoadditive couple stress fluids is presented. The parameters included in the investigations are as follows: journal eccentricity ratio (ε) = 0.1–0.5; couple stress parameter (λ) = 0.05–0.2; dynamic viscosity ratio of bearing/journal surface to base (core) layer (β_{s1}/β_{s3}) = 1–4; bearing/journal surface-adsorbent fluid film layer thickness ratio (Δ_1/Δ_2) = 0.05–0.2; and nanoparticle volume fraction (ψ) = 0.0–0.04. The influence of journal eccentricity ratio (ε), dynamic viscosity ratio of bearing/journal surface to base (core) layer (β_{s1}/β_{s3}), couple stress parameter (λ), and bearing/journal surface-adsorbent fluid film layer thickness ratio (Δ_1/Δ_2) on the nondimensional load capacity and coefficient of friction are analyzed.

Figures 3.2(a–d) show the nondimensional load capacity (W) of a journal bearing with three-layered nanoadditive couple stress fluids. Figures 3.2(a and b) show that the nondimensional load capacity significantly increase with increasing journal eccentricity ratio (ε) and couple stress parameter (λ). The nondimensional load capacity (W) increases (i) substantially with increasing couple stress parameter (λ) from 0.1 to 0.3, and (ii) marginally with increasing nanoparticle volume fraction (ψ) from 0.0 to 0.04. Figures 3.2(c and d) show that the nondimensional load capacity (W) increase with increasing journal eccentricity ratio (ε) and bearing/journal surface-adsorbent fluid film layer thickness ratio (Δ_1/Δ_2). The influence of dynamic viscosity ratio of bearing/journal surface to base (core) layer (β_{s1}/β_{s3}) on the enhancement in the nondimensional load capacity (W) is higher at higher journal eccentricity ratio (ε) and higher bearing/journal surface-adsorbent fluid film layer thickness ratio (Δ_1/Δ_2). However, the bearing/journal surface-adsorbent fluid film layer thickness ratio (Δ_1/Δ_2) configuration with similar values of dynamic viscosity ratio of bearing/journal surface to base (core) layer (β_{s1}/β_{s3}) has an identical influence on the enhancement in the nondimensional load capacity (W).

Figures 3.3(a–d) show the coefficient of friction (C_f) of a journal bearing with three-layered nanoadditive couple stress fluids. Figures 3.3(a and b) show that the coefficient of friction (C_f) decrease significantly with increasing journal eccentricity ratio (ε) and couple stress parameter (λ). Figures 3.3(c and d) show that the coefficient of friction (C_f) decrease with increasing journal eccentricity ratio (ε), bearing/journal surface-adsorbent fluid film layer thickness ratio (Δ_1/Δ_2), and dynamic viscosity ratio of bearing/journal surface to base (core) layer (β_{s1}/β_{s3}). The reduction in the coefficient of friction (C_f) is higher with a higher journal eccentricity ratio (ε) and

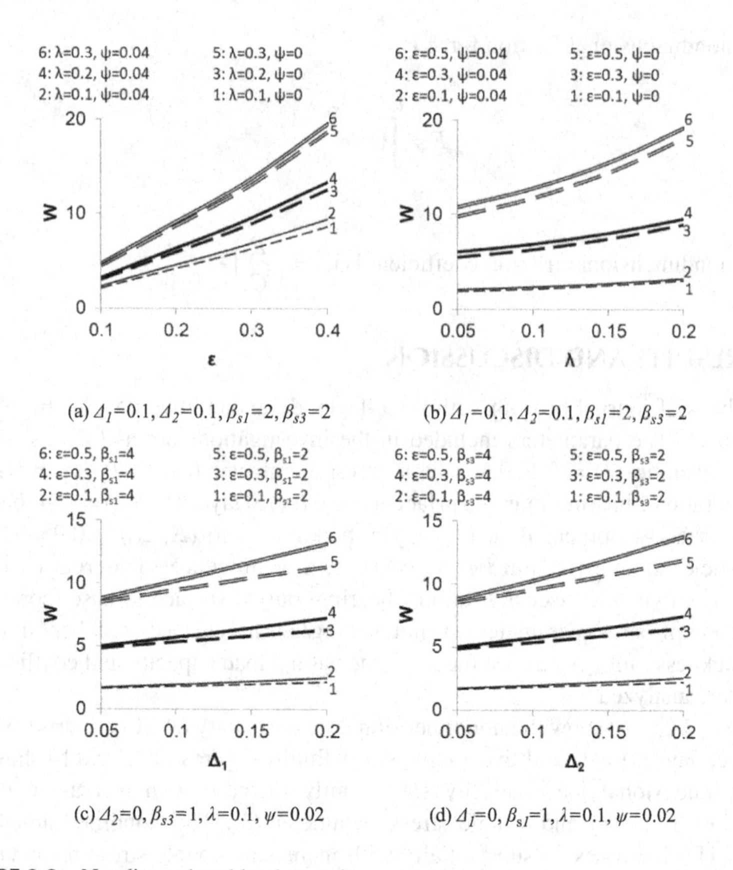

FIGURE 3.2 Nondimensional load capacity.

higher bearing/journal surface-adsorbent fluid film layer thickness ratio (Δ_1/Δ_2). Similar to the nondimensional load capacity (W) characteristics, identical values of the coefficient of friction (C_f) are obtained for the bearing/journal surface-adsorbent fluid film layer thickness ratio (Δ_1/Δ_2) configuration with similar values of dynamic viscosity ratio of bearing/journal surface to base (core) layer (β_{s1}/β_{s3}). The coefficient of friction (C_f) is not influenced by the increase in nanoparticle volume fraction (ψ) from 0.0–0.04. The coefficient of friction (C_f) is not affected by nanoparticle volume fraction (ψ) [29], as both nondimensional pressure and nondimensional shear stress increase by the magnitude of β_n for Newtonian fluids ($\lambda = 0$).

3.4 CONCLUSION

The present study evaluates bionanolubricants for journal bearing with three-layered nanoadditive couple stress fluids. The enhancement in nondimensional load capacity and the reduction in coefficient of friction are presented. The Reynolds boundary conditions are used in the long bearing analysis of journal bearing with three-layered nanoadditive couple stress fluids. The nondimensional load capacity (W) increases

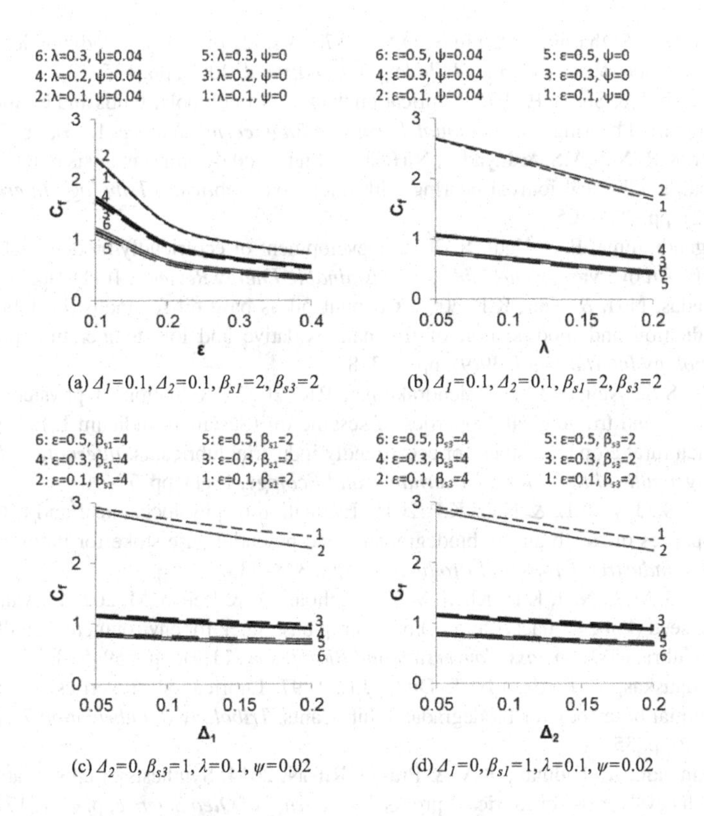

FIGURE 3.3 Coefficient of friction.

(and the coefficient of friction (C_f) decreases) substantially with increasing couple stress parameter (λ), bearing/journal surface-adsorbent fluid film layer thickness ratio (Δ_1/Δ_2), and the dynamic viscosity ratio of bearing/journal surface to base (core) layer (β_{s1}/β_{s3}). The nanoparticle volume fraction (ψ) has a marginal influence on increasing nondimensional load capacity (W). The nanoparticle volume fraction (ψ) has negligible influence on reduction in the coefficient of friction (C_f).

Bionanolubricants for journal bearing replicating three-layered configuration with nanoadditive couple stress fluids has the potential to enhance the load capacity and reduce the coefficient of friction.

REFERENCES

1. Soni, S.C., Sinhasan, R. & Singh, D.V. 1983. Analysis by the finite element method of hydrodynamic bearings operating in the laminar and superlaminar regimes. *Wear*, 84(3), pp. 285–296.
2. Tayal, S.P., Sinhasan, R. & Singh, D.V. 1982. Analysis of hydrodynamic journal bearings having non-Newtonian lubricants. *Tribology International*, 15(1), pp. 17–21.
3. Nair, K.P., Sinhasan, R. & Singh, D.V. 1987a. Elastohydrodynamic effects in elliptical bearings. *Wear*, 118(2), pp. 129–145.

4. Nair, K.P., Sinhasan, R. & Singh, D.V. 1987b. A study of elastohydrodynamic effects in a three-lobe journal bearing. *Tribology International*, 20(3), pp. 125–132.
5. Prakash, J. & Sinha, P. 1975. Lubrication theory for micropolar fluids and its application to a journal bearing. *International Journal of Engineering Science*, 13(3), pp. 217–232.
6. Nair, K.P., Nair, V.S. & Jayadas, N.H. 2007. Static and dynamic analysis of elastohydrodynamic elliptical journal bearing with micropolar lubricant. *Tribology International*, 40(2), pp. 297–305.
7. Nagendramma, P. & Kaul, S. 2012. Development of ecofriendly/biodegradable lubricants: An overview. *Renewable and Sustainable Energy Reviews*, 16(1), pp. 764–774.
8. Jayadas, N.H. & Nair, K.P. 2006. Coconut oil as base oil for industrial lubricants—Evaluation and modification of thermal, oxidative and low temperature properties. *Tribology International*, 39(9), pp. 873–878.
9. Nair, S.S., Nair, K.P. & Rajendrakumar, P.K. 2017. Evaluation of physicochemical, thermal and tribological properties of sesame oil (Sesamum indicum L.): A potential agricultural crop base stock for eco-friendly industrial lubricants. *International Journal of Agricultural Resources, Governance and Ecology*, 13(1), pp. 77–90.
10. Rani, S., Joy, M.L. & Nair, K.P. 2015. Evaluation of physiochemical and tribological properties of rice bran oil–biodegradable and potential base stoke for industrial lubricants. *Industrial Crops and Products*, 65, pp. 328–333.
11. Ahmed, M.S., Nair, K.P., Khan, M.S., Algahtani, A. & Rehan, M. 2021a. Evaluation of date seed (Phoenix dactylifera L.) oil as crop base stock for environment friendly industrial lubricants. *Biomass Conversion and Biorefinery*, 11(2), pp. 559–568.
12. Asadauskas, S., Perez, J.H. & Duda, J.L. 1997. Lubrication properties of castor oil-potential basestock for biodegradable lubricants. *Tribology & Lubrication Technology*, 53(12), p. 35.
13. Sammaiah, A., Padmaja, K.V. & Prasad, R.B.N. 2014. Synthesis of epoxy jatropha oil and its evaluation for lubricant properties. *Journal of Oleo Science*, p. ess13172.
14. Aravind, A., Joy, M.L. & Nair, K.P. 2015. Lubricant properties of biodegradable rubber tree seed (Hevea brasiliensis Muell. Arg) oil. *Industrial Crops and Products*, 74, pp. 14–19.
15. Thottackkad, M.V., Perikinalil, R.K. & Kumarapillai, P.N. 2012. Experimental evaluation on the tribological properties of coconut oil by the addition of CuO nanoparticles. *International Journal of Precision Engineering and Manufacturing*, 13(1), pp. 111–116.
16 Gemsprim, M.S., Babu, N. & Udhayakumar, S. 2020. Tribological evaluation of vegetable oil-based lubricant blends. *Materials Science*, 2214, p. 7853.
17. Nair, S.S., Nair, K.P. & Rajendrakumar, P.K. 2018. Micro and nanoparticles blended sesame oil bio-lubricant: Study of its tribological and rheological properties. *Micro & Nano Letters*, 13(12), pp. 1743–1746.
18. Ahmed, M.S., Nair, K.P., Tirth, V., Elkhaleefa, A. & Rehan, M. 2021b. Tribological evaluation of date seed oil and castor oil blends with halloysite nanotube additives as environment friendly bio-lubricants. *Biomass Conversion and Biorefinery*, https://doi.org/10.1007/s13399-021-02020-9
19. Stokes, V.K. 1966. Couple stresses in fluids. *Physics of Fluids*, 9, pp. 1709–1715.
20. Lin, J.-R. 1997. Effects of couple stresses on the lubrication of finite journal bearings. *Wear*, 206, pp. 171–178.
21. Mokhiamer, U.M., Crosby, W.A. & El-Gamal, H.A. 1999. A study of a journal bearing lubricated by fluids with couple stress considering the elasticity of the liner. *Wear*, 224, pp. 194–201.
22. Li, W.-L. & Chu, H.-M. 2004. Modified Reynolds equation for couple stress fluids – A porous media model. *Acta Mechanica*, 171, pp. 189–202.

23. Brinkman, H.C. 1952. The viscosity of concentrated suspensions and solution. *The Journal of Chemical Physics*, 20, pp. 571–581.
24. Batchelor, G.K. 1977. The effect of Brownian motion on the bulk stress in a suspension of spherical particles. *Journal of Fluid Mechanics*, 83, pp. 97–117.
25. Nair, K.P., Ahmed, M.S. & Al-Qahtani, S.T. 2009. Static and dynamic analysis of hydrodynamic journal bearing operating under nano lubricants. *International Journal of Nanoparticles*, 2, 251–262.
26. Szeri, A.Z. 2010. Composite-film hydrodynamic bearings. *International Journal of Engineering Science*, 48, pp. 1622–1632.
27. Tichy, J.A. 1995. A surface layer model for thin film lubrication. *Tribology Transactions*, 38, pp. 577–582.
28. Meurisse, M.-H. & Espejel, G.M. 2008. Reynolds equation, apparent slip, and viscous friction in a three-layered fluid film. *Proceedings of the Institution of Mechanical Engineers, Part J: Journal of Engineering Tribology*, 222, pp. 369–380.
29. Rao, T.V.V.L.N., Sufian, S. & Mohamed N.M. 2013. Analysis of nanoparticle additive couple stress fluids in three-layered journal bearing. *Journal of Physics: Conference Series*, 431(1), p. 012023.

4 Alternative Industrial Biolubricants
Possible Future for Lubrication

Ponnekanti Nagendramma

CSIR-Indian Institute of Petroleum, Dehradun, India

CONTENTS

DOI: 10.1201/9781003243205-4

4.1 INTRODUCTION

4.1.1 Historical Origin of Industrial Lubricants

Sir Isaac Newton proposed the idea of lubrication in 1687, which served as the foundation for hydrodynamic lubrication theory. Since the discovery of coal oil in the nineteenth century, lubrication has been a significant advance in the industry. The need for industrial lubricants mirrored the industrial revolution, which began in England in the 1700s and expanded across the Atlantic to America. Lubrication systems were first found in ancient Egypt when olive oil was used to transport heavy goods. The first oil well was dug at Titusville, Pennsylvania, in 1859, an event which ushered in the petroleum era. Scientists found how to refine oil and add additives to it to make it lubricate better in the 1920s. In the 1930s, additives that would protect against corrosion, increase pour points, and improve lubricity and viscosity were developed. Synthetic oils were developed in the 1950s and were widely employed in the newly expanding aviation and aerospace sectors [1, 2].

4.1.2 Biodegradable Industrial Lubricants

Severe environmental health issues caused by different lubricants have recently led to an increase in public attention. As a result of increased industrialization and technological innovation, markets are flooded with daily lubricants that make life more comfortable. The lubrication industry is becoming concerned about the massive amount of lubricants being emitted into the environment. The uncontrolled loss of lubricants can damage the environment and deplete natural resources. Lubricants can endanger the environment if they directly contact it due to inappropriate handling, storage, and usage. Over the last 25 years, several governments and large corporations have made enormous efforts to create and find more eco-friendly products and technologies that have reduced the negative impact on our precious environment [3].

Government incentives, ordinances, the implementation of green seals, public awareness, and the imposition of essential limitations have all been used to pressure industries that discharge lubricants into the environment to launch and extend the use of biodegradable lubricants. The first seal, known as the Blue Angel, was issued by Germany. Similarly, a White Swan is Scandinavia's green seal, the Green Cross is America's, and Ecomark is India's and Japan's. These eco-labeling schemes contain varying and constantly changed ecological test criteria, limits, and manufacturer disclosures. Another reason for the use of biodegradable lubricants is the globalization of the lubricant market and the rising use of biodegradable lubricants in ecologically sensitive locations [4].

Biodegradable industrial lubricants include cutting lubricants, turbine lubricants, and hydraulic and metalworking oils. Best lubrication practices are essential for meeting regional quality standards and reducing lubricant usage. Lubricants are becoming increasingly significant as government rules addressing emission limits, natural resource conservation, and disposal increase in stringency. Lubricants must be evaluated based on their ability to withstand higher thermal and mechanical stress and their low usage, biodiversity sustainability, and rapid biodegradability. These factors directly impact lubricant buyers as well as research and development [5].

The search for ecologically sound materials to replace mineral oil in various industrial applications has been identified as an essential priority in the fuel and energy sector. The lubricant industry has spent the last century focused on developing mineral oil-based industrial lubricating fluids. There is ongoing research on sustainable resources like vegetable seed oils in addition to using synthetic esters as viable mineral oil base stock replacements within certain industrial lubricant applications wherein a steady environmental interaction is expected. Another reason to avoid mineral oils is their usage at high temperatures, which increases the risk of fire and incompatibility. Furthermore, natural oils and synthetics offer a number of advantages: these include higher efficiency, fewer deposits, superior radiation resistance, lower maintenance, and longer fluid life. Esters have been discovered to have considerable environmental benefits. These tiny molecules are low in toxicity, generate minimal levels of pollution, and degrade quickly in the environment. Ester-based lubricants are currently widely used in many areas of the globe [6]. The first eco-friendly and biodegradable lubricants were developed in the mid-1970s and early 1980s. It all started with biodegradable lubricants in outboard engines and chain saws [7].

4.1.3 Biodegradable Industrial Lubricants on the Market

Several laws and regulations on petroleum products were imposed in Europe throughout the 1980s, requiring the use of biodegradable lubricants. Few biodegradable lubricants belonging to synthetics and vegetable oils within European market are listed in Table 4.1.

In the 1990s, several American corporations began producing biodegradable products. The Lubrizol Corporation also developed a number of additives and

TABLE 4.1
Synthetic Base Fluid and Their Applications with Trade Names

Manufacturer	Product	Origin	Applications
Castrol	Care lubes HTC	Triglyceride base oil	Hydraulic oils
	Care lubes HES	Ester based	Hydraulic oils
	Care lube GTG	Triglyceride	Gear oils
	Care lube GES	Ester based	Gear oils
Estimol	Esti Chem A/S	Trimethylol propane	Industrial
I.C.I.	Emkarox V.G.W.	Polyol ester	Functional fluids
Chemical &	Emkarox HV	Water-insoluble	Compressor lubricants,
Polymers	Emkarox VG	Polyalkylene glycol	M.W.F., Textile lubricants
	Emkarate	Water-soluble polyalkylene glycol	Fire-resistant hydraulic fluids,
		Water-insoluble polyalkylene glycol	aqueous quench fluids
		Diesters, polyesters, phthalate esters, trimellitate ester, and dimerates	Industrial oils and greases, mould release agents for rubber
Bechem	Hydrostar HEP	Synthetic ester based	Hydraulic fluids
	Hydrostar UWF	Polyethylene glycol-based	Hydraulic fluids
Mobil	Hydraulic E.A.L.	Synthetic ester-based	Hydraulic fluids

lubricants based on sunflower oil. Industrial hydraulic fluids, railway road and mechanical equipment greases, chainsaw oils, industrial gear oils, compressor oils, transformer and coolants are produced using soybean oil. Dual-cycled engine oils, metalworking fluids, as well as more special lubricants are now being tested in the field [8, 9].

Mobil developed high-performance biolubricants, wholly synthetic polyol ester compounds designed solely for the lubrication of refrigeration and air conditioning systems and are used by numerous major compressor and system manufacturers around the world [10, 11].

15 percent of total worldwide lubricant consumption was accounted for and by transformers, rubber, white oils, and printing inks. Other industrial oils include hydraulic fluids (9%), industrial gear oils, turbine oils, and refrigeration fluids (8%), industrial engine oils (7%), metalworking fluids (5%), and greases (3%). Synthetic lubricants will expand at a 4.5% annual pace over the next ten years, although from a tiny foundation [12]. Lubricant industries are performing successful work in eco-friendly synthetic lubricants. Vegetable oil-based lubricants are also gaining in popularity.

4.2 CURRENT STATUS OF THE INDUSTRIAL LUBRICANTS

4.2.1 International Status

According to the most current literature report on biodegradable lubricants, few manufacturers are presently marketing environmentally friendly biodegradable lubricants in the United States, Asia, or Europe. The University of Northern Iowa has developed vegetable oil-based wheel greases, hydraulic multi-grade fluids, transformer electric fluids, and chainsaw oils, as well as rail curve lubricants, under the brand names Soy TRUCK™, BioSOY™, BioTRANS™, SoyLINK™, and SoyTrak™. In addition, the United States of America has demonstrated commercially successful technologies. Since 2002, recyclable soybean-based hydraulic fluid has been utilized in the Statue of Liberty's elevator. Another technical breakthrough is a soy-based hot spinning lubricant used by a large aluminium manufacturing company for hot flat spinning operations on everything from wine bottles to fixed-wing aircraft panels. Bio-based products have long been used successfully in metal working fluids, truck oils, hydraulic and lubricating greases throughout Europe [13–16]. Numerous benefits of recyclable lubes and greases will become evident as income rises, technology progresses, and awareness grows, and the market for biobased lubricants expands.

The Exxon Mobil, Conoco Phillips, Chevron, Valvoline, Hartco, Amsoil from the United States, Nippon Oil, Idemitsu from Japan, Repsol from Spain, Agip from Italy, Lukoil from Russia, Shell from the United Kingdom, Sinopec from China, Total from France, Fuchs from Germany, and Indian oil from India are among the major manufacturers of industrial lubricants [17, 18].

4.2.2 Indian Scenario

The Indian lubricant business is one of the country's fastest-growing retail industries. Since 1998, the Indian Institute of Petroleum (IIP) has had a number of groups working on biolubricant synthesis [19–22] Dehradun, Hindustan Petroleum, Apar

Industries, Bharat Petroleum, Castrol India, Gulf Oil, Indian Oil, Savita Chemicals, and Tide Water are among the nation's lubricant makers. Castrol, Elf Total-Fina, Gulf, and Shell Oil have all established themselves in the market. The presence of private corporations has grown in recent years, as multinational firms have increased their engagement in the Indian market. Engine oil accounts overall for 83 percent of overall sales volumes. Gear oils, transmission fluids, hydraulic braking fluids, and engine coolants have an influence on the overall balance.

4.2.3 World Industrial Lubricant Demand

According to recent market analysis, the industrial lubricant market is predicted to reach $73.3 billion by 2024, having grown at a 3.1% annual pace from $62.8 billion in 2019. Massive industrialization, rising disposable income and rapid urbanization in developing nations are among the market's driving forces. Mineral oil is the industry's most significant section, while the construction and hydraulic fluid segments account for the highest market share in 2019. In terms of volume and value, the Asia-Pacific region remains the largest market for industrial lubricants. The region's growing population and expanding construction expenditures in the emerging economies of China, India, and Indonesia are among the primary factors expected to increase demand for industrial lubricants.

Improved lifestyles, higher employment rates, higher disposable income, and greater foreign investment in different economic sectors are among the key elements that make the Asia-Pacific area appealing to industrial lubricant makers.

Shell, ExxonMobil, Chevron, BP, Total, Petro China Co., Sinopec, Lukoil, Fuchs Petrolub, Indian oil corporation, Idemitsu Kosan Co., and others are among the market's major participants [23].

4.3 ENVIRONMENTALLY ACCEPTABLE INDUSTRIAL LUBRICANTS

Based on the aforementioned environmental factors, the following explanation focuses on the two major types of ester-based products developed at our laboratory.

Lubricant performance, environmental, health, and safety regulations all had an influence on base oil development in the 1990s. As a result, more refined oils, such as poly alpha-olefins, hydrocracked, and esters oils, have become increasingly popular. Edible and inedible seed oils have resurfaced as a result of their rapid biodegradability. Throughout the twenty-first century, the tendency of overall development programme and even higher compatibility will continue [24].

Advanced formulations of industrial lubricants are being tailored to fit the present needs of industrial machinery for productivity increases, efficiency perfection, clean energy, and environmental protection. Mineral oils, vegetable oils, fats, and greases were the only lubricants available in the twentieth century. Solid lubricants, synthetic oils, water-based lubricants, and gas-based lubricants have recently been introduced to the lubricant concept [25, 26].

A lubricant is a material that is used to minimize surface friction and wear. Lubricants can be in the form of a liquid, solid, gas, or a combination of all three states. The primary function of industrial lubricants is to spread a fluid layer over solid surfaces. They are also involved in maintaining morphological structure, thermoregulation, and waste disposal.

Solid lubricants can reduce friction between two moving surfaces at extremely high temperatures and loads. The most common solid lubricants include graphite, molybdenum disulphide, boron nitride, and polytetrafluoroethylene. Semi-solid lubricants are composed of oils and other additives. They may also include solid PTFE, MoS2, and graphite lubricants to avoid corrosion and surface damage. Liquid lubricants such as vegetable, mineral, and synthetic oils are exploited in a wide variety of industrial applications [27].

Mineral oils are principally obtained through the refining of crude oil. These are complex blends of paraffin, aromatic, and naphthenic molecules with carbon numbers ranging from 14 to 40 or higher. These base oils also include trace amounts of heteromolecules such as oxygen, nitrogen, and sulphur. They are classified as paraffinic or naphthenic hydrocarbons based on their hydrocarbon class. White oils, electrical oils, and process oils are among the other base oils. The second type of oil is synthetic oil, composed of ester bonds and additives that alter their properties. Mineral oils do not have the same lubricating characteristics as synthetic oils [28]. In hydrodynamic journal bearings, mineral oils are often utilized. These lubricants are hazardous to the environment because they are poisonous and non-biodegradable. Synthetic and vegetable oil esters have been created as an alternative lubricant since they are biodegradable and non-toxic.

4.3.1 Synthetic Esters

Synthetic lubricants are establishing a position by marketing automotive, industrial, and aviation lubricants. More changes are expected in future decades, as challenges such as volatility, biodegradability, toxicity, and disposability, among others, will have a significant influence on the industry. Phosphate esters, polyalkylene glycols, plant oils, poly alpha olefins, alkylated aromatics, polybutenes, and other synthetics dominate the market. In the twentieth century, synthetic lubricants were made in response to the need for suitable lubrication for moving components under new and extreme conditions [29, 30].

Synthetic lubricants come in a wide range of formulations and applications. They fall within the general ASTM categorization depicted in Table 4.2 [31].

To fulfill the demands of highly efficient lubricants, new ester base fluids were developed. As a result, synthetic esters have become more important as base oil in many commercial and armed forces operations. Machine lubricants, gearbox oils, hydraulic oils, compressed air oils, and turbo oils are a few examples.

4.3.2 Organic Ester Lubricants

Organic esters are the most widely used synthetic lubricant, and their development has paralleled changes in aircraft engine layout in both the United Kingdom and the USA. Synthetic organic esters may be made to function practically in any industry. With the right esters, outstanding oxidation and hydrolytic stability, biodegradability,

TABLE 4.2
Classification of Synthetic Lubricants

Synthesized Hydrocarbons	Other Fluids	Organic Esters
Polybutenes	Perfluoro polyalkyl ethers	Dibasic acid esters
Polyisobutenes		
Olefin oligomers	Polyphenyl ethers	Polyol esters
Poly Alfa olefins	Silicone fluids	
Alkylated aromatics	Silicate esters	
Cycloaliphatics	Metha-acrylate	
	Poly alkylene glycols	
	Phosphate esters	
	Fluoro esters	

lubricity, excellent viscosity index, and good pour point characteristics will be attained [32].

Esters can be generated from mono-, di- or polyfunctional source materials that are linear, branched, saturated and unsaturated. There are thousands of carboxylic acid and alcohol structural components to select from, and the combinations are nearly limitless.

4.3.3 Esterification in the Production of Ester Lubricants

Figure 4.1 displays the fundamental chemical process that produces all esters. Esters are formed when organic acids and alcohols interact. A reversible reaction happens when the water molecule reacts with the ester group and breaks down into its components.

Alkyl groups can have the same chain length and structure, or they can have distinct chain lengths and structures. The acid can be monobasic or dibasic, while the alcohol can have several hydroxyl groups. The molecules are designed for a specific application to maximize stability, reduce volatility, keep pour points low, and give good additive susceptibility. Fluids with good viscosity-temperature properties and natural dispersant/detergent tendencies have the same features.

Despite the fact that components of unsaturated fatty acids must be removed, oleate is often utilized as lubricants. Oleate has several benefits, including lubricity, low volatility, cold flow, biodegradability, renewability, and cheap cost.

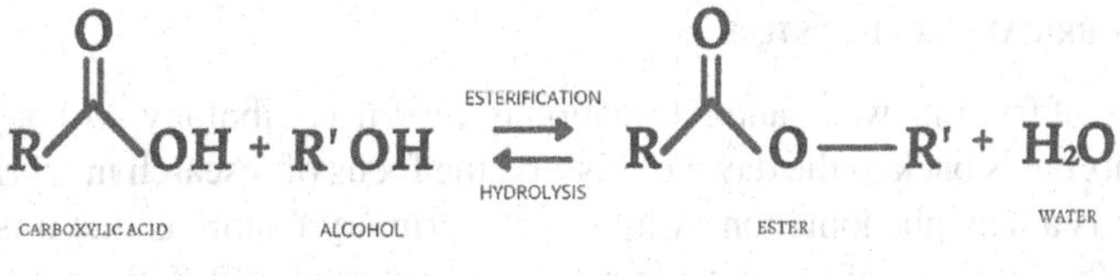

$$R\text{–}C(=O)\text{–}OH + R'OH \underset{\text{HYDROLYSIS}}{\overset{\text{ESTERIFICATION}}{\rightleftharpoons}} R\text{–}C(=O)\text{–}O\text{—}R' + H_2O$$

CARBOXYLIC ACID ALCOHOL ESTER WATER

FIGURE 4.1 Esterification reaction.

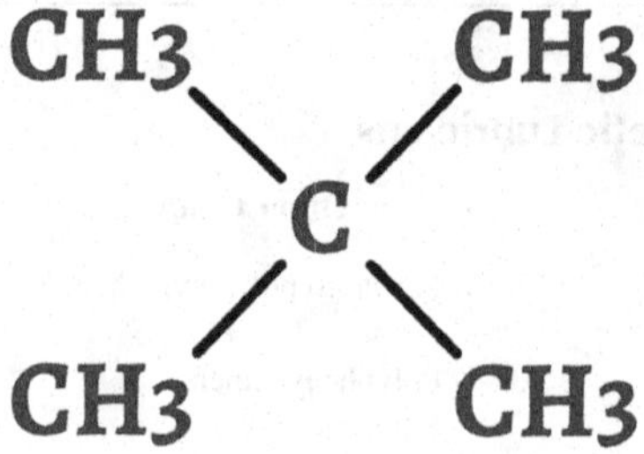

FIGURE 4.2 Neopentane structure.

4.3.4 The Advantages of Neopentane-Structured Polyol Esters

At elevated temperatures, saturated esters are necessary, but there is more to explore. The number and position of hydrogen on a molecule's beta-carbons have a significant impact on oxidative stability at extreme heat. The beta-carbon is the second carbon in the ester group's carbon–oxygen bond. Esters without beta-hydrogen is more thermally stable because beta-hydrogen is extremely reactive to oxygen. These are neopolyol esters, which are named after their structural similarity to neopentane. Because there are no beta-hydrogens in Polyol esters, they all have solid oxidative stability (Figure 4.2). This esters group depends primarily upon a unique five-carbon structure in the polyfunctional alcohol used for esterification with the selected monofunctional acids. This C-structure is neopentane, with a central carbon surrounded by four additional carbon atoms.

Although pure esters are not superior to aliphatic hydrocarbons or mineral oils as boundary lubricants, many esters' hydrolysis or oxidation products are effective, moderate wear preventives and rust inhibitors. As a result, when these esters are used, they develop good boundary-lubricating qualities. Because of their strong solvent qualities, they work well as detergent oils. Because esters are soluble in mineral oils, polyoxoalkylene-type oils, and even silicones, they can be used as coupling and blending ingredients to improve the properties of these oils. Diesters have a far higher thermal stability than mineral oils, with breakdown starting about 500°F and producing acidic decomposition products. It is, however, dependent on the structure of the ester's alcohol component. The oxidation stability of the esters is also better than that of highly refined petroleum fractions [33–35].

Synthetic esters may be made to work at rising temperatures and to evaporate completely before beginning oxidative polymerization, avoiding deposits and varnish formation, by removing the oxidized weak spots.

4.3.5 Lubrication via Esters

The science of friction, wear, and lubrication is known as tribology. Although the use of lubricants dates back to the dawn of history, the focus of research in lubricants and lubrication is a new phenomenon. A lubricant's primary feature is that it is meant to lubricate. The lubricity of a molecule refers to how readily it flows over itself and also to how efficiently it competes to coat the metal surface.

Esters are commonly used as boundary lubricants because they bind with metal surfaces and minimize metal-to-metal contact during sliding motion. The length of the chain, the degree of branching, and the position of ester within the molecule are all structural variables that influence lubricity.

Long carbon chains with fewer branches and higher polarity help in boundary lubrication. Although ester connections are polar, they may be less surface-active if carbon chains shield them. Because synthetic esters are generated from a variety of acid and alcohol feed stocks, the position of the ester groups and the type of carbon chains may be modified.

The lubricity of the ester base stock is influenced by the intensity of the esters with the surface of the metal. Although esters have high lubricity, anti-wear and extreme-pressure additives are required to handle the majority of the load under extreme circumstances. Some believe that esters fight so hard for the metal's surface that essential additives are washed away. On the other hand, several additives are capable of removing an ester from a surface.

It is also critical to select an ester suitable for the purpose. When it comes to boundary lubrication, lubricity is crucial when metal surfaces grind together under pressure.

However, lubricity is less significant if the application includes hydrodynamic lubrication with no metal-to-metal contact. Polyols are suitable for hydrodynamic applications in severe temperatures because they can survive in circumstances that no other lubricant can. Esters may be developed and produced to function in nearly any environment, but this necessitates careful consideration throughout the selection process.

Polyols are increasingly being used in place of mineral oils because of their higher thermal stability. The chemistry of esters is being adjusted in response to environmental pressures to develop compounds with high biocompatibility, non-toxicity, and low engine fumes.

It is difficult to cover all aspects of industrial lubricants in a single chapter. In the literature, several industrial lubricating oils based on mineral resources, plant-based, and synthetic base stocks have been explored. A number of books and articles have also been published on industrial lubricants. However, in this chapter, we will discuss the research conducted at the Indian Institute of Petroleum on the development and tribological investigations of esters, as well as their suitability as a base stock in the production of different industrial lubricants.

4.4 APPLICATIONS OF ESTER OILS IN INDUSTRIAL LUBRICANTS *(WORK PURSUED AT IIP)*

4.4.1 Neopentyl Polyol Ester Oils

The need for biodegradable, environmentally safe lubricants has increased steadily, particularly in eco-conscious areas. There is potential for the production of new eco-friendly safe base oils for a new generation of lubricants. Prior to the incorporation of environmental considerations into the lubricant production process, ester lubricants were used in specialized lubricants for technical reasons [36–37].

Synthetic lubricants of the ester type have been the most suited and widely utilized for modern jet aircraft. Diesters are used in low-speed aircraft, whereas lubricants for high-speed and supersonic transport must tolerate higher temperatures, have a high VI, a low pour point, and other desired qualities. These needs are met significantly by polyol esters reinforced with additives.

The majority of industrial lubricants on the market today are based on mineral oils and have traditional additives to boost performance. Ester oils are more environmentally sustainable and perform much better. Recent trends highlight ester-based fluids with bio-additives. The requirement for high-temperature performance in lubricants is forcing a transition away from mineral oil towards the esters. Industrial lubricating oils are made of base oil and additives [38, 39].

Synthetic esters are utilized in high-performance lubricants as an environmentally friendly base fluid. The selection of base stocks for successful eco-friendly formulations is influenced by the balance of fluidity, volatility, and price. Polyol esters have been proven to be the most cost-effective for biodegradable lubricants.

A number of synthetic polyol-based lube base stocks have been developed. Conventional catalysts are generally used to manufacture esters. These methods use catalysts only once, have disposal problems, produce base oils that require regular monitoring, and result in lower-quality base oils with significant acidity and burnt products. According to published and patented literature, many esters based on neo polyols were produced using conventional catalysts. Nonconventional catalysts have several benefits over conventional catalysts, such as the capacity to be recycled twice without losing reactivity. The method offers benefits in terms of easy processing, lower reaction time, a reduced molar ratio of alcohol to acid, high purity, and cost-effectiveness due to its recyclable nature and yields of 95 percent and above.

All of the esters in this study were produced with an indigenous commercial ion exchange resin catalyst. A simple, cost-effective, efficient process for making synthetic ester base oils using a new catalyst system from indigenous raw materials has been developed for industrial applications.

The acid group in ion exchangers can be chemically bonded to sulfonated polystyrene, allowing for esterification under mild reaction conditions. The final esters are usually highly pure since acid-catalyzed side processes, including dehydration, etherification, and rearrangement, are nearly totally avoided. Such catalysts provide a high-yielding technology that is widely used in industry. The ion exchange is filtered out after the reaction water is removed, and the ester is recovered by distillation.

The developed biodegradable ester lubricants were physico-chemically characterized using several standard ASTM testing techniques, as indicated in Table 4.3.

Then the laboratory-prepared base fluids were tested for tribological behaviour in order to predict their performance in specific applications. Biodegradability and toxicity tests were also performed.

The majority of these products are classed as Group III based on the viscosity index, with the exception of a few polyol esters, which are classified as Group II by the American Petroleum Institute (API) [40]. Synthetic esters are widely utilized in gear oils, marine engines, compressors, hydraulic fluids and grease compositions [41]. These essential fluids can be utilized alone or in blends depending on

TABLE 4.3
Standard Test Methods of Physico-Chemical Characterization

S. No.	Property	Method, ASTM-D
1.	Acid value mg KOH/gm	974
2.	Saponification value, mg/KOH/gm	94
3.	Density d_4^{20} gm/ml	4052
4.	Kinematic viscosity At 100°C cSt At 40°C cSt	445
5.	Viscosity index	2270
6.	Pour point °C	97/96a
7.	Flashpoint °C	92
8.	Noack volatility 250°C	DIN-51581
9.	Copper strip corrosion	BIS:1448:P-15:1976
10.	Oxidation stability	IP-48
11.	Auto ignition temperature °C	BIS:7895
12.	Wear scar dia (mm)	4172-94
13.	Weld load (kgf)	IP-239
14.	Coefficient of friction (μ)	5183
15.	Foaming	BIS. 1448:P-67
16.	Biodegradability	5864

TABLE 4.4
Applications of Ester Oils

S. No.	Neopentyl Polyol Esters	Non Edible Vegetable Oils	Residual Oils
1.	Aluminum cold rolling oils	Neat cutting oils	Multipurpose grease
2.	Mar quenching oils		
3.	Industrial Gear oils		
4.	Automotive Transmission fluids		
5.	Hydraulic fluids fire-resistant		
6.	Gear oils Automotive		
7.	Metalworking fluids		
8.	Steel cold rolling oils		

the product qualities. As per the evaluation, the developed esters are listed in Table 4.4.

4.4.1.1 Aluminium Cold Rolling Oils

Rolling is an industrial method that uses the malleable properties of metals to accomplish shape alteration. Aluminum can be rolled at both hot and cold temperatures. During the hot rolling process, heavy cast aluminum slabs are transformed into hot plates or strips of a specific thickness and then rolled in a cold mill to make sheets and foils of the required finish gauge after cooling to room temperature. In cold

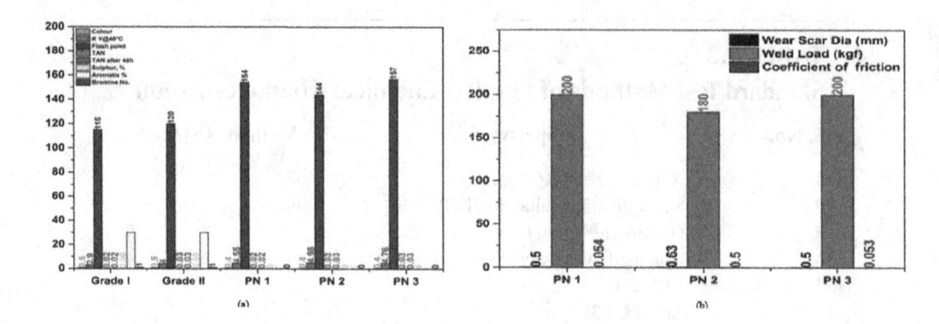

FIGURE 4.3 Aluminium cold rolling oils. (a) Tribo findings for aluminium cold rolling oils.

rolling factories, straight-run light mineral oils with low viscosity are typically utilized for sheet rolling and foil production.

The physicochemical characteristics of the developed biodegradable esters were studied and reported in Figure 4.3. Furthermore, the lubricity, degradability, and toxicity of these esters were investigated in order to establish their feasibility as aluminium cold-rolling oil lubricants (Figure 4.3(a)). According to reports, the materials have a great potential for use in biodegradable aluminium cold rolling oils that meet the I.S.14385-2002 standard [42–44].

4.4.1.2 Mar Quenching Oils

Mar quenching is a steel-hardening technique used to generate high-accuracy and final-quality items from inexpensive low-alloy steels. The oil used for Mar quenching should have a high and consistent quench rate and good steel-hardening properties.

New-generation biodegradable esters were synthesized and studied for their physico-chemical and tribological characteristics. It should have a high flash point and minimal volatility in order to reduce fire dangers. Esters were investigated as a potential enhancement over traditional lubricants. Based on the limited findings, the developed synthetic esters have significant promise for use as biodegradable Mar quenching oils Figure 4.4, as specified by BIS. 4543-2004 [45].

4.4.1.3 Industrial Gear Oils

Gears are machine components that impart motion through tooth engagement. The fundamental function of a gear lubricant is to establish a lubricating layer between the gear mating elements in order to reduce wear and eliminate friction heat generated in sliding rolling contacts. When the danger of scuffing is minor, mineral oils and, in more complicated cases, polyglycols with high load-bearing capacity, exceptionally high viscosity index, and low pour point are used in industrial gears.

Seed oil fatty acids were reacted with petrochemical derivatives to make oleo chemical esters. The physico-chemical features of the synthesized ester and the requirements of commercial EP-type industrial gear oil VG 68 (Figure 4.5) demonstrate that the oleo chemical ester has much promise as a base stock for EP-type industrial gear oil VG 68. Based on the limited research, the synthesized oleo chemical ester has a high potential for application as a biodegradable base stock in

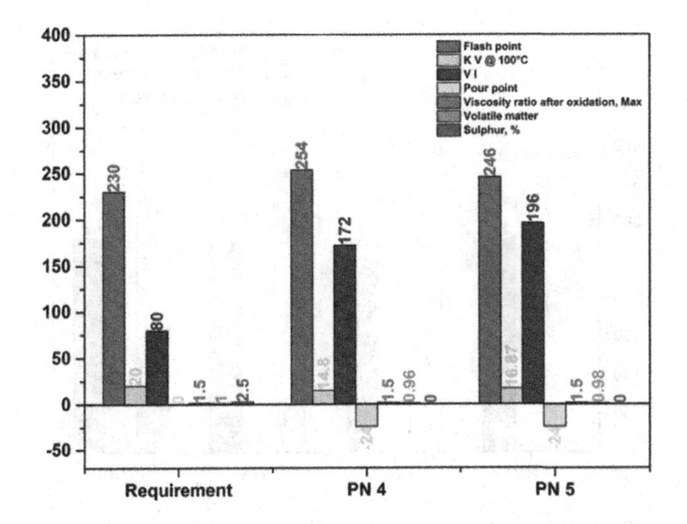

FIGURE 4.4 Mar quenching oils.

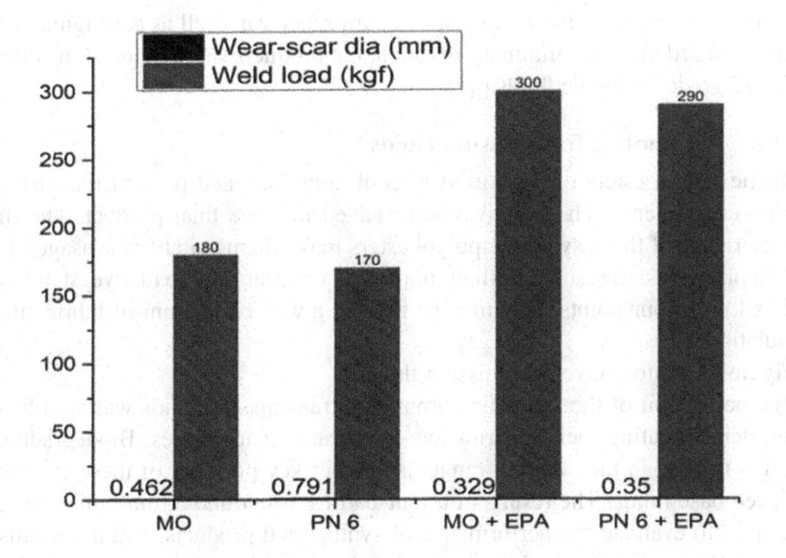

FIGURE 4.5 Tribo findings for Industrial gear oils.

manufacturing EP-type VG68 grade industrial gear oil. For final gear oil formulation production, more research on EP properties on Timken and F.Z.G. Niemann EP tests is required [46].

4.4.1.4 Fire-Resistant Hydraulic Fluids

Synthetic esters appear to be attractive possibilities in areas where environmental protection is critical. The materials were assessed for their lubricating performance and characterized by their physico-chemical properties.

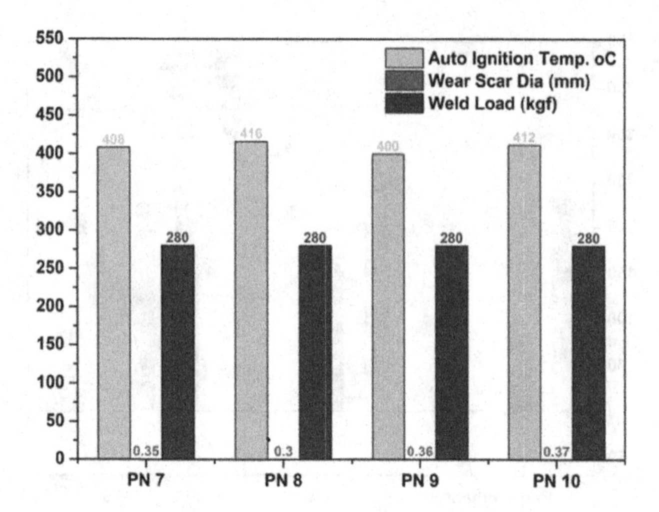

FIGURE 4.6 Tribo findings for fire-resistant hydraulic fluids.

Figure 4.6 depict the improvement in wear property as well as auto-ignition temperature. Based on the preliminary research, the products show tremendous promise as VG-22 grade hydraulic fluids [47].

4.4.1.5 Automotive Transmission Fluids

Synthetic polyol esters oils are used to cool, lubricate, and prevent the rusting of moving components. The additives were added into the final product. The major characteristics of these synthetic polyol esters make them suitable for usage in specific applications. Because of their improved thermal and oxidative stability, as well as lower pour points, they may be used in a wider spectrum of lubricating oil formulations.

Figure 4.7 Automotive transmission fluids

The pour point of the synthetic automobile transmission fluids was significantly lower, demonstrating their superior low-temperature capabilities. Biodegradability, which is usually in the 80–90% range, is another key property of these chemically produced base fluids. The results of a four-ball EP tester and a four-ball wear tester were used to evaluate the performance of synthesized products, and the results are shown in Figure 4.7(a), which indicates that these compounds have greater lubricity properties than mineral oil base stocks. These chemicals can be employed as base stocks for automotive transmission fluids, depending on the specific physicochemical data analysis [48].

4.4.1.6 Metal Working Fluids

The developed polyol esters were tested for tribo properties to use as metalworking fluids. Figure 4.8 shows the load-bearing capacity and lubricity performance of esters. The esters show the best performance with respect to load-bearing capacity along with anti-wear. It appears that the performance of these esters is a function of

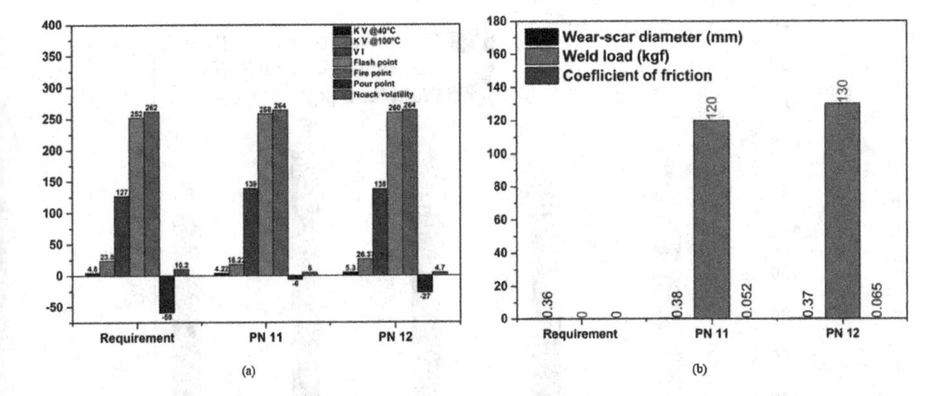

FIGURE 4.7 (a) Tribo findings for automotive transmission fluids.

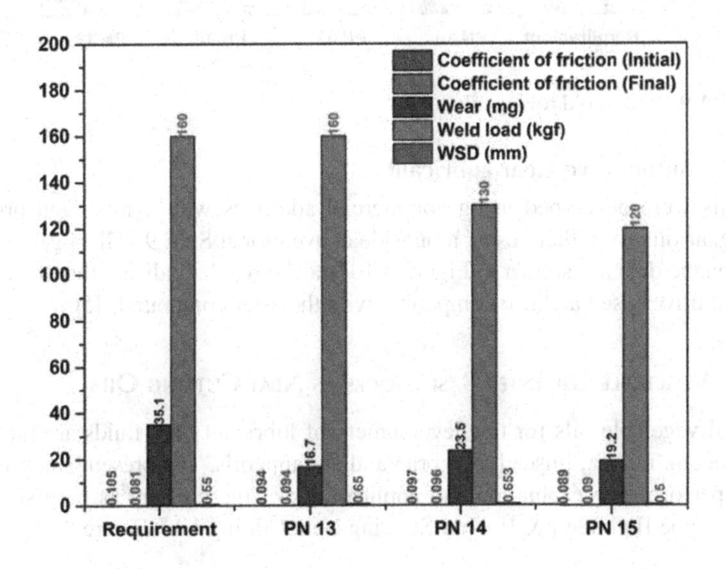

FIGURE 4.8 Tribo findings for metal working fluids.

chemical structure. Comparing the performance of polyol esters with the commercial oils, the esters appear good base oils for the development of eco-friendly metalworking oils.

The developed compounds were tested for their physico-chemical properties and lubricating performance using standard tribo testers. Furthermore, the lubricity properties of these esters were evaluated using a new test established in our lab to see if they might be exploited as cold rolling oils. The CRO should have a load-bearing capacity of moderate to high and good lubricity. Compared to conventional rolling oils, polyol esters appear to be good base oils for producing environmentally friendly cold rolling oils (Figure 4.9) [49].

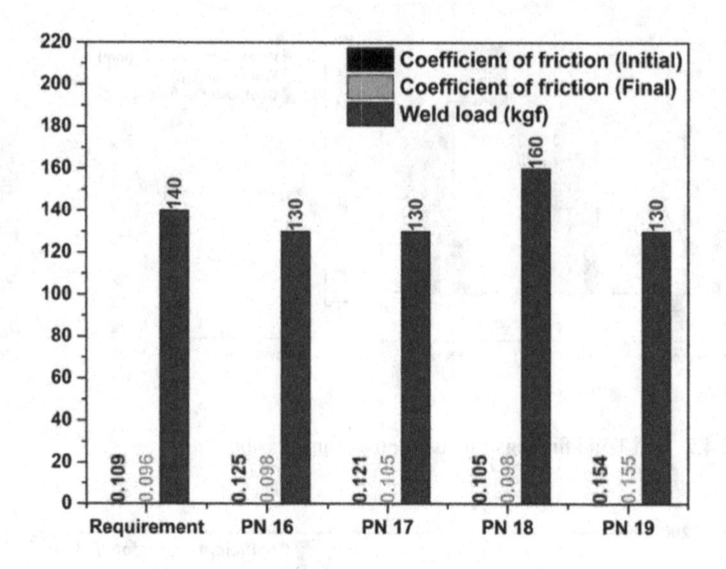

FIGURE 4.9 Steel cold rolling oils.

4.4.1.7 Automotive Gear Lubricants

Gear oils were developed using commercial additives with synthesized products. These gear oils were then tested alongside conventional SAE 90 GL-4 gear oil. The performance data are shown in Figure 4.10(a). The results indicate that the conventional additives used are also compatible with the ester compounds [50].

4.4.2 Vegetable Oil Ester Base Stocks as Neat Cutting Oils

Potential vegetable oils for the development of lubricant base fluids are rape seed, castor, neem, mahua, linseed, rice bran and Karanja oils. The present study focused on the performance evaluation and applicability of these esters as base stocks for IS-3065 Type II Grade I & II, Non-Staining Neat Cutting oils (Figure 4.11).

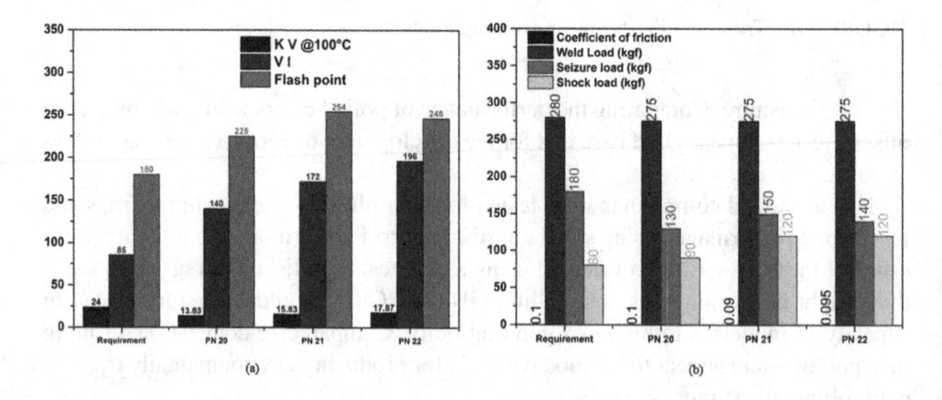

FIGURE 4.10 Automotive gear lubricants. (a) Tribo findings for automotive gear lubricants.

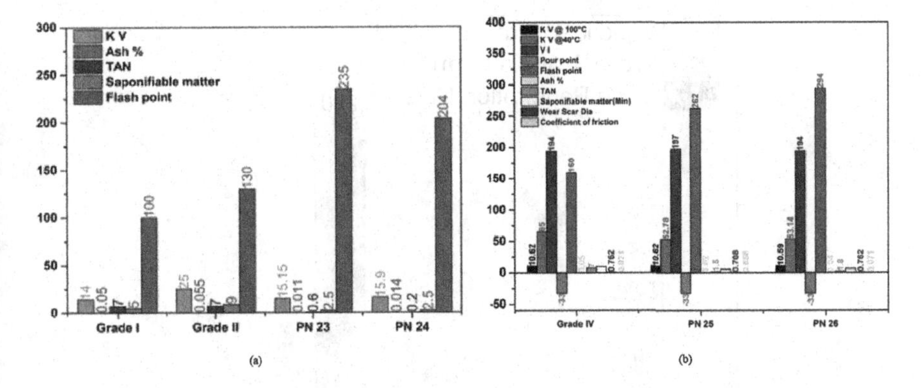

FIGURE 4.11 Neat cutting oils of Grade I & II. (a) Neat cutting oils of Grade IV.

More particularly, the study relates to developing biodegradable base stock as neat cutting oil useful in modern-day cutting operations of turning, milling, drilling, and tapping operations with negligible tool wear. The synthesized products have good antiwear properties, excellent load-bearing capacity, perfect shelf life due to the improved oxidation stability observed for a period of 28 days at 85 ± 2°C, good potential for use as Type II, Grade IV Neat cutting oils. (Figure 4.11(a)) [51–52].

4.4.3 Multipurpose Grease

A biodegradable lubricating grease composition was produced using Jatropha residual oil as the base oil. After that, we acquired multifunctional commercial grease from the Indian market and compared its tribological characteristics to ours. The findings are summarized in Figure 4.12. The synthesized grease values matched the commercial grease in terms of wear and load performance. Furthermore, residual oil, a low-cost byproduct, has led in the manufacture of value-added products. The NLGI grade 2 multifunctional grease developed has exceptional potential as a final product [53].

Further, we granted a patent for the invention of bio reference and gear lubricants for use in enclosed industrial gear systems with tooth pressures and pitch-line velocities that render conventional mineral oils ineffectual. We also published a patent on the production of an eco-friendly and biodegradable neat cutting oil base stocks.

In this chapter, vegetable and polyol ester oils have been developed as degradable industrial lubricants for use in ecologically-sensitive areas. The physico-chemical attributes and tribological performance of the synthesized products were evaluated. Then the developed base oils were compared with mineral oil-based industrial lubricants. Nonconventional ecofriendly indigenous commercial catalysts were used to make the industrial lubricants. The synthesis method used produced high purity esters with low acidity. We developed biodegradable ester oils with and without additives.

The ASTM D-5864-95 test technique was used to examine the biodegradability properties of the synthetic lubricant base stocks [54]. Toxicity was assessed using a

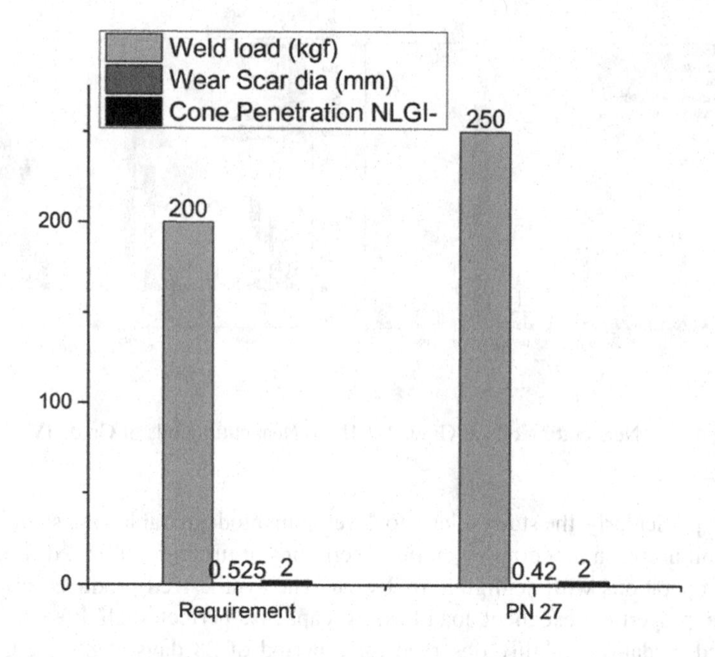

FIGURE 4.12 Tribo findings for multipurpose grease.

modified version of the Algal inhibition test, published in the official Journal of the European Communities under the number L383 A/179–185 (1993) [55]. The standard biodegradability results demonstrates that the ester base stocks developed in the laboratory range in biodegradability from fair to very good, with 80–100 percent biodegradation in 28 days. Mineral-based lubricants, on the other hand, exhibit fair biodegradability, with 20–45 percent biodegradation in 28 days. The toxicity test results show that the samples were non-toxic to sewage microorganisms [56].

The developed new-generation ester lubricants had excellent biodegradability, a high viscosity index, a low pour point, good lubricity, good oxidative stability, excellent wear protection, no evaporation loss, good adherence to metal, corrosion inhibiting properties, and suitability for use with commercial additives. Furthermore, the items are safe for sewage bacteria.

Finally, we conclude that the synthesized vegetable and synthetic polyol esters have promising applications as biodegradable lubricant base stocks. Products are widely used, either alone or as part of formulations and as multifunctional lubricants.

4.4.4 Drawbacks of Ester Oils

Mineral oil-based lubricants are now extensively utilized for a wide variety of relevant applications, but rising green marketing has compelled us to seek alternatives to mineral oils. Alternative industrial lubricant oils include synthetic and vegetable oil esters with good physico-chemical and lubricity properties. Vegetable oils, despite their low cost, have thermo- oxidative stability limitations, as well as irregular suppliers. These esters are generally more versatile in their application utilization than natural esters, with a high degree of biodegradability and minimal aquatic toxicity, due to their far wider viscosity range, increased thermo-oxidative stability, and superior pour point properties. Synthetic fluids are around three to four times the price of regular mineral oils. The lower servicing costs, excellent tribological properties, and longer lifetime of synthetic oils compensate for their higher purchase price. Longer drain intervals and a reduction in total operating expenditures can compensate for the higher initial cost of synthetic ester oils [57–59].

4.5 FUTURE WORK

New and high-performance lubricants are being developed all over the world to meet the demanding requirements of modern technological developments. In this context, the new generation of high-performance biolubricants has proved to possess a broad scope to deliver. Biolubricants have already found their way in a wide range of industrial and automotive applications. Cutting oils, engine oils, gear oils, and hydraulic oils are already extensively accessible in commercial markets. However, there are still a number of scientific and industrial domains where the lubrication and use of biolubricants is a challenge. MEMS are one such area where the use of biolubricants has yet to be researched. We developed and granted a patent for an eco-friendly and biodegradable mineral oil-free lubricant formulation that may be used to lubricate chronometers and other precision components, such as precise bearings, gauges, metres, and clocks in micro electro mechanical system-based devices.

4.6 SUMMARY

For more than a decade, we have been working on biolubricants. We had focused our efforts on new advancements for industrial lubricants made from indigenous raw materials. Industrial lubricant process development and formulations were completed. An analytical center with cutting-edge lubricant testing equipment and a tribological lab for biolubricant performance has been established at our institution. For a variety of industrial applications, lubricants based on neopentyl polyols and non-edible oils have been developed, and various patents have been filed in this area. According to studies, aluminium cold rolling oils, Mar quenching oils, industrial gear oils, fire-resistant hydraulic fluids, automotive transmission fluids, metalworking fluids, steel cold rolling lubricants, automotive gear oils made from polyol esters, neat cutting oils made from non-edible vegetable oils, and vegetable residual oil-based multipurpose greases all have widespread industrial applications, either alone or in combination. More research is required for the final tuning of the products. Attempts have also been made to cover multiple uses with single oil. As a result, few multifunctional lubricants have been developed as a single product for many applications.

ACKNOWLEDGMENTS

The authors would like to acknowledge the Director of CSIR-IIP for granting permission to publish the findings of the study.

REFERENCES

1. MagLube.com, Industrial lubricants, May 22, (2018).
2. D.M. Pirro, M. Webster, *Ekkehard Daschner, Lubrication Fundamentals*, Third Edition. CRC Press. ISBN 978-1-4987-5291-6, (2016).
3. A ready reference for lubricant and fuel performance – Includes detailed. Information on API, ILSAC, and ACEA specifications and test procedures. http://www.lubrizol.com/referencelibrary/lubtheory/base.htm September 17, (1998).
4. P. Nagendramma, Development of eco-friendly/biodegradable synthetic (polyol & complex) ester lube base stocks. Thesis. Srinagar, India: H.N.B. Garhwal University, (2004).
5. J.M. Perez, E.E. Klaus, Dibasic Acid and Polyol Esters, *C.R.C. Handbook of Lubrication and Tribology*, e-book, -CRC press, Volume III, Edited by E.R. Booser, December 21, (1993) 237–252.
6. S.J. Randles, P.M. Stroud, R.M. Mortier, S.T. Orszulik, T.J. Hoyes, M. Brown, *Synthetic Base Fluids, Chemistry and Technology of Lubricants*, Second Edition, Edited by R.M. Mortier, S.T. Orszulik, Springer, (1992) 32–61.
7. R.L. Goyan, R.E. Melley, P.A. Wisner, W.C. Ong, Biodegradable lubricants, *Lubrication Engineering*, 54(7), (1998) 10–17.
8. L. Tocci, Getting your money's worth, *Lubes-n-Greases*, 2(3), (1996) 18.
9. D. Theodori, R.J. Saft, H. Krop, P. Van Broekhuizen, Development of criteria for the award of the European eco-label to lubricants. *IVAM Research and Consultancy on Sustainability*. www.ivam.uva.nl, November 27, (2003).
10. Gosalia A, Another brick in the wall. *Lubes-n-Greases*, 15(1) (2009) 14–18.
11. T. Mang, Lubricants and Their Market, Chapter 1, *Lubricants and Lubrication* Second Edition, Edited by Th. Mang, W. Dresel, Wiley-VCH, (2007) 1–6.
12. Product review biodegradable fluids and lubricants, *Industrial Lubrication and Tribology*, 48(2), (1996) 17–26.
13. L.A. Honorary, Recent advances in bio-based lubricants and greases in the United States, 11th NLGI, Lubricating Greases – Emerging Trends, Mussoorie February 19-21 (2009).
14. C. Kajdas, *Industrial Lubricants, Chemistry and Technology of Lubricants*, Second Edition, Edited by R.M. Mortier, S.T. Orszulik, Springer (1997)228–263.
15. L.A.T. Honorary, Biodegradable/biobased lubricants and greases, *Machinery Lubrication*, August, (2010).
16. J. Van Rensselaer Biobased lubricants: Gearing up for a green world. *Tribology and Lubrication Technology* 66(1) January, (2010) 32–48.
17. M. Fuchs, The world lubricants market current situation and outlook, *12th International Colloquium Tribology*, Esslingen, (2000).
18. G. Lingg, A. Gosalia, The automotive and industrial lubricant market, Automotive and Industrial lubrication, *15th International Colloquium on Tribology*, Technische Akademie Esslingen 17–19 January, (2006).
19. P. Nagendramma, T.D. Gananath, A. Neeraj, B. Babita, Development of new generation bio reference samples and gear lubricants for enclosed industrial gear drives. Indian Patent Grant Number 359671, 26 February (2021).
20. S. Ponnekanti Nagendramma, A. Goyal, J. Ray, N. Atreys Singh, Studies on the sustainability of vegetable oil as a biocompatible multipurpose green lubes and additives, *Indian Journal of Chemical Technology* 26, (2019) 454–457.
21. P. Nagendramma, A. Ray, G.D. Thakre, N. Atray, Ecofriendly and biodegradable lubricant formulation and process for preparation thereof, Application Number: 16/760, 963, 11142718, 12 October (2021).

22. P. Nagendramma, R.P.S. Savita Kaul Bisht and M.R. Tyagi. Development of biodegradable lubricating oil formulations from non-edible vegetable oils. *101st AOCS Annual Meeting & Expo*. Arizona, USA, May 16–19, (2010) 75.
23. www.marketsandmarkets.com, Industrial lubricant market to surpass $73B by 2024, *Noria news wires*, (2021).
24. R.P.S. Bisht, S. Singhal, A laboratory technique for the evaluation of automotive gear oils of API GL-4 *Tribotest Journal*, 6(1) 69-77 (1999).
25. https://interflon.com, The Importance of Industrial Lubrication, 29 May (2019).
26. H. Spikes, The history and Mechanisms of ZDDP, *Tribology Letters*, 17(3), 1 October, (2004) 469–489.
27. M.M. Hussaina, A.P. Pratap, V.R. Gavala, Tribology in Industry, www.tribology.rs Study of Vegetable Oil-Based Biolubricants and Its Hydrodynamic *Journal Bearing Application: A Review*, 10 October, (2002).
28. R.K. Hew Stone, Environmental health aspects of lubricant additives. *Science of the Total Environment*, 156(3), (1994) 243–254.
29. M. Brown J.D. Fotheringham, T.J. Hoyes, R.M. Mortier, S.T. Orszulik, S.T. Randles, P.M. Stroud, Synthetic Base Fluids, Chapter 2, *Chemistry and Technology of Lubricants*. Edited by R.M. Moritier, S.T. Orszulik (1997) 61.
30. R.E. Hatton, *Synthetic Lubricants*, Edited by Gunderson, R.C. and Hart, A.W. Reinhold Publishing Corporation, New York, Chapter 8 Silicate esters 323-360 (1962).
31. W.E. McTurk. Synthetic lubricants. WADC Technical Report, (1953) 53–88.
32. Synthetic Esters: Engineered to Perform, Tyler Housel, Inolex Chemical Co. Machinery Lubrication 4, (2014).
33. D.F. Smith, G.O. Hawk, P.L. Golden, The mechanism of the formation of higher hydrocarbons from water gas, *Journal of the American Chemical Society*, 52, 3221 (1930).
34. G. Cohen, C.M. Murphy, J.G. Rear, H. Ravner, W.A. Zisman, NRL Report 4066, (1952).
35. MIL-L-9236B, U.S. Air force, March (1960).
36. F. Gaining Max, F. Mike, A.A. Reglitzky, D. Plomer Piet, Environmental needs and new automotive technologies drive lubricants quality, *Proc. 14th World Petroleum Congress*, vol. 3, (1994) 99.
37. D.A. Lauer, Industrial Applications, *C.R.C. Handbook of Lubrication and Tribology*, Volume III, Edited by E. Richard Booser, December 21 (1993).
38. F.T.G. Smith, *Neopentyl Polyol Esters*, Chapter 10, Edited by R.C. Gunderson, W.A. Hart, Reinhold Publishing Corporation, New York, (1962).
39. G. Vander Waal, D. Ken Beck, Testing, application and future, development of environmentally friendly ester base fluids. *Journal of Synthetic Lubrication*, 10 (1993) 67–83.
40. Product Review, *Industrial Lubrication and Tribology* 49(2) (1997) 78.
41. P. Nagendramma, S. Kaul, Development of ecofriendly/biodegradable lubricants: An overview, *Renewable and Sustainable Energy Reviews*, 16(1) January (2012)764–774.
42. P. Nagendramma, B.M. Shukla, D.K. Adhikari, Synthesis, characterization and tribological evaluation of new generation materials for aluminum cold rolling oils, *Lubricants* 4(23), 28 June (2016) 1–10.
43. Bureau of Indian Standards, *Aluminium Cold Rolling Oils Specification*; IS 14385; Bureau of Indian Standards, New Delhi, India, 1996, Reaffirmed (2002).
44. Bio-Aluminum Cutting Oil, Food Grade Stabilized Renewable Lubricants. Available online: http://www.renewablelube.com, accessed on February 9, (2016).
45. P. Nagendramma, N. Atray, D.K. Adhikari, Development of new generation biodegradable base stocks for Mar quenching oils, *9th International Symposium on Fuels & Lubricants, A019.IOCL*, Faridabad, India, April 15–17, (2014).

46. P. Nagendramma, Study of pentaerythritol tetraoleate ester as industrial gear oil, *Lubrication Science* 23 (2011) 355–362.
47. P. Nagendramma, S. Kaul, Study of synthesized eco-friendly biodegradable esters: Fire resistance and lubricating properties. *Journal of Synthetic Lubrication* 22 January (2010) 103–110.
48. R.P.S. Bisht, S. Kaul, P. Nagendramma, V.K. Bhatia, A.K. Gupta, Eco-friendly base fluids for lubricating oil formulations. *Journal of Synthetic Lubrication*, 3, 19 October (2002), 243–248.
49. B.M. Ponnekanti Nagendramma R.P.S. Shukla S.K. Bisht, V.R.K. Sastry, Study of synthetic esters as ecofriendly base fluids for steel cold rolling oils. *3rd International Conference on Industrial Tribology*, Jamshedpur, India, 7.1–7.4, April (2001) 8–11.
50. P. Nagendramma, S. Kaul, Study of synthetic complex esters as automotive gear lubricants. *Journal of Synthetic Lubrication* 25 May (2008), 131–136.
51. P. Nagendramma, K. Savita, T.G. Doulat, R.P.S. Bisht, M.R. Tyagi, Development of new biodegradable base stocks as neat cutting oil P.NO. 3483 Del 2012, 15 August (2014).
52. P. Nagendramma, S. Kaul, Studies of polyol ester lube base stocks as such for neat cutting oils. *8th International Symposium on Fuels & Lubricants, Advances in Fuels, Lubricants and Alternatives, Indian Habitat Centre*, New Delhi, India. March 7–9, (2012).
53. P. Nagendramma, P. Kumar, Eco-friendly multipurpose lubricating greases from vegetable residual oils, *Lubricants* 3, (2015) 628–636.
54. ASTM, ASTM-D-5864-95, Standard Test Method for Determining Aerobic Aquatic Biodegradation of Lubricants or Their Components, American National Standards Institute, New York, NY (1996).
55. K. Van Miert Toxicity, Methods for the determination of ecotoxicity, C3—Algal inhibition test, *Official Journal of the European Communities* 383A (1993) 179–185.
56. P. Nagendramma, S. Kaul, D.K. Adhikari, Biodegradability and toxicity studies of synthesized polyol ester lube base stocks. Greasetech India, *A Quarterly Journal of NLGI-India Chapter*, XV(1), July 1–September 30, (2012).
57. B.R. Hohn, K. Michaelis, R. Dobereiner, Load carrying capacity properties of fast biodegradable gear lubricants. *Journal of the STLE Lubrication Engineering*, 55(11), (1999) 15–38.
58. J. Eastwood, A. Swallow, S. Colmery, Selection criteria of esters in environmentally acceptable hydraulics, UNIQEMA, NCFP I05-4.2, (2007).
59. T. Bartels, Gear Lubrication Oils, Chapter 10, *Lubricants and Lubrication*, Second Edition, Edited by T. Mang, W. Dresel, Wiley-VCH Verlag GmbH & Co. KGaA, Weinheim (2007) 230–273.

5 Tribological Performance Involving DOE of an Additively Manufactured Cu-Ni Alloy

R. Rajesh
Global Academy of Technology, Bangalore, India

Mithun V. Kulkarni
Vijaya Vittal Institute of Technology, Bangalore, India

P. Sampathkumaran
Sambhram Institute of Technology, Bangalore, India

S. Anand Kumar
Indian Institute of Technology, Jammu, India

S. Seetharamu
RV College of Engineering, Bangalore, India

Jitendra Kumar Katiyar
SRM Institute of Science and Technology, Chennai, India

CONTENTS

DOI: 10.1201/9781003243205-5

5.1 INTRODUCTION

The metal parts are made primarily from powder metallurgy, casting, machining, and metal injection moulding. However, additive manufacturing involving 3D printing and the layer-by-layer building up of objects is a relatively new method employed in industries such as medical, automotive, space, and aircraft manufacture. In the 3D printing technology, both the DMLS and the SLS processes are popular and widely used [1]. A reasonable quantum of research and developmental work in 3D printing, also known as rapid prototyping, is reported from the point of exploring the mechanical, metallurgical and tribological characteristics [2–7]. Here, the adoption of different input settings such as the layer thickness, laser scan speed and layer orientation, and types of powders [8] used are the essential parameters, which have been optimized and reported [9–11].

The 3D printing of non-ferrous materials such as aluminium (Al), steel, titanium (Ti) has been reported to assess the mechanical and wear properties. The information on copper (Cu)-based alloys for mechanical and wear evaluations is limited as they possess excellent mechanical properties, good wear resistance and corrosion properties apart from good weldability and castability. The high thermal conductivity of Cu alloys makes the alloy more suitable for many engineering applications. The other issues relating to hardness, strength, surface roughness, impurity levels, porosity, and residual stresses dictate the properties. Copper (Cu) and nickel (Ni) alloys are isomorphous and the mixture is completely soluble due to the face-centred cubic crystal structure. Cu-Ni alloys need to be studied under sliding wear, galling and pitting corrosion conditions despite exhibiting good mechanical and thermal characteristics and they are being used to produce parts such as bearings, wear plates, etc. Thus, the wear behaviour becomes a crucial, important, and challenging issue. The wear phenomenon affects equipment life, primarily operated under sliding and corrosive environmental conditions.

Tribological aspects relating to engineering materials have been addressed through the laser processing routes, such as laser hardening, laser peening, laser cutting, laser melting etc. In laser hardening process, higher power laser source is made to bombard on the surface of a sample resulting in heating of the localized areas of the sample. It is done from the point of enhancing the strength and durability of a component. The laser peening is another process wherein pulsed laser system (shock waves) is made to impinge on the surface of a sample resulting in modification of the surface with improved properties. The laser cutting technology makes use of laser beam to vaporize a sample and results in a cut edge. This is intended for industrial manufacturing applications. In the laser melting process, the melted particles are made to deposit on a bed powder with a heat source and mainly used in metal additive manufacturing method to produce near net shape parts. In a work reported by Anand Kumar et al. [7, 12] on the surface treatment using laser peening on nickel-based super alloy 718 in one case [7] and Ti-6A-V in the other case [12], wherein it is reported that the processes adopted have contributed to grain refinement due to the formation of nano crystals, increased surface hardness and the development of higher compressive stresses and more specifically improved fretting wear behaviour with alumina & steel as counter bodies. It is seen from this work that the wear volume losses have increased with the increase in the application of load. This work is reported [12] from the viewpoint of laser treatment (peening) & laser melting (DMLS) of different materials subjected to tribological tests influenced by load as load is the key parameter affecting the wear loss. Further, the tribological behaviour of 3D printed samples, especially Cu-Ni alloys, has been less often reported and hence this current work aims to study the wear and friction aspects of DMLS processed Cu-Ni alloys. In this context there is a discussion of some of the literature concerning the effect of input parameters of DMLS process affecting the wear. The wear and friction behaviour of bronze nickel alloys produced by sand casting was studied in detail by Ilangovan et al. [13]. As an alternative to this, bronze nickel samples fabricated using the DMLS process would be a more recent method of producing 3D samples. Hussain et al. [14] studied the SS316-based metal matrix composites processed by the DMLS method to study the effect of process variables, such as laser scan speed, laser power and powder mixing ratio on the wear resistance, and it is reported that laser power is one of the significant factors in determining the wear behaviour. A study was carried out by Massimo Lorusso et al. [15] on the wear and friction behaviour on Alsi10Mg-TiB_2 prepared by DMLS and casting processes. As reported [15], higher Cof and lower wear rates were obtained for the DMLS part compared to cast samples of the same compositions. Their findings were based on the analysis of microstructure, grain size and hardness. Similar work was conducted by Rahul Kumar et al. [16] on 3D-printed steel parts' wear behaviour, wherein the results obtained were compared with the conventional processed Hardox 400 steel and AISI 316 stainless steel. It is reported that the additive manufactured alloy steel and composite showed a lower wear rate than that of Hardox 400 steel and AISI 316 stainless steel. Hongling Qin et al. [10] conducted a detailed literature review on the additive manufactured parts, in which it is reported that the proper selection of process parameters has significantly improved the density and hardness and, in turn, improved the wear resistance.

Often, the number of operating parameters increases and, hence, one has to apply a statistical technique which could determine a predominant parameter and its optimization. In this regard, a rational, robust and effective process control technique is very much required. Complex problems involving multiple factors are efficiently handled and investigated through factorial experiments. The usage of appropriate scientific methods for experimental planning, conduction and investigation and analysis could significantly reduce the extensive time duration for conducting the experiments and the high cost involved. The Taguchi's experimental design is the commonly employed method to analyze the diverse nature of problems encountered in research, development and manufacturing. The foremost objective of the design of experiment (DOE) is to achieve the solution for the problem at a lower cost, simultaneously securing quick results and obtaining more useable data. The practical limitations existing in machine tools could be effectively overcome by employing the DOE, which would bring out the experimental plan leading to results generation, which is not likely to obtain by any other means or at expensive cost. So, DOE is framed to obtain quick results at a lower cost and generate results by mathematical formulation and equations for more straight forward interpretation.

The repeated heating and cooling cycles adopted during the layer-by-layer building up process results in the generation of residual stress. The influence of residual stress on the mechanical performance causes part distortion and dimensional instability. In this context, numerous investigations have been reported [17] to focus on the beneficial and detrimental effects of the induced residual stress in a component. When a component is subjected to the wear process, if compressive stress gets generated, it would inhibit the microcrack initiation and prevent cracks propagation, while the induction of tensile stress would result in catastrophic failure. Thus, the wear rate decreases due to the control and generation of compressive stress in a component. The information regarding the effect of residual stress on the wear resistance of engineering components is rather limited. The uniqueness or novelty of the work lies in investigating the wear & friction behaviour of Cu-Ni alloys prepared using 3D printing (DMLS process) as this aspect could not be cited in the literature & forms first of its kind.

In view of the above points, evolving correlation of tribological characteristics with residual stress has been addressed for the DMLS-processed Cu-Ni alloys. The present study focuses on the development of the alloy followed by density, hardness, surface roughness measurements and analysis as well as wear damage assessment using a Scanning Electron Microscope (SEM).

5.2 MATERIALS AND METHODS

5.2.1 Materials

A mixture of bronze and nickel materials (sourced from EOS GmbH, Germany), having an average particle size of about 20 μm, has been employed to produce DMLS samples. The nominal compositions of the DMLS processed sample are 76.6% Cu, 14.6% I, 6.7% Sn, and 1.6 P.

5.2.2 Experimental Methods

5.2.2.1 DMLS Procedure

The machine used for developing the cylindrical samples of 4 mm length and 10 mm diameter by the DMLS method was the EOS-INT M250 facility. CATIA R19 V5 3D CAD/CAM software was used to model the sample and then saved as an .STL file. Further, the file was converted to the ".bdf" format involving EOS RP tool software. Following this, the sample is sliced into several layers of different thicknesses 20 μm, 40 μm and 60 μm. The final dimensions of the sample obtained from the bed size 250 mm × 250 mm × 22 mm had 4 mm thickness & 10 mm diameter involving the EDM wire cutting method to separate the sample from the bed. The porosity determination of the samples were done theoretically. The Vickers hardness measurements have been made on the test samples as per ASTM standard at a load of 15 kg with a dwell time of 15 s using a square diamond indentor with the allowable standard deviation of ±1 VHN. The measurements have been carried out on 3 representative samples and the average is repeated.

5.2.2.2 Fabricated Specimens with Varied Process Parameters

Table 5.1 gives details of the input parameters selected for the work. The photographs of the samples prepared in the laboratory is shown in Figure 5.1.

5.2.2.3 Sliding Wear Behaviour and Friction Measurement

The wear and friction properties of DMLS samples were assessed using a pin on disc machine. Figure 5.2 gives a photograph of the Pin-on-Disc wear tester (Ducom Make, model TR-20) used in this work.

The wear tests were conducted as per ASTM G-99 standard for 30 N, 40 N, and 50 N loads with a sliding speed of 1.04 m/s (200 rpm) and a sliding distance of 376 m for the pin geometry of length 4 mm and diameter 10 mm. A total of three measurements were made and the average value has been reported. The coefficient of variation is determined based on the standard deviation obtained as per the ASTM G65 and should lie below 15%. The wear loss is expressed in terms of specific wear rate (mm³/Nm) obtained by converting weight loss into volume loss and dividing it by sliding distance(m) and applied load (*N*) as per the Equation (5.1) given below.

$$Ws = \frac{V}{F_N * L}\left[\text{mm}^3/(\text{N}-\text{m})\right] \tag{5.1}$$

TABLE 5.1
DMLS Process Parameters Selected

Sl. No	Input Parameters	Selected Values
1	Layer thickness	20,40,60 μm
2	Laser scan speed	100,200,300 mm/s
3	Laser Power	240 W
4	Hatch spacing	0.2 mm
5	Laser beam diameter	0.4 mm

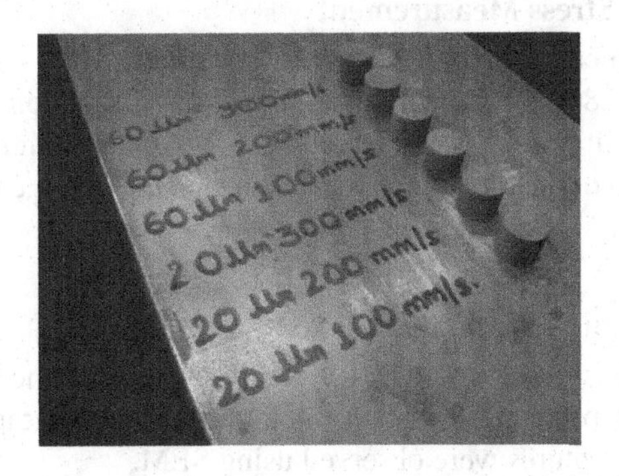

FIGURE 5.1 Fabricated specimens.

FIGURE 5.2 Wear test on pin on disc.

where ΔV is the loss in volume, F_N is the applied normal load, and L is the covered sliding distance.

The COF is determined using Equation (5.2) as given below.

$$COF = F_T / F_N \quad (5.2)$$

Where F_T = Frictional load & F_N = applied normal load

5.2.2.4 Surface Roughness

The surface roughness measurements were carried out using a portable surface roughness tester (Mitutoyo SJ-210 make) by setting the probe-travelling speed at 0.5 mm/s to find out the arithmetic roughness (R_a) value. In the case of SR measurements, a minimum of three readings are recorded and the average value is reported. The standard deviation allowed is ±1% of the surface roughness value as per the standard practice.

5.2.2.5 Residual Stress Measurement

The residual stress measurements (make PROTO, model MGR 40P) were conducted as per the ASTM E 2860-12 standard using the X-ray diffraction technique (Bragg's law) for the test conditions are X-ray target – Cr_K- Alpha, Current –4 mA, Voltage – 20 KV). A minimum of three readings was taken, and the average value was reported. The standard deviation allowed is ±30 MPa.

5.2.2.6 Microscopic Examination

The microstructural assessment of DMLS samples was performed based on the standard metallographic principle to obtain the light microphotographs. The worn surface morphological features were observed using SEM.

TABLE 5.2
Factor Levels for the Experiment

Factors	Level 1	Level 2	Level 3
Laser scanning speed-LSS (mm/s)	100	200	300
Layer thickness-LT (micron)	20	40	60
Normal load-NL (N)	30	40	50

5.2.3 Design of Experiments of Wear Tests

Taguchi's experimental design, being a powerful technique, can identify the dominant factors, their influence on the mechanical properties, and their interactions. The choice of the area of experiment and the identification of the suitable process parameters and their levels are based on screening of the experiments. These process parameters are listed in Table 5.2.

In the present investigation, the influence of three factors, viz., scan speed, layer thickness and normal load each at three levels, was studied using an L_{27} (3^3) full-factorial orthogonal design and it is shown in Table 5.2. Minitab 19 software is used for the DOE and statistical model analysis.

5.3 RESULTS AND DISCUSSION

The values of density, porosity, hardness, surface roughness and residual stress are shown in Table 5.3. From Table 5.3, it can be seen that the density of the samples decreased with an increase in laser scanning speed and layer thickness. The hardness values showed a similar trend. i.e., hardness decreased with an increase in layer thickness as well as speed. The observed lower hardness and higher porosity for higher scan speed conditions could be due to the partial melting of the powder [18]. Similar observations were reported by various researchers [19–21].

TABLE 5.3
Porosity, Microhardness, Surface Roughness and Residual Stress of LPBF Processed Samples

Alloy Designation	Laser Scan Speed (mm/s)	Layer Thickness	Density (g/cm³)	Porosity (%)	Hardness (VHN)	Surface Roughness (R_a) µm	Residual Stress (MPa)
DS 20/100	100	20 µm	8.05	3.7	146	12.8	−100
DS 20/200	200		7.85	6.3	111	14.1	−83
DS 20/300	300		7.78	6.5	79	15.4	−65
DS 40/100	100	40 µm	7.98	4.8	96	15.0	−71
DS 40/200	200		7.79	6.9	78	16.3	−57
DS 40/300	300		7.76	7.1	66	17.1	−45
DS 60/100	100	60 µm	7.91	5.1	91	16.8	−47
DS 60/200	200		7.78	7.0	75	17.9	−39
DS 60/300	300		7.75	7.2	65	19.2	−35

Interestingly, the shape, size and morphology of the powder play a vital role in accomplishing the desired porosity during the LPBF technique. The powder size is in the range of 15 to 50 μm with an average particle size of 30 μm is employed in the present study. The tap density of the particle used is crucial during LPBF printing, since it relates the particle size distribution. For instance, powder size distribution near the upper limit (~40 to 50 μm) can lead to larger porosity volume fraction during LPBF printing. It is due to the interlayers and particle arrangement may generate a further noteworthy porosity content. Conversely, powder size distribution nearing the lower limit (~1 to 10 μm) can maximize the opportunity for lack of melting and thus escalates the porosity volume fraction during LPBF printing [22]. Therefore, powder size distribution considerably impacts the porosity content of the LPBF processed parts. Because of the above, it is comprehensible that the particle shape, size and morphology influences the porosity and thus the mechanical properties such as hardness of the LPBF fabricated parts as shown in Figures 5.3 and 5.4.

Figure 5.5 show the effect of laser scanning speed and layer thickness on the surface roughness development. It is observed that the surface roughness increased with the increase in the laser scan speed and layer thickness. The larger magnitude of laser scan speed will lead to a shorter time interval for the complete melting of the powder particle; therefore unmelted and fused particles contribute towards the creation of a rougher LPBF part surface. It was observed that, in the case of layer thickness of 20 μm with a twofold increase in scanning speed from 100 mm/s to 300 mm/s has led to an increase of R_a value of about ~20%. Similarly, for both 40 μm and 60 μm layer thicknesses, it was found to be around ~14% and 14.5%, respectively. Hence, it is understood that for complete melting to occur a sufficient time should be allowed for laser-powder interaction; therefore, lower scan speeds delivers the improved smoother surface. A similar trend was reported by other researchers [1, 23–25]. In addition, it is noted that the R_a value rises for higher layer thickness value. It was

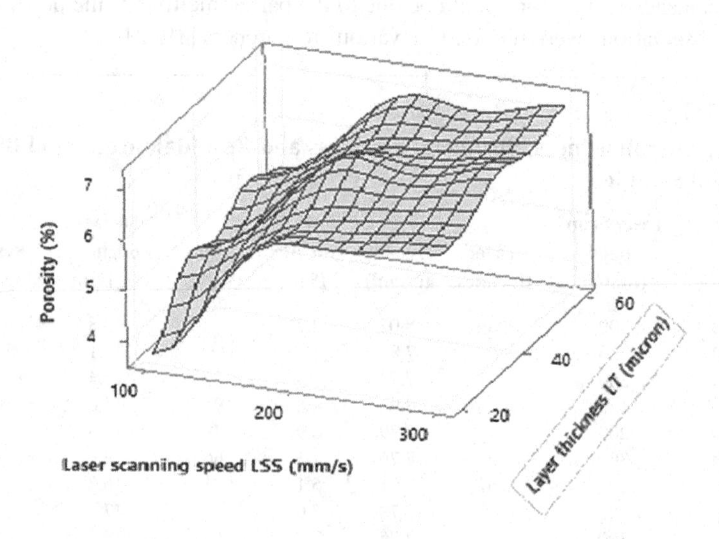

FIGURE 5.3 Porosity plot.

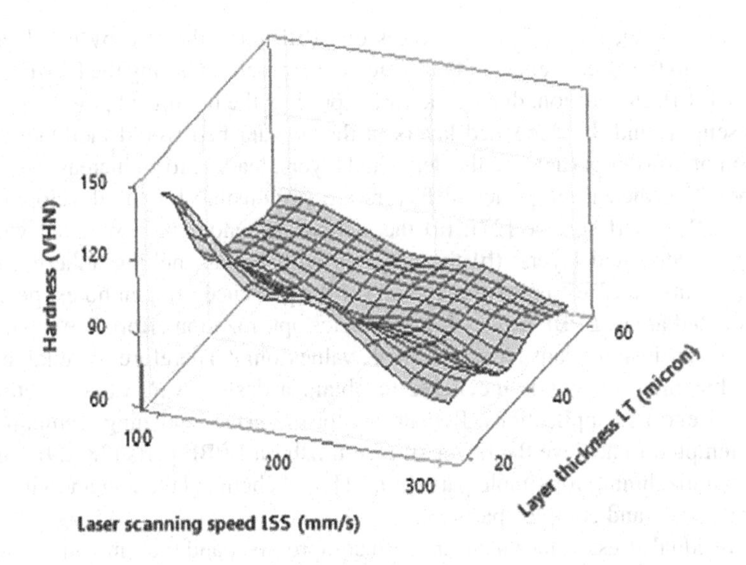

FIGURE 5.4 Hardness plot.

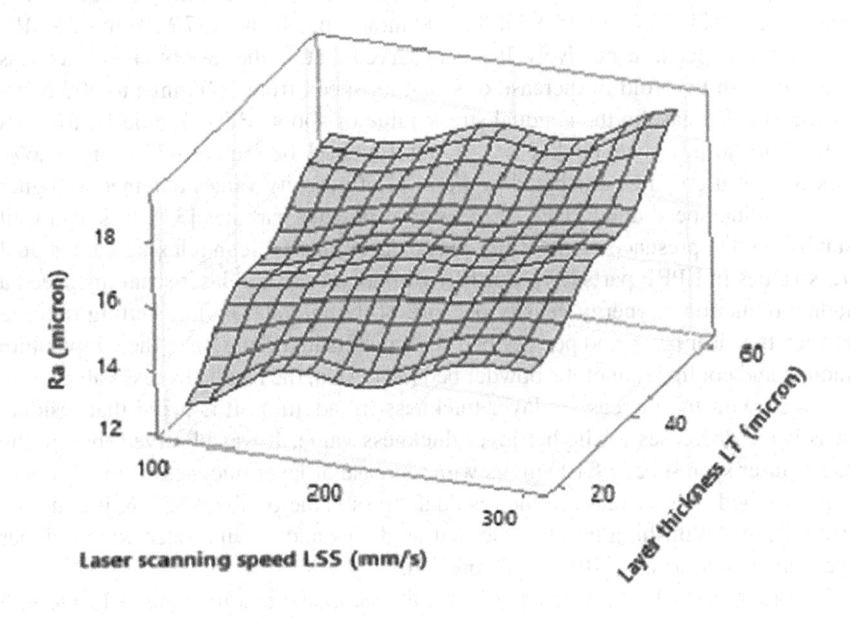

FIGURE 5.5 Surface roughness plot.

observed that, in the case of laser scan speed of 100 mm/s with an increase of layer thickness from 20 μm to 60 μm has led to an increase in the R_a value of about ~31%. In the case of lower-layer thickness samples, powder being spread on the powder bed requires lesser effort to fuse particles compared to the samples with higher thickness samples. Given the above findings, it is understood that both laser scan speed and layer thickness optimized for minimum values for achieving lower surface roughness values. In general, the surface roughness of LPBF parts dictated by the degree of oxidation and the effectiveness of complete powder melting during the LPBF process (Li et al. [26]). In addition, during the LPBF process the unutilized powder particles are present around the deposited layers in the powder bed would facilitate further adhesion of powder particles to the deposited layers, leading to an increase in surface roughness. Further, three principal factors are responsible for the development of rougher LPBF part surface [27]: (a) the staircase phenomena associated with the number of deposited layers; (b) the powders with partial melting adhering to the deposited outside LPBF part surface; and (c) the presence of open holes, pores and partly melted areas. LPBF processing parameter optimization efforts are intended to address to an insignificant reduction in R_a values only. Therefore, a suitable post-surface finishing process is inevitable to obtain a desirable R_a value meeting the functional end-use applications. Recently, a lot of surface finishing techniques has been attempted to achieve the better surface finish on LPBF parts [28, 29]. Further, traditional machining for simple parts [30, 31], and chemical etching and vibra-honing for intricate and complex parts [32].

The residual stress values measured are compressive, and their magnitudes are in the range of 35 to 100 MPa. From Table 5.3, it is noticed that the residual stress decreases with an increase in the scan speed, i.e., from −100 MPa to −65 MPa for 100 mm/s speed, −71 MPa to −45 MPa for 200 mm/s speed, and −47 MPa to −35 MPa for 300 mm/s speed, respectively. It was observed that, in the case of layer thickness of 20 μm with two-fold of increase of scanning speed from 100 mm/s to 300 mm/s has led to a decrease in the residual stress value of about ~35%. Similarly, for both 40 μm and 60 μm layer thicknesses, it was found to be around ~37% and ~26%, respectively. It could be attributed to the level of porosity values obtained at higher laser scanning speed due to lack of melting of powder particles [33]. It is also well established that presence of porosities plays a prominent role in relaxing the residual stress values in LPBF parts [33]. Further, with an increase in laser scanning speed a gradual reduction in energy density because of the interaction time getting reduced between the laser beam and powder bed. This results in a decrease in the temperature gradient and cooling rate of the powder bed [34]. Also, the residual stress values have decreased with the increase in layer thickness. In addition, it is noted that residual stress value decreases for higher layer thickness value. It was observed that, in the case of laser scan speed of 100 mm/s with increase of layer thickness from 20 μm to 60 μm has led to a decrease in the residual stress value of about ~53%. It could be attributed to diminishing trend of thermal gradients and cooling rates due to higher layer thickness during LPBF deposition [35].

Overall, the residual stress obtained for the samples lie in the range −110 to −35 Mpa, the Vickers hardness lies in the range 146 to 65, a substantial variation seen in both of these cases.

The role of compressive residual stresses generated due to layer thickness and scan speed have been established in the preset work. Further, higher the hardness value obtained, higher is the compressive stress getting induced in the sample. During hardness indentation measurement, the indentor load/stress is directed perpendicular to the sample surface. Subsequently, the direction of contact shear stress underneath the indenter is the opposite with the compressive residual stress direction. Hence, the compressive residual stress is confined to the amount of shear stress required for

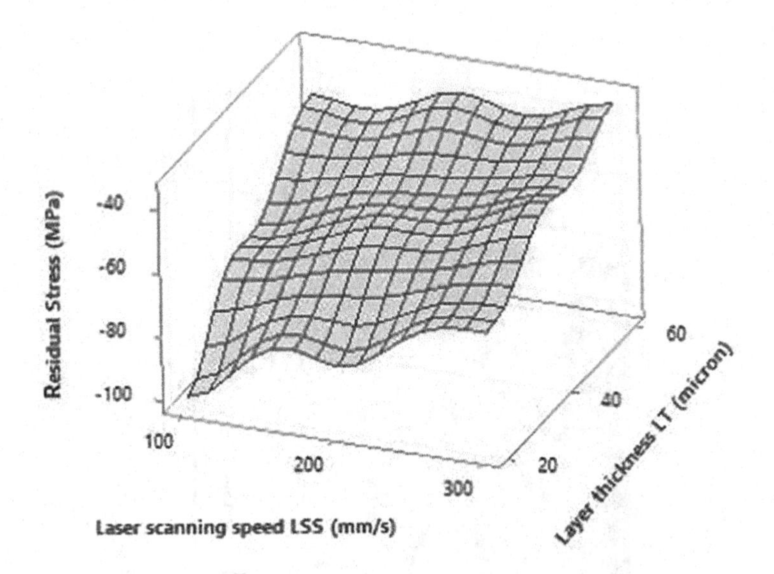

FIGURE 5.6 Surface plots of residual stress.

surface plastic deformation. The limited shear stress value can offer more resistance to the plastic deformation behavior due to indentation [36] (Figure 5.6).

5.3.1 Sliding Wear Behaviour and CoF of DMLS Processed Samples

Figure 5.7 show the specific wear rate for various load applications of 30 N, 40 N, and 50 N and a scan speed of 100, 200 and 300 mm/s in respect of layer thickness of 20 μm, 40 μm and 60 μm respectively at 1.04 m/s sliding speed.

With the increase in load, the specific wear rates irrespective of the layer thicknesses have increased. The trends observed with increase in thickness affecting density, hardness and surface roughness are understandable, even for the increase in scan speed as they get full support from the literature points [37; 38].

A work carried out by Ramesh et al. [37] revealed that the wear behaviour of the DMLS-processed alloy have reported that the wear losses increased with the increase in the layer thickness as well as with the increase in scan speed. The other properties, namely, hardness, tensile stress and percentage elongation have decreased with laser scanning speed.

A similar attempt by Ramesh et al. [38] focused on Ni-coated Iron silicon carbide composites to develop components by the DMLS method for laser speed of 50 mm/s to 150 mm/s. It is reported in the work that an increase in load from 10 N to 80 N and in speed from 75 mm/s to 100 mm/s, the specific wear rates have decreased. From their findings, it is seen that lower the scan speed, higher is the hardness & lower is the wear rate and Vis- Vis.

This type of trend is noticed in published papers [38], where an increase in the load contributing to lower wear rate mainly due to higher plastic deformation resulting in a higher probability of shearing of asperity at the junctions taken place. As regards the sliding speed, the different trends are noticed for iron and SiC

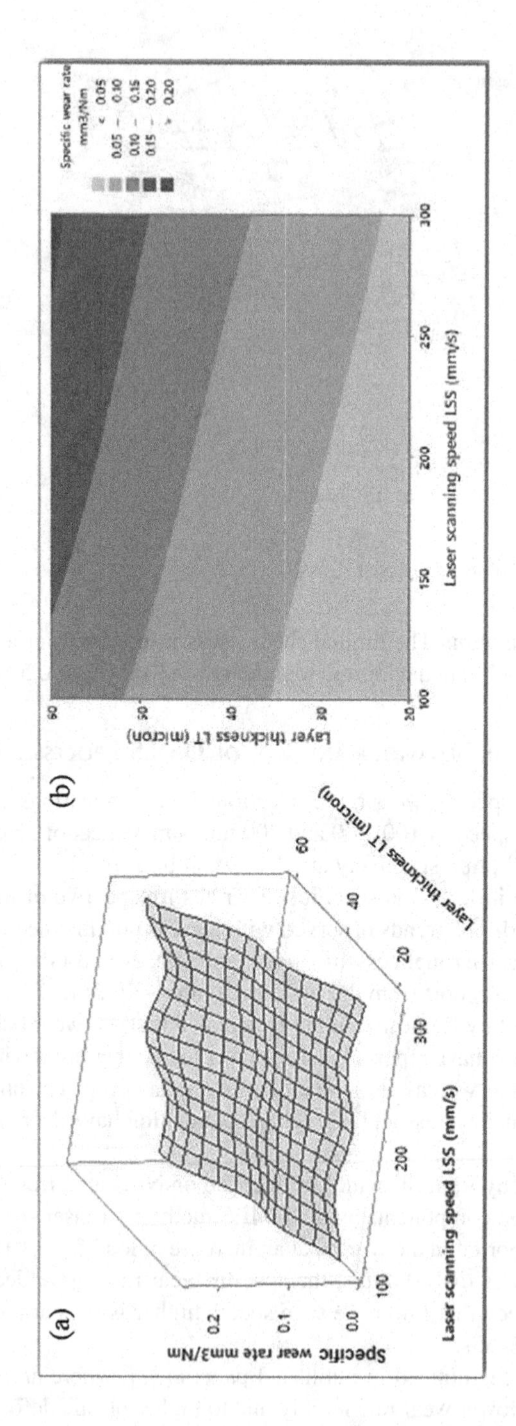

FIGURE 5.7 Surface and contour plots-wear.

components. The trends reported [35, 38] and the results obtained in the present work on hardness & sliding wear resistance are in very broad agreement.

Further, it is well documented [39, 40] that higher the density and hardness, the wear rate decreased. The reason for this was attributed to lower porosity levels obtained for the lower-layer thickness samples. These are on the expected and the established trends.

Another work, by Cody Ingenthron et al. [41], reported the effects of layer thickness on dry-sliding wear of additively manufactured stainless steel and bronze composite samples, i.e. the thinner the layer, the higher is the wear resistance.

Figure 5.8 (a) plot pertaining to the mean of means of specific wear rate due to the change in the layer thickness, laser scan speed and load. Here it is evident that specific wear rate is quite significant for the layer thickness increase compared to either laser scanning speed or the normal load of LPBF samples. In addition, Figure 5.8(b) depicts the interaction plots and is advantageous to assess collaboration among the factors. Observing the non-parallel nature of lines in the interaction plot suggests that there is an interaction among the parameters. It is observed that the lines are not parallel, which indicates that all the three parameters, laser scanning speed, layer thickness and normal load, have a considerable influence on the specific wear rate.

Figure 5.9 shows that the COF values increased with increased layer thickness and an increase in scan speed. A similar observation was reported by Naiju [42] on the direct metal deposition-processed layer of H13 tool steel, of varied thicknesses of 0.5, 1.0 and 1.5 mm on mild steel sample have reported that for 0.5 mm layer thickness, the average COF obtained was 0.159 and for 1.5 mm layer thickness the average COF recorded was 0.18. This demonstrates that CoF increases with an increase in layer thickness. Thus, the reported investigations [42, 43] give evidence to the present work. The porosity data results support this finding. The CoF behaves with a linear inverse relationship with the hardness. The friction force at the tribo-pair interface is reflected due to adhesion, deformation and intervention caused by a third body [44]. As per well-established mechanical attraction friction theory [45] the CoF value is the summation of the deformation (CoF_d) and adhesive (CoF_a) constituents as given by Equation (5.3).

$$CoF = CoF_a + CoF_d \tag{5.3}$$

Moreover, the adhesive component of the friction force is minimized by enhancement in the hardness of the sample. Samples processed under lower-layer thickness (20 μm) had lowered the adhesive component of friction force due to high hardness, resulting in lesser total friction force during sliding wear testing under all conditions. Hence lesser COF values are observed, leading to smoother worn surfaces (see Figures 5.11(a) and 5.13(a)). Sample processed with a layer thickness (40 μm and 60 μm), owing to its lower hardness, has a higher interface area ensuing in higher friction force than samples processed under lower layer thickness (20 μm) during sliding wear testing. Therefore, the overall COF values are significantly higher, leading to rough and irregular worn surfaces (see Figures 5.12 and 5.14).

Figure 5.10(a) shows the mean of means plot for the CoF behaviour of LPBF processed samples. It is noted that the effect of layer thickness on COF is significant

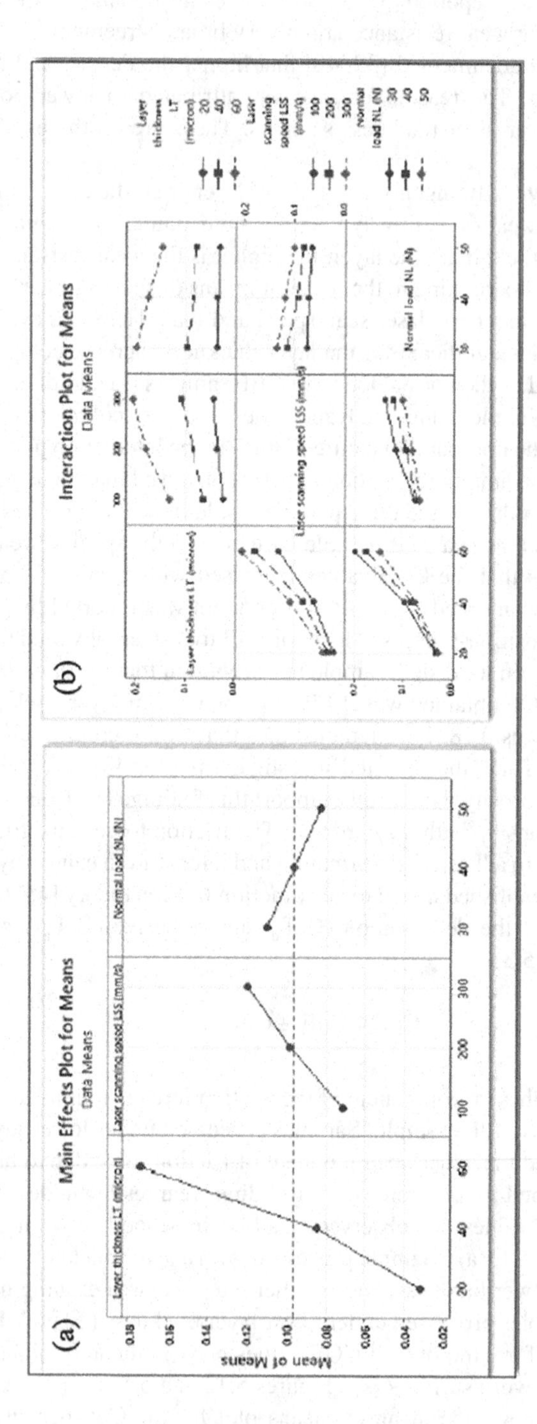

FIGURE 5.8 Specific wear rate of LPBF samples: (a) mean of means plot; (b) interaction plot.

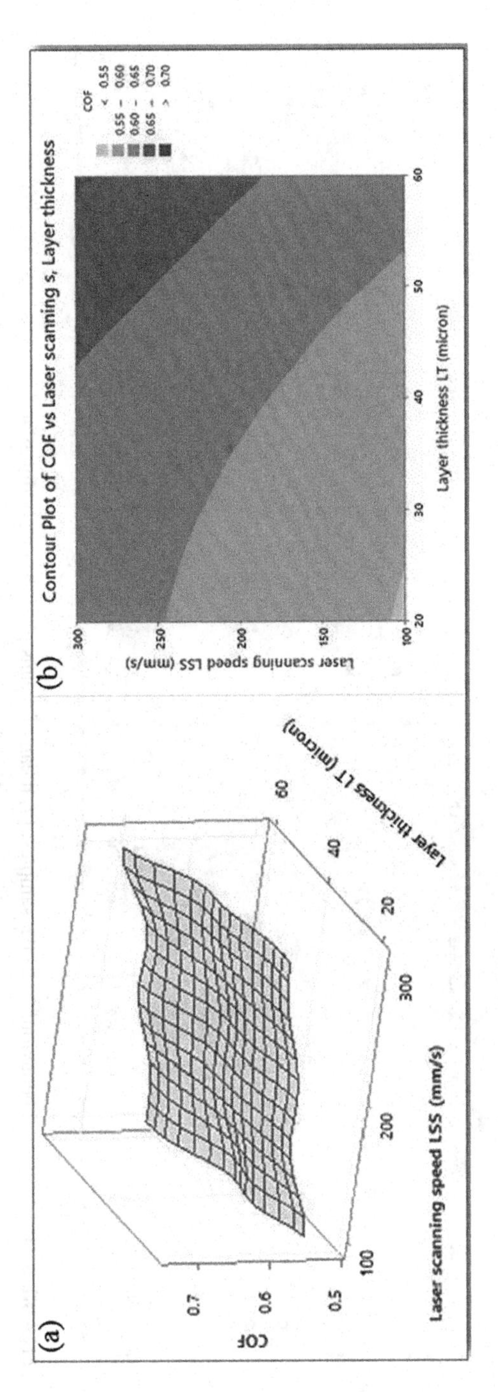

FIGURE 5.9 CoF behaviour: (a) surface plot; (b) contour plot.

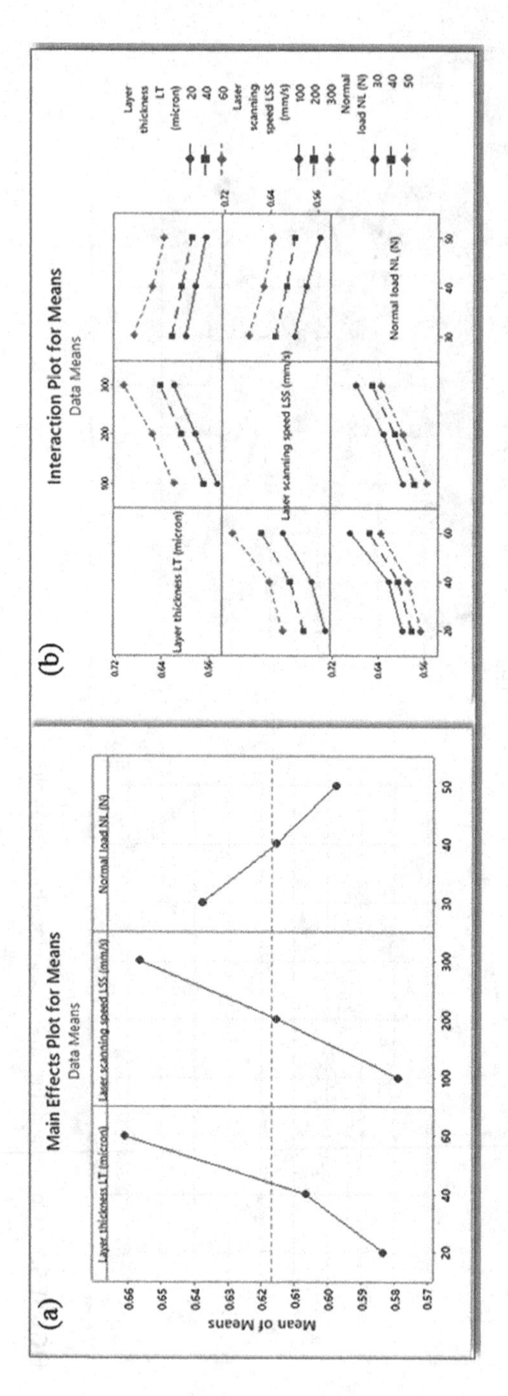

FIGURE 5.10 Frictional behaviour of LPBF samples: (a) mean of means plot; (b) interaction plot.

compared to laser scanning speed and the normal load employed. Further, Figure 5.10(b) depicts the interaction plots. It illustrates that the lines are not parallel, which indicates that all the three parameters, laser scanning speed, layer thickness and normal load, have a considerable influence on the COF behaviour.

5.3.2 Worn Surface Examination

Figure 5.11 shows the SEM pictures of the Cu-Ni alloy's worn surfaces after sliding under a load variation of 30 N and 50 N. The sliding wear mechanism operates in accordance with Archard's adhesive theory of wear. The worn-out asperities on the surface penetrate the sample surface, as they act as abrasive media and plough a series of grooves in it, thus causing three-body wear. It is evident from Figure 5.11(b) for the sample DS20/300 that increased debris formation, together with higher deformation pattern in the form of parallel, continuous, and deep grooves, are observed at 30 N load compared to sample DS20/100 (Figure 5.11(a)), which exhibits significantly less debris formation and the number of continuous grooves noticed are less involving the delamination process. Hence, lower specific wear rate and lower COF values are observed for samples (DS20/100) under a load of 30 N. It is noticed that the wear damage for the sample DS 20/300 is more significant (Figure 5.11(b)) due to the ploughing mechanism involving abrasive wear, which is responsible for higher COF values. On the other hand, Figure 5.12(a and b), for the samples DS60/100 and DS60/300 for 30 N load, the wear damage assessment using SEM reveals that DS 60/300 shows higher damage features (Figure 5.12(b)) such as higher degree of deep grooves, increased debris formation, and ploughing features along with the sliding directions. In contrast, DS 60/100 (Figure 5.12(a)) has lesser shallow groves formation, and lesser debris has formed in the material, indicating a lesser degree of damage. The DS 20/100 and DS 20/300 for 50 N load are shown in Figures 5.13(a and b). Figure 5.13(a) indicates shallow groves with a debris formation in the matrix, perhaps plastic deformation involving the delamination process. On the other hand, DS20/300 (Figure 5.13(b)) exhibits deep

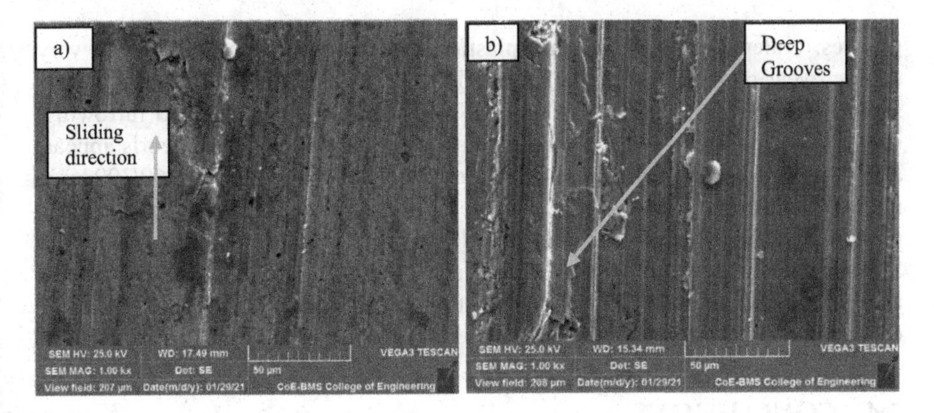

FIGURE 5.11 SEM of DMLS samples having 20 μm layer thickness for scanning speeds of (a) 100 mm/s, (b) 300 mm/s. at a load of 30 N.

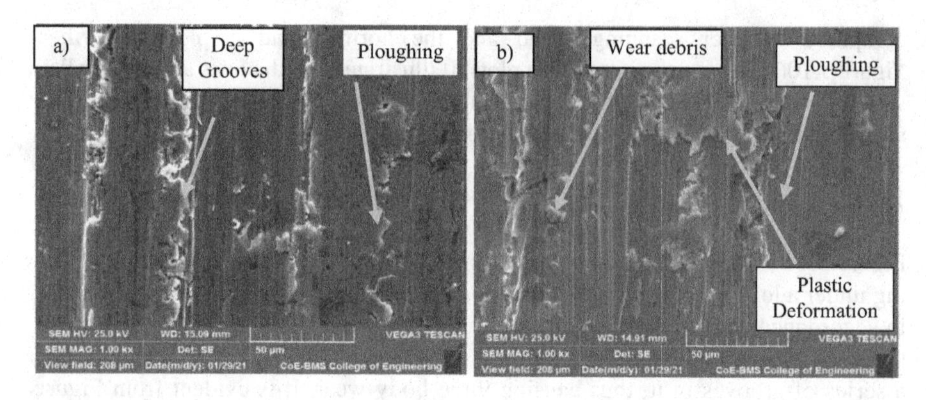

FIGURE 5.12 SEM of DMLS samples having 60 μm layer thickness for scanning speeds of (a) 100 mm/s, (b) 300 mm/s. at a load of 30 N.

FIGURE 5.13 SEM of DMLS samples having 20 μm layer thickness for scanning speeds of (a) 100 mm/s, (b) 300 mm/s. at a load of 50 N.

grooves, craters, voids and higher plastic deformation zones. They reveal very distinct wear damage features regarding DS 60/100 and DS 60/300 at 50 N load shown in Figures 5.14(a and b). The DS 60/100 have contributed to furrowing, deeply formed grooves on the surface, craters formation, moderate debris appearance, and flaky type sheet features of delamination mechanism. In DS 60/300 large number flakes/sheet formation due to shear of asperities, intense grooves, due to ploughing action and an overall higher degree of damage compared to the other SEM feature observed in Figures 5.11(a and b), 5.12(a and b) and 5.13(a and b) as well as Figure 5.14(a). Hence contributing to the higher COF values among all the samples.

5.4 CONCLUSIONS

The key points that emerged from the present experimental study are as follows:

FIGURE 5.14 SEM of DMLS samples having 60 μm layer thickness for scanning speeds of (a) 100 mm/s, (b) 300 mm/s. at a load of 50 N.

- The Cu-Ni alloy samples with varied thicknesses of 20 μm, 40 μm and 60 μm were produced by the novel LPBF process involving three different scan speeds, namely 100 mm/s, 200 mm/s and 300 mm/s
- The samples processed under lower-layer thickness and laser scanning speed exhibited higher Vickers hardness (~45%), compressive residual stress (~35%), and lower surface roughness (~20%) and porosity content (~43%). It was attributed to the powder and laser interaction phenomena during the LPBF process.
- Regarding the sliding wear behaviour, the sample with 20 μm layer thickness showed the least specific wear rate (at least one order of magnitude) than the sample with 60 μm layer thickness, at a laser scanning speed of 100 mm/s under all normal loads.
- As regards the frictional behaviour, the sample with 20 μm layer thickness showed the least COF (at least ~10%) than the sample with 60 μm layer thickness, at a laser scanning speed of 100 mm/s under all normal loads.
- The superior sliding wear performance is attributed to the better mechanical properties of the LPBF sample processed under lower laser scanning speed and thinnest layer thickness.
- The scanning electron microscopic studies on the worn surface morphologies corroborate well with the sliding wear data obtained, thus demonstrating the culmination of various wear mechanisms.

The work may be summarized that the LPBF parts, made of bronze-nickel alloy with varied frictional and wear responses, may be suitable for damping and bearing pad applications.

ACKNOWLEDGMENTS

The authors gratefully acknowledge Central Manufacturing Technology Institute (CMTI) Management for providing the facilities to prepare the 3D printed samples. The authors wish to acknowledge the support and help rendered by the Management of the Global Academy of Technology, the Vijaya Vittal Institute of Technology, the BMS Institute of Technology & Management, the Sambhram Institute of Technology, and the RV College of Engineering. The authors are very much thankful to the management of the Ambedkar Institute of Technology and the Vemana Institute of Technology Bengaluru for providing the necessary facilities to conduct the supporting tests.

REFERENCES

1. F. Calignano, D. Manfredi, E. P. Ambrosio, L. Iuliano, and P. Fino, "Influence of process parameters on surface roughness of aluminum parts produced by DMLS," *Int. J. Adv. Manuf. Technol.*, vol. 67, no. 9–12, pp. 2743–2751, 2013, https://doi.org/10.1007/s00170-012-4688-9.
2. A. Shrivastava, S. Anand Kumar, and S. Rao, "A numerical modelling approach for prediction of distortion in LPBF processed Inconel 718," *Mater. Today: Proc.*, vol. 44, pp. 4233–4238, 2020. https://doi.org/10.1016/j.matpr.2020.10.538.
3. B. K. Nagesha, S. A. Kumar, K. Vinodh, A. Pathania, and S. Barad, "A thermo-mechanical modelling approach on the residual stress prediction of SLM processed HPNGV aero-engine part," *Mater. Today: Proc.*, 2021. https://doi.org/10.1016/j.matpr.2020.12.940
4. V. Rajkumar, B. K. Nagesha, A. K. Tigga, S. Barad, and T. N. Suresh, "Single crystal metal deposition using laser additive manufacturing technology for repair of aero-engine components," *Mater. Today: Proc.*, 2021. https://doi.org/10.1016/j.matpr.2021.02.083
5. A. Pathania, S. A. Kumar, B. K. Nagesha, S. Barad, and T. N. Suresh, "Reclamation of titanium alloy based aerospace parts using laser based metal deposition methodology," *Mater. Today: Proc.*, 2021. https://doi.org/10.1016/j.matpr.2021.01.354
6. A. Shrivastava, S. Rao, B. K. Nagesha, S. Barad, and T. N. Suresh, "Remanufacturing of nickel-based aero-engine components using metal additive manufacturing," *Mater. Today: Proc.*, 2021.
7. S. A. Kumar, et al., "Fretting wear behavior of laser peened Ti-6Al-4V," *Tribol. Trans.*, vol. 55, no. 5, pp. 615–623, 2012. https://doi.org/10.1080/10402004.2012.686087.
8. F. Attarzadeh, B. Fotovvati, M. Fitzmire, and E. Asadi, "Surface roughness and densification correlation for direct metal laser sintering," *Int. J. Adv. Manuf. Technol.*, vol. 107, no. 5–6, pp. 2833–2842, 2020. https://doi.org/10.1007/s00170-020-05194-0.
9. H. Hassanin, A. Elshaer, R. Benhadj-Djilali, F. Modica, and I. Fassi, "Surface finish improvement of additive manufactured metal parts," *Micro Precis. Manuf.*, pp. 145–164, 2018. https://doi.org/10.1007/978-3-319-68801-5_7.
10. H. Qin, R. Xu, P. Lan, J. Wang, and W. Lu, "Wear performance of metal materials fabricated by powder bed fusion: A literature review," *Metals (Basel)*, vol. 10, no. 3, 2020. https://doi.org/10.3390/met10030304.
11. B. Zhang, Y. Li, and Q. Bai, "Defect formation mechanisms in selective laser melting: A review," *Chinese J. Mech. Eng. (English Ed.)*, vol. 30, no. 3, pp. 515–527, 2017. https://doi.org/10.1007/s10033-017-0121-5.
12. S. Anand Kumar, et al., "Effects of laser peening on fretting wear behaviour of alloy 718 fretted against two different counterbody materials," *Proc. Inst. Mech. Eng., Part J: J. Eng. Tribol.*, vol. 231, pp. 1276–1288, 2017. https://doi.org/10.1177/1350650117692707.
13. S. J. S. Ilangovan, "Dry sliding wear behaviour of sand cast Cu-11Ni-6Sn alloy," *Int. J. Res. Eng. Technol.*, vol. 3, no. 7, pp. 28–32, 2014. https://doi.org/10.15623/ijret.2014.0307005.
14. M. Hussain, V. Mandal, P. K. Singh, P. Kumar, V. Kumar, and A. K. Das, "Experimental study of microstructure, mechanical and tribological properties of cBN particulates SS316 alloy based MMCs fabricated by DMLS technique," *J. Mech. Sci. Technol.*, vol. 31, no. 6, pp. 2729–2737, 2017. https://doi.org/10.1007/s12206-017-0516-3.

15. M. Lorusso et al., "Tribological behavior of aluminum alloy AlSi10Mg-TiB2 composites produced by direct metal laser sintering (DMLS)," *J. Mater. Eng. Perform.*, vol. 25, no. 8, pp. 3152–3160, 2016. https://doi.org/10.1007/s11665-016-2190-5.
16. R. Kumar, M. Antonov, U. Beste, and D. Goljandin, "Assessment of 3D printed steels and composites intended for wear applications in abrasive, dry or slurry erosive conditions," *Int. J. Refract. Met. Hard Mater.*, vol. 86, p. 105126, 2020. https://doi.org/10.1016/j.ijrmhm.2019.105126.
17. A. N. Jinoop, S. K. Subbu, C. P. Paul, and I. A. Palani, "Post-processing of laser additive manufactured inconel 718 using laser shock peening," *Int. J. Precis. Eng. Manuf.*, vol. 20, no. 9, pp. 1621–1628, 2019. https://doi.org/10.1007/s12541-019-00147-4.
18. C. U. Brown et al., *The Effects of Laser Powder Bed Fusion Process Parameters on Material Hardness and Density for Nickel Alloy 625, NIST Advanced Manufacturing Series*, pp. 100–119, 2018. https://doi.org/10.6028/NIST.AMS.100-19
19. K. Kempen et al., "Microstructure and mechanical properties of selective laser melted 18Ni-300 steel," *Phys. Procedia*, vol. 12, pp. 255–263, 2011.
20. S. D. Nath, G. Gupta, M. Kearns, O. Gulsoy, and S. V. Atre, "Effects of layer thickness in laser-powder bed fusion of 420 stainless steel," *Rapid Prototyp. J.*, vol. 26, no. 7, pp. 1197–1208, 2020. https://doi.org/10.1108/RPJ-10-2019-0279.
21. N. Khanna et al., "Investigations on density and surface roughness characteristics during selective laser sintering of Invar-36 alloy," *Mater. Res. Express*, vol. 6, p. 086541, 2019.
22. A. K. Subramaniyan, S. R. Anigani, S. Mathias, A. Pathania, P. Raghupatruni, and S. S. Yadav, "Influence of heat treatment on microstructure, mechanical and wear properties of maraging steel fabricated using direct metal laser sintering (DMLS) technique," *J. Mater.: Des. Appl.*, 2021. https://doi.org/10.1177/14644207211037342.
23. B. Liu, R. Wildman, C. Tuck, I. Ashcroft, and R. Hague, "Investigaztion the effect of particle size distribution on processing parameters optimisation in selective laser melting process," in *22nd Annual International Solid Freeform Fabrication Symposium – An Additive Manufacturing Conference SFF 2011*, pp. 227–238, 2011.
24. H. M. Gajera, K. G. Dave, V. P. Darji, and K. Abhishek, "Optimization of process parameters of direct metal laser sintering process using fuzzy-based desirability function approach," *J. Brazilian Soc. Mech. Sci. Eng.*, vol. 41, no. 3, pp. 1–23, 2019. https://doi.org/10.1007/s40430-019-1621-2.
25. M. Resch, A. F. H. Kaplan, and D. Schuoecker, "Laser-assisted generating of three, dimensional parts by the blown powder process," in *XIII International Symposium on Gas Flow and Chemical Lasers and High-Power Laser Conference*, vol. 4184, pp. 555–558, 2001.
26. Y. Li, H. Yang, X. Lin, W. Huang, J. Li, and Y. Zhou, "The influences of processing parameters on forming characterizations during laser rapid forming," *Mater. Sci. Eng. A*, vol. 360, no. 1, pp. 18–25, 2003.
27. P. Li, D. H. Warner, A. Fatemi, and N. Phan, "Critical assessment of the fatigue performance of additively manufactured Ti–6Al–4V and perspective for future research," *Int. J. Fatigue*, vol. 85, pp. 130–143, 2016.
28. S. Anand Kumar, A. Sudarshan Reddy, S. Mathias, A. Shrivastava, and P. Raghupatruni, "Investigation on pulsed electrolytically polished AlSi10Mg alloy processed via selective laser melting technique," *Proc. Inst. Mech. Eng., Part L: J. Mater.: Des. Appl.*, 2021. https://doi.org/10.1177/14644207211045301
29. A. Shrivastava, S. A. Kumar, B. K. Nagesha, and T. N. Suresh, "Electropolishing of Inconel 718 manufactured by laser powder bed fusion: Effect of heat treatment on hardness, 3D surface topography and material ratio curve," *Opt. & Laser Technol.*, vol. 144, p. 107448, 2021.

30. A. A. Shapiro, et al., "Additive manufacturing for aerospace flight applications," *J. Spacecr. Rocket.*, vol. 53, no. 5, pp. 952–959, 2016.
31. C. de Formanoir, S. Michotte, O. Rigo, L. Germain, and S. Godet, "Electron beam melted Ti–6Al–4V: Microstructure, texture and mechanical behavior of the as-built and heat-treated material," *Mater. Sci. Eng. A*, 652, pp. 105–119, 2016.
32. J. H. Shaikh, N. K. Jain, and V. C. Venkatesh, "Precision finishing of bevel gears by electrochemical honing," *Mater. Manuf. Process.*, vol. 28, no. 10, pp. 1117–1123, 2013.
33. L. Mugwagwa, I. Yadroitsev, and S. Matope, "Effect of process parameters on residual stresses, distortions, and porosity in selective laser melting of maraging steel 300," *Metals (Basel)*, vol. 9, no. 10, 2019. https://doi.org/10.3390/met9101042.
34. B. K. Panda and S. Sahoo, "Thermo-mechanical modeling and validation of stress field during laser powder bed fusion of AlSi10Mg built part," *Results Phys.*, vol. 12, no. January, pp. 1372–1381, 2019. https://doi.org/10.1016/j.rinp.2019.01.002.
35. H. Ali, H. Ghadbeigi, and K. Mumtaz, "Processing parameter effects on residual stress and mechanical properties of selective laser melted Ti6Al4V," *J. Mater. Eng. Perform.*, vol. 27, no. 8, pp. 4059–4068, 2018. https://doi.org/10.1007/s11665-018-3477-5.
36. S. Anand Kumar, et al., "Effects of laser peening and counterbody material on fretting wear behaviour of alloy 718," *J. Eng. Tribol.*, vol. 233, pp. 1276–1288, 2017.
37. C. S. Ramesh, C. K. Srinivas, and K. Srinivas, "Friction and wear behaviour of rapid prototype parts by direct metal laser sintering," *Tribol. - Mater. Surfaces Interfaces*, vol. 1, no. 2, pp. 73–79, 2007. https://doi.org/10.1179/175158407x231330.
38. C. S. Ramesh and C. K. Srinivas, "Friction and wear behavior of laser-sintered iron-silicon carbide composites," *J. Mater. Process. Technol.*, vol. 209, no. 14, pp. 5429–5436, 2009. https://doi.org/10.1016/j.jmatprotec.2009.04.018.
39. Y. Zhu, J. Zou, and H. Yang, "Wear performance of metal parts fabricated by selective laser melting: A literature review," *J. Zhejiang Univ. Sci. A*, vol. 19, no. 2, pp. 95–110, 2018. https://doi.org/10.1631/jzus.A1700328.
40. B. Sagbas, "Post-processing effects on surface properties of direct metal laser sintered AlSi10Mg parts," *Met. Mater. Int.*, vol. 26, no. 1, pp. 143–153, 2020. https://doi.org/10.1007/s12540-019-00375-3.
41. C. Ingenthron et al., "Wear studies in binder jet additive manufactured stainless steel-bronze composite," in *Proceedings of Solid Freeform Fabrication Symposium*, pp. 732–744, 2015.
42. C. D. Naiju and P. M. Anil, "Influence of operating parameters on the reciprocating sliding wear of direct metal deposition (DMD) components using taguchi method," *Procedia Eng.*, vol. 174, pp. 1016–1027, 2017. https://doi.org/10.1016/j.proeng.2017.01.254.
43. Ş. Kasman and I. E. Saklakoğlu, "Surface roughness and wear behavior of laser machined AISI H13 tool steel," *Adv. Mat. Res.*, vol. 264–265, pp. 1264–1269, 2011. https://doi.org/10.4028/www.scientific.net/AMR.264-265.1264.
44. E. Rabinowicz, *Friction and Wear of Materials*. New York: John Wiley & Sons Inc, 1995.
45. I. V. Kragelskii, *Friction and Wear*. Moscow: Mashinostroenie, 1968.

6 Tribology of Graphene Oxide-Based Multilayered Coatings for Hydrogen Applications

Prabakaran Saravanan

BITS-Pilani Hyderabad Campus, Hyderabad, Telegana, India

CONTENTS

6.1 INTRODUCTION

The tribology at different scales such as nano-, micro- and macro-scales needs further understanding. The energy efficiency and life span of devices with moving components increases with decrease in friction. The scientific estimation is that one-third of the fuel consumed in automobiles used simply to tackle the friction [1]. The mitigation of energy losses due to friction may contribute significantly to the efficiency and lifespan of machines, as well as helping to fight against climate change by cutting anthropogenic CO_2 emissions.

DOI: 10.1201/9781003243205-6

Fossil fuel-based oils are the common lubricants used to reduce friction and wear. Although, in many other critical applications, such as in the aerospace industry, the usage of conventional liquid lubricants is restricted, particularly in extremes of temperature, pressure and environments (i.e., reactive/corrosive) [2, 3]. Therefore, the alternative solid lubricants are of great interest. State-of-the-art solid lubricants include molybdenum disulfide (MoS_2), graphite, boron nitride (BN), tungsten disulfide (WS_2), magnesium silicate (e.g. $Mg_3Si_4O_{10}(OH)_2$, talc), and polymers such as polytetrafluoroethylene (PTFE, Teflon®) [3, 4]. These common lubricants can be applied as surface coatings or as fillers in self-lubricating composite materials.

Solid lubricant coatings on sliding contacts most often produce a thin layer of transfer material from the coating to the counterface. This newly transferred layer of material may be called as tribofilm or transfer layer or tribo layer. Due to such transfers, different chemistry or microstructure may be observed on wear surfaces because of surface chemical reactions occurring with the surrounding environment. Henceforth, it is a concern with solid lubricant coatings that it may reduce friction or wear in one environment but fail quickly in a different one [4]. Due to the controls of the shear resistance in the interface zone, the incorporation of solid lubricants could also improve the tribological behavior in polymer and its composites [5].

Graphene and graphene oxide (GO) are emerging solid lubricants which have been investigated for sliding applications in various environments [6]. The majority of these works investigated graphene (and related materials) as potential additives for traditional fossil fuel-based oil lubricants. Dou et al. recently reported an enhancement of the lubrication properties of polyalphaolefin (PAO) oil blended with crumpled micron-sized graphene balls, with a particular focus on particle morphology [7]. Recent studies on solid lubrication have demonstrated that few-layer graphene onto 440C stainless steel using solvent casting technique can provide a reduction in the coefficient of friction (COF) from ~1.0 to ~0.2 and a significant increase in wear durability by three orders of magnitude [8, 9]. The macroscopic tribology of chemical vapor deposition (CVD) grown monolayer graphene transferred onto 440C steel has also been investigated. The COF was ~0.2, and in hydrogen (900 mbar) the wear life was around 40 times longer than in nitrogen (900 mbar) [10]. The excessive damage to the graphene layer prevented by hydrogen passivation of defects is attributed to be reason for this excellent frictional behavior in hydrogen.

The adhesion of graphene onto different solid substrates remains a challenge. For instance, Liang et al. have reported pure GO film, deposited on silicon substrates via the electrophoretic deposition technique, achieved the COF values as low as 0.05 and suggested that GO can serve as a common MEMS parts solid lubrication material [11]. Recently, the layer-by-layer (LbL) self-assembly technique has been utilized to build a GO multilayered composite. In this technique, the electrostatic attraction between the negative and positive charges from GO and polymers, respectively, is used for fabrication. In this way, extremely delicately defined multilayers of alternating positively and negatively charged materials can be built up simply by dip-coating, spin-coating, printing, or spray-coating. Generally, the further transformation of such electrostatic assemblies to covalent bonding can be performed using UV irradiation or heat treatment [12]. Solid merits of this coating technique include simplicity, a high degree of thickness control (within several nanometers), facile deposition onto complex geometries, low cost, and scalability [13]. In tribology, thin films of poly(sodium 4-styrenesulfonate)-mediated graphene assembled by LbL with polyethylenimine (PEI) were measured to have a COF of ~0.2. A minimal increase in wear life was reported with the thickness increase [14, 15]. Very few articles in the literatures have investigated the environmental effect on the frictional behavior of graphene and GO solid lubricants. The vast majority report a COF of ~0.2 for graphene or/graphene oxide coatings on steel [9, 10]. However, macroscale superlubricity was observed for graphene sliding against a diamond-like carbon (DLC) counterface (COF ~0.004) [16]. The formation of nanoscrolls at the contact area between the surfaces is demonstrated as the reason for this observed superlubricity.

In this chapter, we report the interesting tribological behavior of polyethylenimine/graphene oxide $(PEI/GO)_{15}$ LbL solid lubricant coatings in air, vacuum, nitrogen, and hydrogen gas environments [17]. The basic motivation for using graphene oxide was its high strength, processability (water dispersible), and the fact that it is a monolayer. PEI was chosen specifically to "glue" graphene oxide layers together via the LbL approach in order to create thicker coatings of GO and improve adhesion to the steel surface. The most related conventional solid lubricant – graphite – has low COFs in humid environments, however fail in dry one, therefore any new system that improves on this is desirable. In our work, extraordinarily low COFs and long wear lifetimes are reported, depending on the thickness and the environment. Detailed microstructural analysis of wear surfaces and debris and computer simulations are carried out to gain understanding into the mechanism behind these substantial improvements.

6.2 EXPERIMENTAL METHODS

6.2.1 Coating Deposition

The 0.2 wt% of polyethyleneimine (PEI) solution was prepared by dissolving the branched polyethyleneimine (PEI) (Mw ~25000 g mol^{-1}, purity >99%, Sigma-Aldrich) in water. The pH of the PEI solution was measured as 10.3 (basic) by digital pH meter (SevenEasy, Mettler Toledo). To make a 0.1 mg ml^{-1} graphene oxide (GO) aqueous solution, as bought GO aqueous solution (4 mg ml^{-1}, Sigma-Aldrich, monolayer content >95%) was diluted in water and the pH of the final solution was measured as 3.75 (Acidic). The oxygen content in the GO is >36% (according to the datasheet). The deionized pure water (18.3 MΩ cm^{-1}) used for solutions and washing purposes was prepared using reverse osmosis, then ion exchange and filtration (Millipore, Direct-QTM).

Figure 6.1 shows a schematic diagram of the layer-by-layer $(PEI/GO)_{n=15}$ coating deposition. Electrostatic LBL deposition was performed by the absorption of oppositely charged molecular species alternatively onto different substrates (i.e. steel, quartz glass, silicon) [17, 18]. The substrates were rinsed in ethanol and dried in nitrogen, prior to film deposition. Subsequently, the substrates were subjected to oxygen plasma treatment using a plasma-etching system (FA-1, SAMCO, Japan) to introduce a negatively charged surface for absorption of the first layer of cationic PEI. Plasma treatment conditions are as follows: RF power: 55 W, and an O_2 flow:

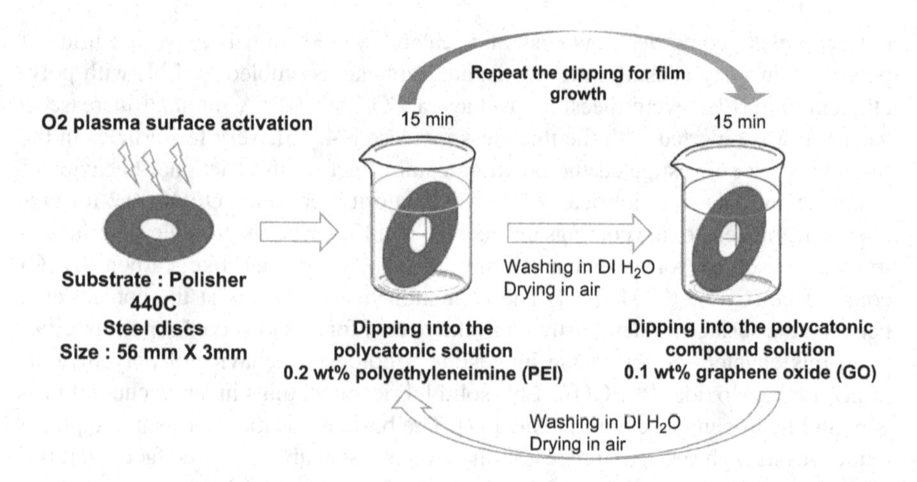

FIGURE 6.1 Procedure of the layer-by-layer $(PEI/GO)_{n=15}$ coating deposition.(Reproduced with permission from Saravanan, et al. [17].)

10 ml min^{-1} and the treatment time: 5 min for the steel substrates, respectively. After O_2 plasma etching, the substrates were alternatively immersed (for 15 min) into aqueous solutions of PEI (0.2 wt%) and GO (100 mg/L), as shown in Figure 6.1. Rising and drying the substrates in water and in nitrogen gas, respectively, strictly followed after each immersion step to avoid the physisorption.

To study the effect of the coating thickness on tribology, three distinct coatings with 5, 10 and 15 bilayers of PEI/GO were fabricated, respectively. The fabricated films are abbreviated as $(PEI/GO)_n$, where n is the number of bilayers deposited and n = 15 in this study. In each case, the outermost layer was GO.

6.2.2 Characterizations of $(PEI/GO)_{15}$ Coatings

The LbL $(PEI/GO)_{15}$ coating deposition was controlled using a quartz crystal microbalance (QCM), and characterized using Raman spectroscopies, UV-Vis, FT-IR-RAS and contact angle measurements. The optical microscopy, scanning electron microscopy (SEM) and transmission electron microscopy (TEM) were used for coatings, worn surface and wear debris morphology characterization.

QCM on electrodes with a frequency of 9 MHz (USI System, Fukuoka, Japan) were used for monitoring the adsorption of the PEI and GO layers after each deposition cycles. The frequency change before after deposition was used to estimate the mass change using the modified Sauerbrey equation [19].

The $(PEI/GO)_{15}$ coatings on quartz glass substrates were subjected to UV-vis absorption analysis, immediately after each deposition cycle at room temperature using a UV-vis-NIR V-670 spectrophotometer.

Raman characterization was done on fresh and worn surfaces (after tribology tests) of (PEI/GO) coating by a DXR™ Raman Microscope (Thermo Scientific, USA). The wavelength and a maximum laser power are 532 ± 1 nm and 10 mW, respectively. The spectras were acquired in the range 0 to 3500 cm^{-1}, with an exposure time and laser power of 20 s and 1 mW, respectively. The numerical aperture of the 20X lens was 50 μm and the laser beam diameter was 1 mm.

To study the nature of the surface (hydrophobic/hydrophilic), the water contact angle was carried at the end of each deposition cycle, using a CA-W automatic contact angle meter (Kyowa Interface Science, Japan) equipped with an AD-31 auto-dispenser. The reported data was averaged from 10 separate measurements on the same coating.

Morphological investigation of the coating surfaces, cross-sections and wear debris after the sliding tests was conducted using field emission scanning electron microscopy (5 kV, FESEM Hitachi S-5200). The samples, fractured in liquid nitrogen and dried under vacuum, were used for cross-sectional imaging. All samples were coated with a thin platinum layer to avoid charging due to electron beam. Finally, high-resolution transmission electron microscopy (HRTEM, JEM-2010, JEOL, Japan) with an operating voltage of 200 kV was utilized to characterize the wear debris collected after the sliding tests done in different environments.

6.2.3 Tribological Testing

The sliding tests were carried out in air (~140 ppm of H_2O), vacuum, dry hydrogen (H_2) and dry nitrogen (N_2) environments using an in-house made multi-environmental tribotester. Figure 6.2 shows the graphical illustration and a digital photograph of the tribotester setup, and reported in our previous works [20–23]. All tests were done at room temperature (25°C). The substrate and counterface are polished AISI 304 stainless steel (hardness ≈ 60 HRC, Ra. 0.01 μm) and SUS steel balls (8 mm diameter), respectively. The normal load was 5 N; the sliding speed was 0.47 m/s (250 rpm); and the sliding distance was ~1600 m. The hardness of (PEI/GO)15 coating on steel: ~0.12–0.16 GPa and the corresponding contact pressure was 0.1 GPa. The

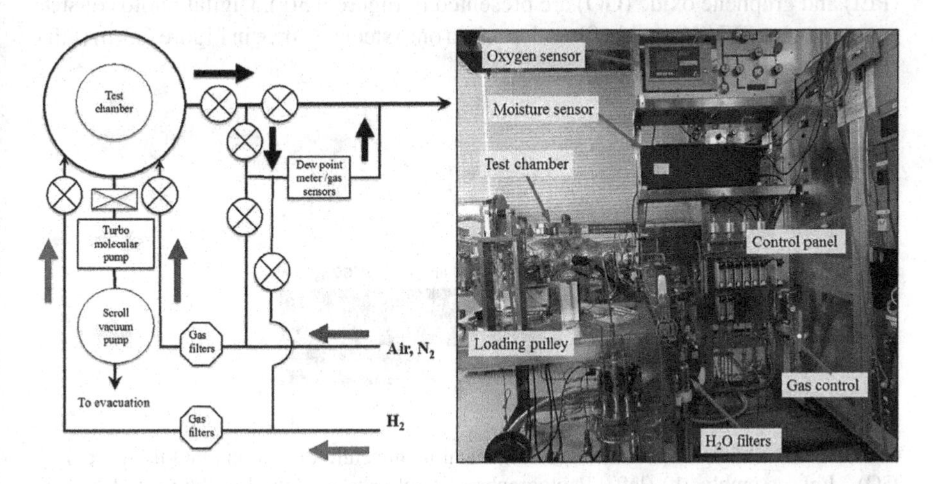

FIGURE 6.2 (Left) Digital photographs of the custom built tribometer setup a, overall view, b, detailed test chamber view. (Right). Schematic illustration of the custom-built tribometer setup.(Reproduced with permission from Saravanan, et al. [17].)

steady-state coefficient of friction (COF) was estimated by averaging from the point where steady-state behavior was first observed until the end of the test, or until the failure of the coating. When COF is higher than 0.4 or any abnormal fluctuation in COF is termed as failure of coating.

Our observation is such that when a COF is higher than 0.4, a sudden increase in COF to much higher values (e.g. 0.8 or 1) with significant fluctuations in COF values is often observed. Average COF values calculated from three repeated trails are reported for all cases. Similar definitions for COF have been used elsewhere [24–26] and in our previous works ([27–29]).

6.2.4 Density Functional Theory (DFT) Simulations

The insight into the interaction between GO layers under pressure in different gas environments was studied by DFT calculations. The influence of PEI was not taken into account at this point for simplicity. In addition, the amount of PEI in the LbL films is negligible. The H_2O, hydrogen and nitrogen molecules were intercalated between GO bilayers to simulate the experimental results. The calculations were done with the periodic plane wave DFT method, implemented in the Vienna Ab-initio Software Package (VASP) [30–32]. The general gradient approximation (GGA) was employed, and PAW pseudo potentials were used [33]. The cut off energy was set to 400 eV, and the k-point sampling was 2 × 2 × 1.

6.3 RESULTS AND DISCUSSION

6.3.1 Coating Characterization Results: Morphology and Growth

The multilayer $(PEI/GO)_{15}$ thin films were coated onto 440C steel substrates using LbL electrostatic self-assembly. The chemical architectures of polyethyleneimine (PEI) and graphene oxide (GO) are presented in Figure 6.3(a). Digital photos of steel substrate and $(PEI/GO)_{15}$ thin films deposited onto steel is shown in Figure 6.3(b) (left).

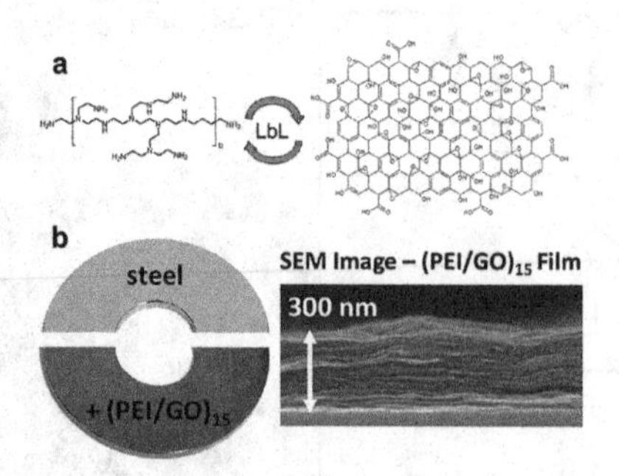

FIGURE 6.3 Sample specimens. (a) Chemical architecture of PEI and GO used for $(PEI/GO)_{15}$ LbL assembly. (b) (left), Photographs of steel substrates used in the macrotribological tests, coated with $(PEI/GO)_{15}$ bilayers. (b) (right), Cross-sectional SEM images of the $(PEI/GO)_{15}$ coatings fabricated on silicon (Si) wafers showing the approximate thicknesses. (Standard Deviation: ±5% for 15 bi-layers.)

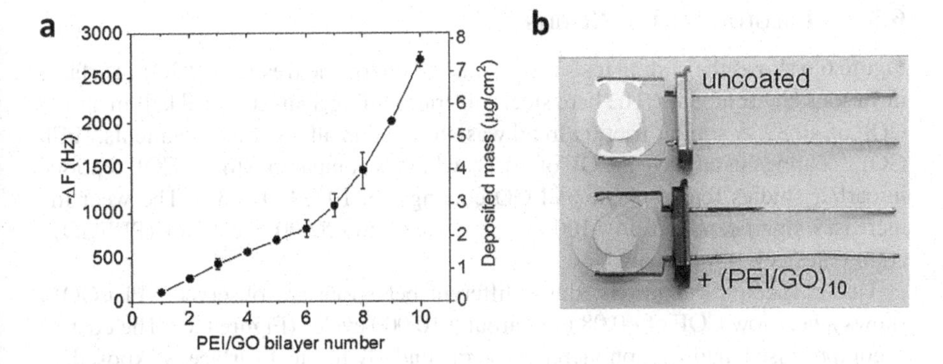

FIGURE 6.4 QCM measurements. (a) Change in QCM frequency of oscillation due to the deposition of $(PEI/GO)_n$ films. (b) Photograph of uncoated and (PEI/GO)10 -coated QCM sensors.(Reproduced with permission from Saravanan, et al. [17].)

The SEM images captured at cross-section of 15 bilayer films deposited on Si wafers (Figure 6.3(b)-right) clearly reveals the consistent layered structure of the coatings. Figure 6.4 shows the quartz crystal microbalance (QCM) measurements, performed to characterize the coating growth after each deposition cycle. Quantitative gravimetric data shown in Figure 6.5(a), confirms that the coating is mainly made of GO and much less PEI. The estimated density was *ca.* 0.5 g/cm³, which is similar that of dry GO powders. Cross-sectional SEM image in Figure 6.3(b) (right) was used to measure the thicknesses and also affirms the layered structure.

The roughness of the $(PEI/GO)_{15}$ film is 253 Å, measured using a profilometer and that is very consistent with the observed surface SEM images, i.e. the roughness increases with the number of LbL layers. Taking into account these values, the very minor influence of roughness on the friction result in the macroscale lubricity testing was expected as the values are few tens nanometers that is quite small considering the significantly large contact area between ball and counterface.

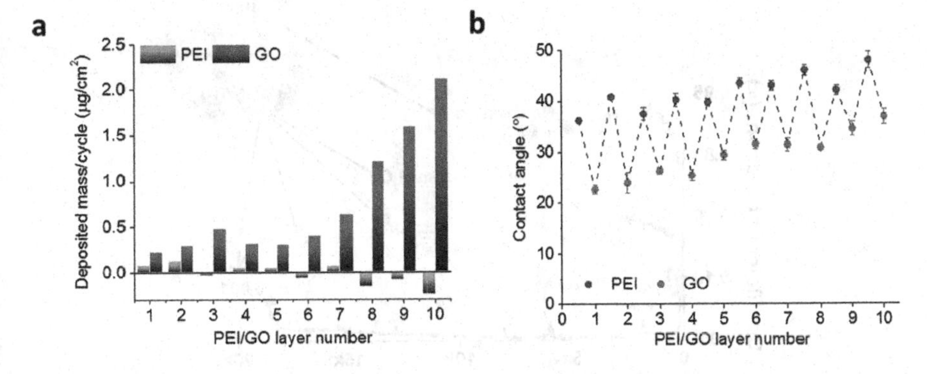

FIGURE 6.5 (a) Mass increase at each deposition step (measured by QCM) for PEI and GO, demonstrating faster GO mass increase after the 6th cycle. (b) Water contact angle measurements after each deposition step confirming the deposition of PEI despite the slight mass decrease at later cycles. (Reproduced with permission from [17].)

6.3.2 Tribological Test Results

Figure 6.6 shows the sliding tests results carried out on the fresh $(PEI/GO)_{15}$ surfaces in various environments. The bare steel substrate sliding against a steel ball in air has COF of ~0.89. A similar friction trend was observed in all other environments. $(PEI/GO)_{15}$ coatings in air have a COF of ~0.17, which is comparable to the COF reported in earlier studies for graphene and GO coatings [6, 10, 11, 14, 34]. The wear life increases significantly from ~100 cycles for steel to ~5,000 cycles for $(PEI/GO)_{15}$ coting on steel, respectively.

Under vacuum conditions, quite different behaviors are observed. $(PEI/GO)_{15}$ shows a very low COF of ~0.08 until around 10,000 cycles (Figure 6.6). The coating is compromised, and friction increases as the underlying steel surface is exposed. In dry H_2 with 2 ppm of water, $(PEI/GO)_{15}$ showed that a much lower COF of ~0.04 is observed up to ~16,000 cycles. However, in dry N_2, an extremely low COF value of < 0.02 is observed after a "running-in" period (approx. < 500 cycles), up until ~18,000 cycles. Significantly stable ultra-low COF is seen in N_2 for $(PEI/GO)_{15}$ over a greater time duration [17]. A specific self-healing behavior is observed in a hydrogen atmosphere as there was a COF increase for a short duration of a few thousand cycles before there was a return to the low COF again. This may be attributed to the hydrogen passivation effect postulated in earlier studies Berman et al. [10].

In brief, steel shows a COF ~ 0.9 in all environments. $(PEI/GO)_{15}$ coatings enhanced the friction and wear behavior and expanded the wear life in all gas atmospheres. Drier conditions lead to better tribological properties and, in particular, the best results are achieved in dry nitrogen. The reasons behind the lower friction coefficients (COF < 0.01) are explained briefly below.

Presumably, this is the very first experimental observation of macroscale superlubricity for GO-based solid coating on steel. In a prior study, superlubricity between a DLC-coated ball and SiO_2 was demonstrated by the inclusion of nanodiamonds (NDs) to solution-casted graphene layers [16]. The main finding is that the

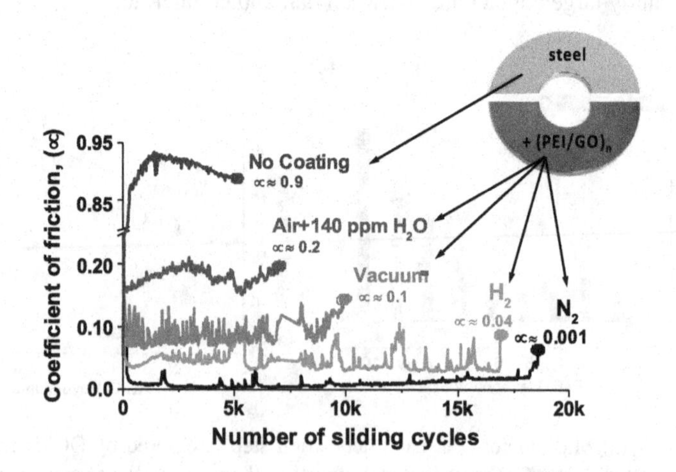

FIGURE 6.6 Typical coefficient of friction (COF) Vs. Number of cycles plot, showing friction behavior of bare steel and $(PEI/GO)_{15}$ coating in different environments.

nanodiamond particles promoted the transformation of graphene sheets into nanoscrolls in dry environments. A similar observation of conceding the lowest friction in the drier environments (H_2, N_2) is also seen here. Hence, as in [16]), the main possibility is that the observed ultra-low friction has its origin at the microstructural level. The environment-induced transfer layer formation and regeneration of transfer layers on the counterface material are among the other possibilities. Therefore, the worn surfaces and debris are thoroughly examined hereafter to gain an insight into the underlying mechanisms.

6.3.3 Worn/Wear Debris Microstructural Characterization

Figure 6.7 shows the optical micrographs of the tested samples, which demonstrates the condition of coating and severity of wear, the transfer film's existence and the size of contact area between the ball and surface. The apparent contact area was larger for the balls tested in air (Figure 6.7, air), in contrast with the ones tested in other gas environments. In summary, micrographs indicate that the reduction in contact area correlates well with observed frictional behavior.

SEM observation shows that the fragmentation of $(PEI/GO)_{15}$ coating into micron/sub-micron-scale particles does occur in all testing conditions, suggesting that the coating does not remain undamaged. Figure 6.8(a) shows that the wear debris after tested in air has more agglomerates with larger particles. Typical features are round-edged particles with an amorphous appearance (Figure 6.8(a)). While the in-dry conditions (vacuum, H_2, and N_2), the debris are typically smaller in size with a retention of the GO-layered structure. In addition, many smaller platelets do exist in the images (Figure 6.8(b–d)). The layered structure is much more obvious in N_2 than in vacuum and hydrogen.

TEM images of the wear debris collected after testing in air (Figure 6.8(e)) indicates typical microstructure of GO and similar to the particles observed in SEM (Figure 6.8(e)). By contrast, in drier environments (Figure 6.8(f)) the larger particles (μm sized) are covered with an ample amount of nano-featured particles such as hollow nanoparticles, and nanotubes/nanoscrolls. The wall thickness of these

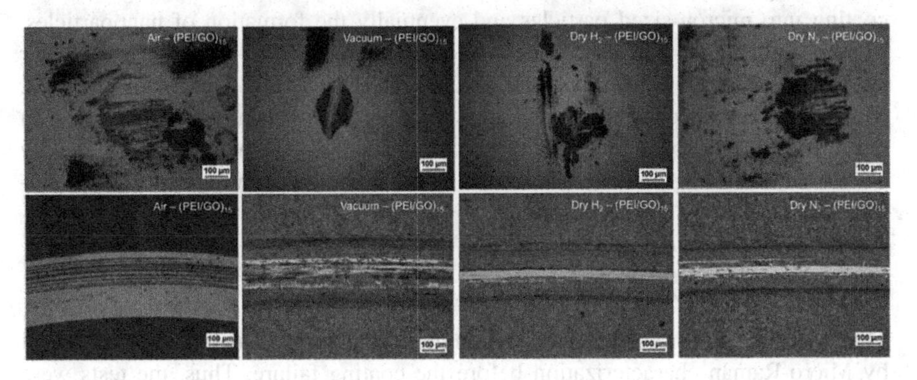

FIGURE 6.7 Optical microphotographs of counterface ball and worn surfaces after the sliding tests in in various environments.

FIGURE 6.8 Scanning electron microscopy (SEM) images of wear debris collected after sliding tests in various environments. a. air; b. vacuum; c. dry H_2; and d. dry N_2. Transmission electron microscopy (HR-TEM) image of the wear debris collected after the friction test in, e. air; f. N_2 (similar microstructure was observed in vacuum and H_2).(Reproduced with permission from Saravanan, et al. [17].)

nano-featured particles is about 2–10 nm, indicating that layers of $(PEI/GO)_{15}$ turned into these particles by the sliding induced scrolling and rolling process.

The tribology (i.e. friction, wear)-induced development of these nanostructures is extremely common, as shown by Berman et al. [16]. The energy supplied by a sliding ball on GO layers removes them from coating. Then this kinetic energy facilitates the scroll development through the promotion of energy imbalance in the system. This is believed to be responsible for the obtained superlubricity by means of behaving as nanoscale bearings. During the "running-in" period in N_2, the significant decrease in COF during the initial few hundred cycles may correspond to the milling up of the coating into micron-sized particles and eventually the formation of nanoparticles. These nanofeatured particles are considerably durable and strong, because their structures (i.e. hollow form) remain undamaged even after 25,000 sliding cycles. Moreover, there is considerable similarity between nanostructures observed here in our work to the earlier observation communicated by Berman et al., where dry conditions assisted graphene nanoscrolls formation resulted in macroscale superlubricity [16].

6.3.4 Raman Characterization Results

The local chemical structure of the GO on the sliding surfaces is further investigated by Micro-Raman characterization before the coating failure. Thus, the tests were stopped after around 5,000 cycles. The much lighter wear tracks were observed for dry environments (Figure 6.9(a)). The typical 'D' and 'G' bands, corresponding to

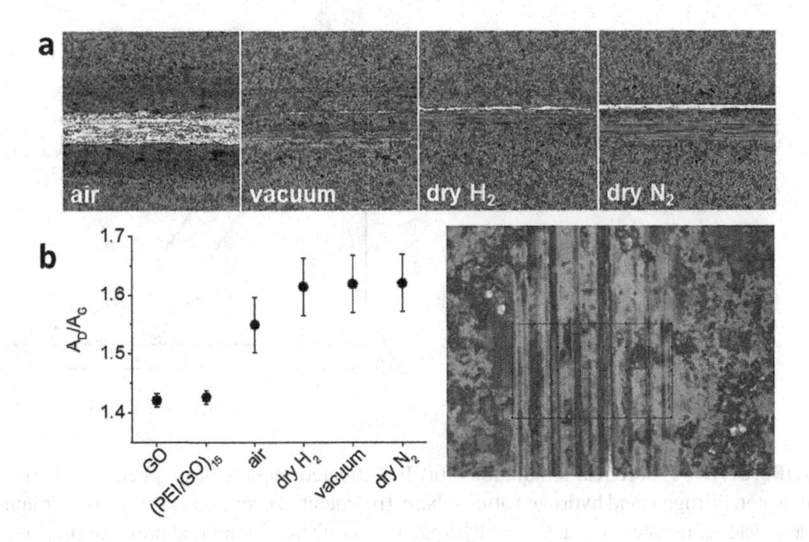

FIGURE 6.9 (a) Wear track images after 5,000 cycles in different environments. (b) A_D/A_G ratios calculated from averaged Raman spectras acquired on GO, fresh surface of $(PEI/GO)_{15}$, and wear tracks after sliding tests. Optical microscopy image shows the typical place of spectra acquisition from wear track. (Reproduced with permission from [17].)

GO, were seen for fresh surfaces [35, 36]. The (A_D/A_G) ratio for the fresh surface of all $(PEI/GO)_{15}$ coatings and pristine GO was about 1.42, which is suggesting that that no noticeable alterations occurred in the structure of GO due to LbL activity [37, 38]. However, the (A_D/A_G) ratios on the wear track after the test in air and dry environments (H_2, N_2 and vacuum) were increased to around 1.52 and 1.61, respectively. There are some reports in the literature on 'D' band intensity surges due to milling induced friction and sliding processes. This is a direct indication of an increase in the number of defects [37, 39, 40]. This indicates that significant alterations do occur in the chemical structure of GO in dry conditions than in air (humid), caused by sliding tests. The Raman results are in good accordance with the earlier observation of friction-induced formation of nanoparticles in dry environments.

6.3.5 DFT Simulations

In order to investigate further into the atomistic scales of this system, density functional theory (DFT) simulations were carried out on graphene oxide bilayers intercalated with molecules of H_2O, H_2 and N_2. A GO model was developed by introducing the epoxy, oxygen atoms, carboxyl and hydroxyl groups to pure graphene. The interlayer spacing between layers without any intercalated species was adjusted to 5.8 Å. The optimized interlayer distances with intercalated gas molecules for water, hydrogen, and nitrogen, were 6.3 Å, 7.1 Å and 7.0 Å, respectively (Figure 6.10). The contact pressure endured by the films in tribological tests was replicated by decreasing the cell parameter gradually, normal to the plane of the GO sheets with a 0.2 Å step size. At each step, the atomic coordinates were fully relaxed, but the cell parameters in all spatial directions were fixed. Figure 6.10(b) shows that the potential energy of

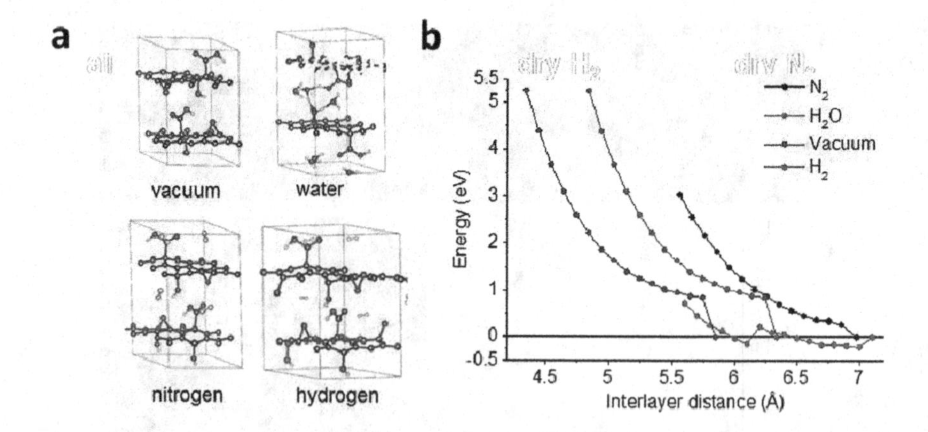

FIGURE 6.10 Theoretical simulations. (a) Investigated models of graphene oxide in vacuum, water, nitrogen, and hydrogen atmosphere. (b) Potential energy surface scans of graphene oxide in vacuum, water, nitrogen, and hydrogen atmosphere, for applied pressure perpendicular to the graphene oxide layers.(Reproduced with permission from Saravanan, et al. [17].)

surfaces for vacuum, hydrogen and nitrogen are reasonably similar without water, with an increase in repulsive force when the pressure is applied. Whilst in the presence of water, multiple local minima indicates the hydrogen bonding occurs and also a sequence of geometrical reorientations take place for of the water to minimize energy.

6.4 DISCUSSION

To discuss and interpret further the above results, the effect of various environments on the tribological behavior of $(PEI/GO)_{15}$ films can be elucidated by the observation such as film microstructure, nanostructures formation and interlayer interactions. In humid air, the film crumbles into much larger chunks of coating under the action of sliding, which resulted in reasonably low COF of ~0.2. Similarly, the hydrophilic nature of both PEI and GO allows the humidity (water) in the air to be absorbed. Based on DFT simulations, the humidity in the air leads to strong hydrogen bonding between the layers under pressure. Consequently, this hydrogen bonding hinders the breaking up of the sheets and milling them into smaller micro- and nanostructures, as ascertained by TEM observation.

The amount of humidity present in the nitrogen gas environment is negligible, compared to air. As claimed by the DFT simulations, the strong electrostatic repulsion between the GO layers promotes the separation and breaking up of the individual GO layers into flakes and nanostructures, as shown by the SEM and TEM images. Eventually, this results in a smaller apparent contact area and, subsequently, low COF and wear. Same particles were also observed in a vacuum.

The DFT simulations indicated that that the situation of hydrogen should be similar to nitrogen. By contrast, the tribological behavior in hydrogen was somewhat poorer than in nitrogen, with fewer nanostructures, as observed by TEM. The formation of water via reaction between H_2 molecules and O_2-based functional groups on GO is thought to account for the observed inferior behavior in H_2 [41].

The typical solid lubricant graphite has very contrasting and interesting friction and wear behavior than these LbL $(PEI/GO)_{15}$ coatings. Graphite shows high friction and superlubricity, in dry and humid environments, respectively [42–45]. By contrast, GO has superlubricity and high friction, in dry and humid environments, respectively. One postulation for observing low friction in a humid environment is due to the fact the weakening of the binding force between the basal planes of graphite near surface by intercalation. Another finding is that the chemisorbed water molecules increases the interlayer spacing between bulk and near surface basal planes [42, 46]. However, in GO-based coatings, intercalated water between two GO layers leads to the opposite effect due to the generation of more bonds between basal planes. The highly hydrophilic nature of GO is to be attributed for this behavior in contrasted with the hydrophobic nature of graphite. While in dry atmospheres, hydrophobic gas molecules in the GO coating behave the same as water in graphite, specifically expanding the distance between basal planes.

The frictional behavior of $(PEI/GO)_{15}$ coatings is compared with common solid lubricant diamond-like carbon (DLC) coatings. The friction coefficient of $(PEI/GO)_{15}$ coatings in dry nitrogen and hydrogen is comparable to hydrogenated amorphous DLC coatings in dry N_2 (COF ~ 0.06) [47, 48] and hydrogen (COF ~ 0.01) [49]. However, DLC is much more durable than $(PEI/GO)_{15}$ coatings. This may be attributed to the poor alignment and cross-linking of GO flakes. Further work is necessary to elucidate and validate the mentioned postulations.

Another important comparison of our findings should be done with the microscale friction tests on modified graphene. Several works revealed that force of friction increases upon the hydrogenation, fluorination or oxidation of graphene [50]. In the recent study, Lee et al. elucidated that low friction happens in the sp^2-rich GO subdomains [51]. The important difference of our work and mentioned nanoscale studies is that low friction in our case is considered to take place between the layers of GO and is enhanced depending on gaseous atmosphere. Nanoscale studies measure the friction between the AFM tip and usually single layer of (modified) graphene. In our case we have a combined effect: steel ball may actually have higher friction against the top layer of GO, but the GO layer has very low friction against the subsequent layer, therefore resulting in a total macroscale friction reduction.

Overall, based upon the DFT outcomes, it may be suggested that the repulsion between the GO layers in various environments has a great significance on tribological properties. Observed superlubricity appears to be the product of formation of micro- and nanoparticles, as well as the interlayer repulsion in various atmospheres. Both criteria are met only in dry N_2, in which excellent friction and wear behavior were witnessed for $(PEI/GO)_{15}$ coatings. The results obtained from the DFT calculations should be taken with care to the scale difference between the built computational model and the size and complexity of real coating/environment. However, as the model for DFT calculations represent an infinitely repeating unit cell of the real system and therefore do indeed represent a macroscopic object.

6.5 CONCLUSIONS

In summary, the tribological behavior (i.e. friction and wear) of $(PEI/GO)_{n=15}$ coatings on a steel substrate was explored in four different environments (humid air, vacuum, hydrogen and nitrogen). Overall, the coating provided a noticeable reduction in COF and enhancement in wear lives in all dry environments. $(PEI/GO)_{n=15}$ coatings in dry environments had a considerably superior tribological performance than the performance in a humid atmosphere. Superlubricity of COF ~ 0.02, obtained in dry atmospheres, is attributed to the formation of specific nanostructures under dry conditions, leading to a huge reduction in the contact area, friction and wear. DFT simulation suggested that in the presence of intercalated water under pressure, a strong hydrogen bonding network forms, preventing the separation of GO sheets to form such nanostructures. In dry environments, inert gases such as N_2 result in better tribological performance than reactive gases such as H_2. This is due to the formation of nanoparticles in dry environments, which are not observed in humid environments. These results open up the field for durable and almost frictionless graphene oxide solid lubricants for mechanical engineering applications.

ACKNOWLEDGMENTS

This research was supported by the World Premier International Research Center Initiative (WPI), MEXT, Japan, the Japan Science and Technology Agency (JST), JST-CREST program with Grant Number: JPMJCR13C1, Japan and, the JSPS Kakenhi start-up research grant (Grant No: UFG5H06471). The views expressed herein are those of the authors and are not necessarily those of the funding agencies. The names of the supplier companies mentioned in the paper are for reference only, and we believe same results may be obtained by the products and chemicals supplied by other companies.

REFERENCES

1. Holmberg, K., Andersson, P. & Erdemir, A., 2012. Global energy consumption due to friction in passenger cars. *Tribology International*, 47, pp. 221–234. Available at: https://doi.org/10.1016%2Fj.triboint.2011.11.022.
2. Donnet, C. & Erdemir, A., 2004. Historical developments and new trends in tribological and solid lubricant coatings. *Surface and Coatings Technology*, 180–181, pp. 76–84.
3. Mang, T. & Dresel, W. eds., 2006. *Lubricants and Lubrication*, Wiley-Verlag GmbH & Co. KGaA. Available at: https://doi.org/10.1002%2F9783527610341.
4. Scharf, T.W. & Prasad, S. V., 2012. Solid lubricants: A review. *Journal of Materials Science*, 48(2), pp. 511–531. Available at: https://doi.org/10.1007%2Fs10853-012-7038-2.
5. Shalwan, A. & Yousif, B.F., 2013. In state of art: Mechanical and tribological behaviour of polymeric composites based on natural fibres. *Materials & Design*, 48, pp. 14–24. Available at: https://doi.org/10.1016%2Fj.matdes.2012.07.014.
6. Berman, D., Erdemir, A. & Sumant, A.V., 2014a Graphene: A new emerging lubricant. *Materials Today*, 17(1), pp. 31–42. Available at: https://doi.org/10.1016/j.mattod.2013.12.003.

7. Dou, X. et al., 2016. Self-dispersed crumpled graphene balls in oil for friction and wear reduction. *Proceedings of the National Academy of Sciences*, 113, pp. 1528–1533. Available at: http://www.pnas.org/lookup/doi/10.1073/pnas.1520994113.
8. Berman, D., Erdemir, A. & Sumant, A. V., 2013a. Few layer graphene to reduce wear and friction on sliding steel surfaces. *Carbon*, 54, pp. 454–459. Available at: https://doi.org/10.1016%2Fj.carbon.2012.11.061.
9. Berman, D., Erdemir, A. & Sumant, A.V., 2013b. Reduced wear and friction enabled by graphene layers on sliding steel surfaces in dry nitrogen. *Carbon*, 59, pp. 167–175. Available at: http://linkinghub.elsevier.com/retrieve/pii/S0008622313002108.
10. Berman, D. et al., 2014b Extraordinary macroscale wear resistance of one atom thick graphene layer. *Advanced Functional Materials*, 24(42), pp. 6640–6646. Available at: www.afm-journal.de.
11. Liang, H. et al., 2013. Graphene oxide film as solid lubricant. *ACS applied materials & interfaces*, 5, pp. 6369–75. Available at: http://www.ncbi.nlm.nih.gov/pubmed/23786494.
12. Wei, Z., Barlow, D.E. & Sheehan, P.E., 2008. The assembly of single-layer graphene oxide and graphene using molecular templates. *Nano Letters*, 8(10), pp. 3141–3145.
13. Decher, G., 1997. Fuzzy nanoassemblies: Toward layered polymeric multicomposites. *Science*, 277(5330), pp. 1232–1237. Available at: https://doi.org/10.1126%2Fscience.277.5330.1232.
14. Liu, S. et al., 2012. Layer-by-layer assembly and tribological property of multilayer ultrathin films constructed by modified graphene sheets and polyethyleneimine. *Applied Surface Science*, 258(7), pp. 2231–2236. Available at: http://linkinghub.elsevier.com/retrieve/pii/S0169433211013912.
15. Ou, J. et al., 2012. Fabrication and tribological investigation of a novel hydrophobic polydopamine/graphene oxide multilayer film. *Tribology Letters*, 48(3), pp. 407–415. Available at: http://link.springer.com/10.1007/s11249-012-0021-x.
16. Berman, D. et al., 2015. Macroscale superlubricity enabled by graphene nanoscroll formation. *Science*, 348(6239), pp. 1118–1122. Available at: https://doi.org/10.1126%2Fscience.1262024.
17. Saravanan, P. et al., 2016. Macroscale superlubricity of multilayer polyethylenimine / graphene oxide coatings in different gas environments. *ACS Applied Materials & Interfaces*, 8(40), pp. 27179 – 27187. Available at: https://doi.org/10.1021/acsami.6b06779.
18. Saravanan, P. et al., 2017. Ultra-low friction between polymers and graphene oxide multilayers in nitrogen atmosphere, mediated by stable transfer film formation. *Carbon*, 122, pp. 395–403. Available at: http://linkinghub.elsevier.com/retrieve/pii/S0008622317306681.
19. Selyanchyn, R. et al., 2015. A nano-thin film-based prototype QCM sensor array for monitoring human breath and respiratory patterns. *Sensors (Switzerland)*, 15(8), pp. 18834–18850.
20. Fukuda, K., Hashimoto, M. & Sugimura, J., 2011. Friction and wear of ferrous materials in a hydrogen gas environment. *Tribology Online*, 6(2), pp. 142–147. Available at: https://doi.org/10.2474%2Ftrol.6.142.
21. Fukuda, K. & Sugimura, J., 2013. Influences of trace water in a hydrogen environment on the tribological properties of pure iron. *Tribology Online*, 8(1), pp. 22–27. Available at: https://doi.org/10.2474%2Ftrol.8.22.
22. Fukuda, K. et al., 2010. Influence of trace water and oxygen in a hydrogen environment on pure fe friction and wear. *Tribology Online*, 5(2), pp. 80–86. Available at: http://joi.jlc.jst.go.jp/JST.JSTAGE/trol/5.80?from=CrossRef.

23. Tanaka, H. et al., 2009. New experiment system for sliding tests in hydrogen and surface analysis with transfer vessel. *Tribology Online*, 4(4), pp. 82–87. Available at: http://joi.jlc.jst.go.jp/JST.JSTAGE/trol/4.82?from=CrossRef.
24. Hu, T., Zhang, Y. & Hu, L., 2012. Tribological investigation of MoS 2 coatings deposited on the laser textured surface. *Wear*, 278–279, pp. 77–82. Available at: https://doi.org/10.1016/j.wear.2012.01.001.
25. Neidhardt, J. et al., 2004. Structural, mechanical and tribological behavior of fullerene-like and amorphous carbon nitride coatings. *Diamond and Related Materials*, 13(10), pp. 1882–1888. Available at: https://doi.org/10.1016%2Fj.diamond.2004.05.012.
26. Sugimoto, I. & Miyake, S., 1988. Solid lubricating fluorine-containing polymer film synthesized by perfluoropolyether sputtering. *Thin Solid Films*, 158(1), pp. 51–60.
27. Saravanan, P. et al., 2014. The role of functional end groups of perfluoropolyether (Z-dol and Z-03) lubricants in augmenting the tribology of SU-8 composites. *Tribology Letters*, 56(3), pp. 423–434. Available at: https://doi.org/10.1007/s11249-014-0419-8.
28. Saravanan, P., Satyanarayana, N. & Sinha, S.K., 2013a. Self-lubricating SU-8 nanocomposites for microelectromechanical systems applications. *Tribology Letters*, 49(1), pp. 169–178. Available at: https://doi.org/10.1007/s11249-012-0055-0.
29. Saravanan, P., Satyanarayana, N. & Sinha, S.K., 2013b. Wear durability study on self-lubricating SU-8 composites with perfluoropolyther, multiply-alkylated cyclopentane and base oil as the fillers. *Tribology International*, 64, pp. 103–115. Available at: http://linkinghub.elsevier.com/retrieve/pii/S0301679X13001357.
30. Kresse, G. & Furthmüller, J., 1996a. Efficient iterative schemes for ab initio total-energy calculations using a plane-wave basis set. *Physical Review B*, 54(16), pp. 11169–11186. Available at: https://doi.org/10.1103%2Fphysrevb.54.11169.
31. Kresse, G. & Furthmüller, J., 1996b. Efficiency of ab-initio total energy calculations for metals and semiconductors using a plane-wave basis set. *Computational Materials Science*, 6(1), pp. 15–50.
32. Kresse, G. & Hafner, J., 1993. Ab initio molecular dynamics for liquid metals. *Physical Review B*, 47(1), pp. 558–561. Available at: https://doi.org/10.1103%2Fphysrevb.47.558.
33. Blöchl, P.E., 1994. Projector augmented-wave method. *Physical Review B*, 50(24), pp. 17953–17979.
34. Ou, J. et al., 2010. Tribology study of reduced graphene oxide sheets on silicon substrate synthesized via covalent assembly. *Langmuir*, 26(20), pp. 15830–15836. Available at: http://pubs.acs.org/doi/abs/10.1021/la102862d.
35. Bayer, T. et al., 2014. Characterization of a graphene oxide membrane fuel cell. *Journal of Power Sources*, 272, pp. 239–247. Available at: http://www.sciencedirect.com/science/article/pii/S0378775314013366.
36. Cançado, L.G. et al., 2011. Quantifying defects in graphene via raman spectroscopy at different excitation energies. *Nano Letters*, 11(8), pp. 3190–3196. Available at: https://doi.org/10.1021%2Fnl201432g.
37. Munir, K.S. et al., 2015. Quantitative analyses of MWCNT-Ti powder mixtures using raman spectroscopy: The influence of milling parameters on nanostructural evolution. *Advanced Engineering Materials*, 17(11), pp. 1660–1669. Available at: http://doi.wiley.com/10.1002/adem.201500142.
38. Sui, Z.-Y. et al., 2013. Preparation of three-dimensional graphene oxide–polyethylenimine porous materials as dye and gas adsorbents. *{ACS} Applied Materials & Interfaces*, 5(18), pp. 9172–9179. Available at: https://doi.org/10.1021%2Fam402661t.
39. Dresselhaus, M.S. et al., 2010. Perspectives on carbon nanotubes and graphene raman spectroscopy. *Nano Letters*, 10(3), pp. 751–758. Available at: http://pubs.acs.org/doi/abs/10.1021/nl904286r.

40. Mondal, O. et al., 2015. Reduced graphene oxide synthesis by high energy ball milling. *Materials Chemistry and Physics*, 161, pp. 123–129. Available at: https://doi.org/10.1016%2Fj.matchemphys.2015.05.023.
41. Pei, S. & Cheng, H.M., 2012. The reduction of graphene oxide. *Carbon*, 50(9), pp. 3210–3228.
42. Dienwiebel, M. et al., 2004. Superlubricity of graphite. *Physical Review Letters*, 92(12). 126101. Available at: https://doi.org/10.1103%2Fphysrevlett.92.126101.
43. Savage, R.H., 1948. Graphite lubrication. *Journal of Applied Physics*, 19(1), pp. 1–10. Available at: https://doi.org/10.1063%2F1.1697867.
44. Skinner, J., Gane, N. & Tabor, D., 1971. Micro-friction of Graphite. *Nature Physical Science*, 232(35), pp. 195–196. Available at: https://doi.org/10.1038%2Fphysci232195a0.
45. Yen, B.K., Schwickert, B.E. & Toney, M.F., 2004. Origin of low-friction behavior in graphite investigated by surface x-ray diffraction. *Applied Physics Letters*, 84(23), pp. 4702–4704.
46. Lancaster, J.K. & Pritchard, J.R., 1981. The influence of environment and pressure on the transition to dusting wear of graphite. *Journal of Physics D: Applied Physics*, 14(4), p. 747. Available at: http://stacks.iop.org/0022-3727/14/i=4/a=027.
47. Andersson, J., Erck, R.A. & Erdemir, A., 2003. Friction of diamond-like carbon films in different atmospheres. *Wear*, 254(11), pp. 1070–1075. Available at: https://doi.org/10.1016%2Fs0043-1648%2803%2900336-3.
48. Erdemir, A., 2001. The role of hydrogen in tribological properties of diamond-like carbon films. *Surface and Coatings Technology*, 146–147, pp. 292–297. Available at: https://doi.org/10.1016%2Fs0257-8972%2801%2901417-7.
49. Okubo, H. et al., 2015. Effects of hydrogen on frictional properties of {DLC} films. *Tribology Online*, 10(6), pp. 397–403. Available at: https://doi.org/10.2474%2Ftrol.10.397.
50. Ko, J.-H. et al., 2013. Nanotribological properties of fluorinated, hydrogenated, and oxidized graphenes. *Tribology Letters*, 50(2), pp. 137–144. Available at: https://doi.org/10.1007%2Fs11249-012-0099-1.
51. Lee, H. et al., 2016. Friction and conductance imaging of sp2- and sp3-hybridized subdomains on single-layer graphene oxide. *Nanoscale*, 8(7), pp. 4063–4069. Available at: https://doi.org/10.1039%2Fc5nr06469d.

7 Analysis of Frictional Stress Variations along Tooth Contact of Spur and Helical Gears

Santosh S. Patil
Manipal University Jaipur, Jaipur, India

Saravanan Karuppanan
Universiti Teknologi PETRONAS, Bandar Seri Iskandar, Malaysia

CONTENTS

7.1 INTRODUCTION

H. Hertz [1] in 1881 first solved the contact problem of two elastic cylinder bodies in contact and this has subsequently been the base of contact mechanics theory. All modern-day contact mechanics problems are interpreted and solved using this

DOI: 10.1201/9781003243205-7

classical theory. Archard [2] was the principal researcher after Hertz to have experimented on the contact pressure between two contacting and deforming bodies. The contact-related problems have been increasing and causing problems such as pitting and scuffing on gears, which has been the focus of much research interest [3]. According to the Hertz theory, it can be observed that the contact stresses depend on curvature radius and material properties of contacting bodies unlike for bending which depends on the shape and geometry of gear tooth [4]. In the past two or three decades, many researchers worked with different methodologies to evaluate the gear contact stress variations, i.e. through analytical/mathematical models, experimental and Finite Element (FE) techniques. Initial investigations on FE stress variation estimation of gear tooth without including friction were conducted by Ramamurti and Rao [5], Sundarajan and Young [6], Rao and Muthuveerappan [7], Chen and Tsay [8], Gosselin et al. [9] and Vijayakar et al. [10]. Other works with FE models presented by Chen et al. [11], Barone et al. [12], Chacon et al. [13], Parker and Vijayakar [14] and Atanasovska [15] used analytical calculations to validate the FE model for helical gears. They argued that these models are well suited and can be employed effectively. The frictional inclusion in gear contact stress evaluation is one of the important areas too. A literature review by Patil et al. [16] on contact stress evaluation, both with and without friction, explains different methodologies. It is also evident that Finite Element Analysis (FEA) is one of the better estimates of contact stresses in most of the cases. However, other works of Patil et al. [17–19] on frictional contact evaluation have not discussed the stress variations along the contact path (line of action), except one in which spur gear stress variations have been shown. Few other works of Hsieh et al. [20] and Atanasovska [21] also used a single tooth model, which may have a few limitations in appropriate estimation of stresses.

Therefore, the present chapter makes an attempt to discuss the detailed contact stress variation along the contact path of spur and different helical gears. This estimation will bring out the percentage variation of contact stresses along the contact path with respect to the pitch point of contact. Also, it will reveal the percentage increase in the stresses at different points due to frictional variation between 0 to 0.3 friction coefficient.

7.2 METHODOLOGY

The gear contact stresses have been evaluated using three approaches, analytical approach, Finite Element Method (FEM) and Romax Software. The flow of the current study is represented in Figure 7.1. It shows that the frictionless contact stresses calculated using the AGMA formula were used to validate frictionless FEM formulation. Later the validated FEM and RomaxDESIGNER were used to calculate the contact stresses along the contact path and contact stresses variation with friction.

$$\sigma_H = C_p \sqrt{\frac{F_n}{bdI}\left(\frac{\cos\beta}{0.95CR}\right)K_v K_o\left(0.93K_m\right)} \tag{7.1}$$

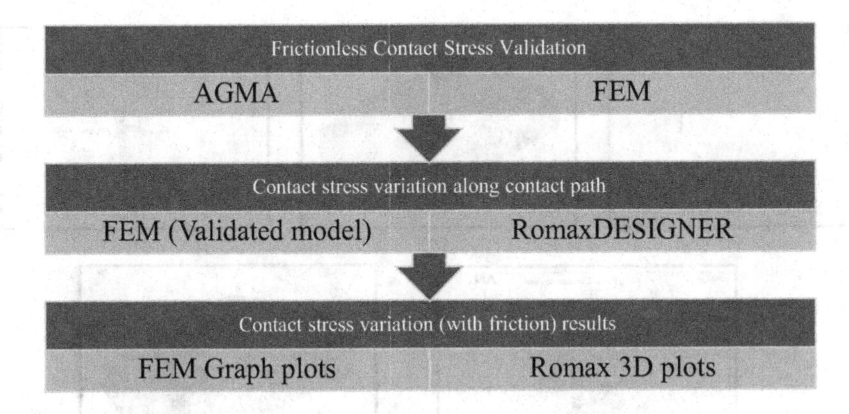

FIGURE 7.1 Flow chart of the study.

Analytical calculations were carried out using the AGMA equation represented by Equation 7.1 [17]. AGMA contact stress calculations were performed for the gears and comparison was done with FEM. The detailed specifications of the spur and helical gears are shown in Table 7.1. The FE methodology used to determine the contact stress distribution on the spur and helical gears pairs with frictionless contact conditions is summarized in Figure 7.2. A precise APDL program [22] was used to create involute geometries with no errors, but this requires the user to be skilled in APDL programming. The gear pairs were made of structural steel and their properties are mentioned as reported by the gear manufacturer. Figure 7.2 shows the steps which were followed to generate the 3D gear segment, where both the pinion and gear were correctly assembled to form the gear pair model. The same modelling procedure, right from the involute profile generation was followed to model all other

TABLE 7.1
Specifications of Finite Element Gear Sets

Analysis Gears	Spur Gear		5-Degree Helical Gear Pair		15-Degree Helical Gear Pair		25-Degree Helical Gear Pair		35-Degree Helical Gear Pair	
Parameter	**Pinion**	**Gear**	**Pinion**	**Gear**	**Pinion**	**Gear**	**Pinion**	**Gear**	**Pinion**	**Gear**
No. of Teeth	20	20	20	20	20	20	20	20	20	20
Normal Module	4.5	4.5	4.5	4.5	4.5	4.5	4.5	4.5	4.5	4.5
Normal Pressure Angle (degree)	20	20	20	20	20	20	20	20	20	20
Helix Angle (degree)	0	0	5	5	15	15	25	25	35	35
Pitch Diameter (mm)	91.5	91.5	91.84	91.84	94.72	94.72	100.95	100.95	111.69	111.69
Face Width (mm)	20	20	20	20	20	20	20	20	20	20
Centre Distance (mm)	91.5		91.84		94.72		100.95		111.69	
Contact Ratio	1.55		1.54		1.52		1.51		1.43	
Torque (Nm)					302					
Speed (rpm)					1000					

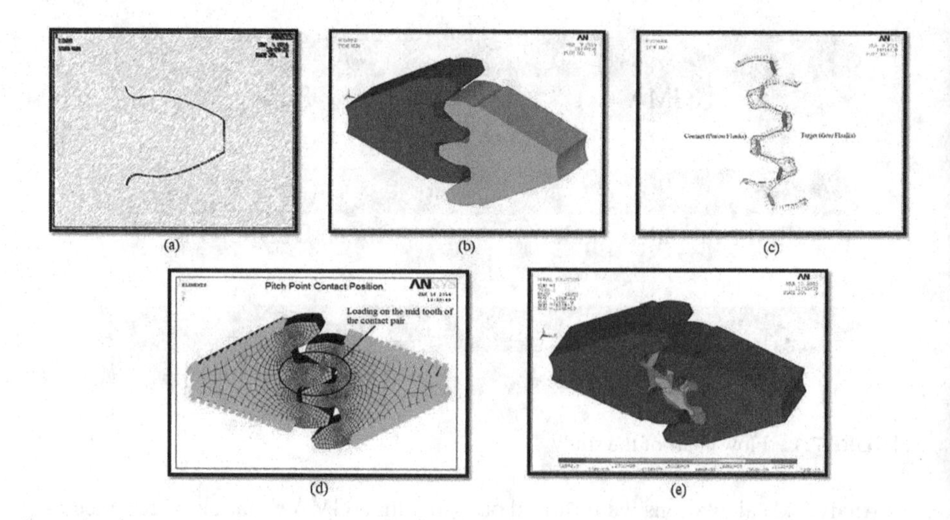

FIGURE 7.2 ANSYS FEA flow process (a) Involute profile of a spur gear, (b) Meshed segment of pinion and gear, (c) TARGET and CONTACT region on meshed gear pair, (d) Boundary conditions and loading on gear pair, (e) Sample stress contour of a gear pair.

helical gear sets. In the next step, the models are assembled, constrained and applied with boundary conditions prior to frictional contact analysis via FEM. The analytical results of the test gears (m = 4.5) were used for comparison with FEM frictional results.

The analysis results of the finite element method and the RomaxDESIGNER were then obtained and compared. Necessary improvements on the mesh refinement and solver settings were performed when the results did not converge and/or results were not within the allowable limits. The mesh refinement was performed and, based on the convergence test, a refined mesh was selected for further Finite Element Analyses. The effect of friction on contact stresses, along the contact path for spur and helical gears was evaluated and discussed.

7.3 VALIDATION OF FRICTIONLESS SPUR AND HELICAL GEAR MODELS

In this section, comparison of the tooth contact stresses obtained from the three-tooth section 3D model using ANSYS was done with the results obtained using AGMA standards. Equation 7.1 is recommended by the AGMA and the values of the parameters used in this equation were selected appropriately. First, the spur gear model was analysed. The specifications of spur gear were referred to Table 7.1.

$$\begin{aligned}\sigma_H &= 191\sqrt{\frac{6601.25}{0.02\times0.0915\times0.0803}\left(\frac{\cos 0}{0.95\times1.55}\right)1.18\times1\times(0.93\times1.2)} \\ &= 1210\,\text{MPa}\end{aligned} \tag{7.2}$$

Detailed investigations on the effect of different helix angles on the tooth contact stress were then carried out. The analysis was extended to a 5-degree helical gear pair. The calculation of AGMA contact stress for the meshed 5-degree helical gear (specification in Table 7.1) is given below:

$$\sigma_H = 191\sqrt{\frac{6601.25}{0.02\times0.09184\times0.0806}\left(\frac{\cos 5}{0.95\times1.54}\right)1.18\times1\times(0.93\times1.2)}$$
$$= 1208\,\text{MPa}$$

Similarly, the contact stress calculations for 15-, 25- and 35-degree gear pairs are shown below;

$$\sigma_H = 191\sqrt{\frac{6601.25}{0.02\times0.09472\times0.0806}\left(\frac{\cos 15}{0.95\times1.52}\right)1.18\times1\times(0.93\times1.2)}$$
$$= 1165\,\text{MPa}$$

$$\sigma_H = 191\sqrt{\frac{6601.25}{0.02\times0.10095\times0.0806}\left(\frac{\cos 25}{0.95\times1.51}\right)1.18\times1\times(0.93\times1.2)}$$
$$= 1071\,\text{MPa}$$

$$\sigma_H = 191\sqrt{\frac{6601.25}{0.02\times0.11169\times0.0806}\left(\frac{\cos 35}{0.95\times1.43}\right)1.18\times1\times(0.93\times1.2)}$$
$$= 961\,\text{MPa}$$

The calculation of the von Mises stresses at the contact region of gears for the above cases were repeated in ANSYS APDL. The FEA stress distribution results for the frictionless cases are summarized in Figure 7.3. The comparison of the AGMA results and the FEA results from ANSYS showed acceptable match. The results and the percentage differences between the two analyses approaches are included in Table 7.2.

The maximum values of the tooth contact stress obtained by the ANSYS method were considered and presented. For the spur gear pair, the ANSYS results differed by 1.0% of the values obtained by the AGMA. Similarly, for 5-, 15-, 25- and 35-degree helical gear pairs, the ANSYS results differed by 1.5%, 6.5%, 4.8% and 4.1%, respectively, of the values obtained by AGMA calculations. From these results, it was found that all cases gave a close approximation to the values obtained by the AGMA calculations. The observation in all the cases also shows that the AGMA stress values are greater than those obtained from the Finite Element Method. It can be justified that AGMA being the conservative method, which includes all types of corrective factors other than the friction factor, is expected to give higher stress values compared to FEM. Overall, in all of the cases of gear pairs, the FEM frictionless analysis

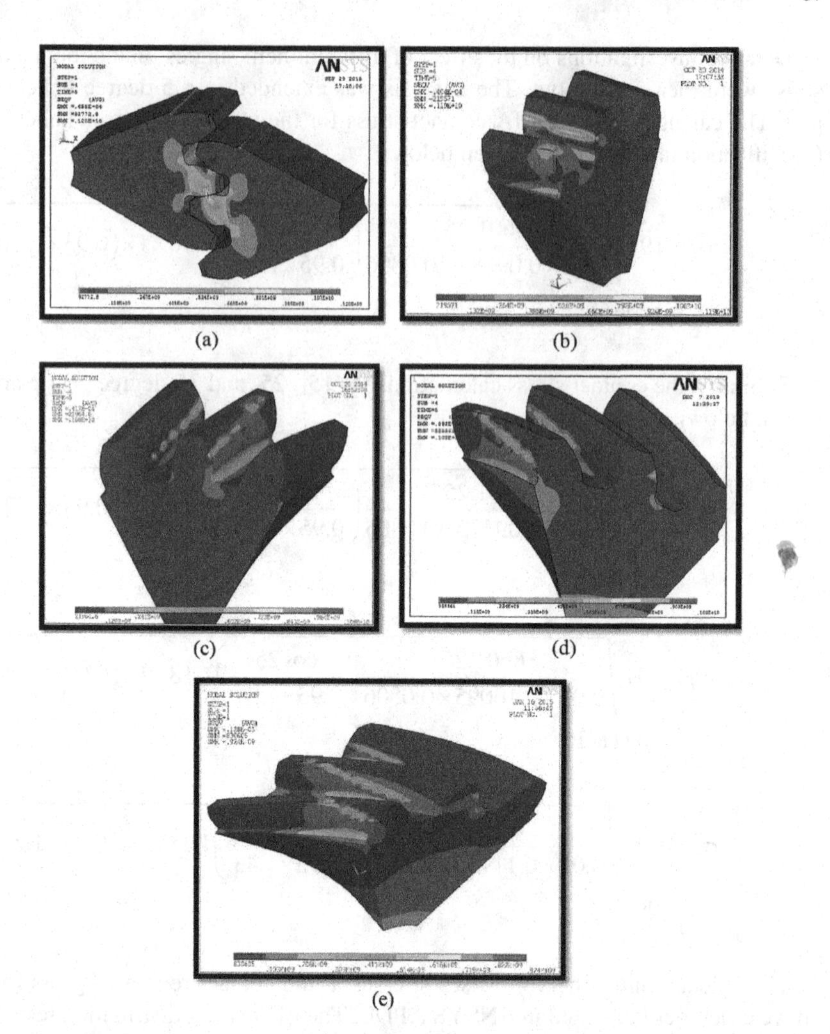

FIGURE 7.3 Frictionless contact von Mises stress distributions (a) Spur gear; (b) 5-degree helical gear; (c) 15-degree helical gear; (d) 25-degree helical gear; (e) 35-degree helical gear.

TABLE 7.2
Frictionless Contact Stress Comparison between AGMA and Finite Element Analysis

Gear Sets	Gear Contact Stress, MPa		Percentage Difference (%)
	AGMA Calculated Results	FEA (ANSYS) Results	
Spur gear	1210	1200	1.0
5-degree helical gear	1208	1190	1.5
15-degree helical gear	1165	1090	6.5
25-degree helical gear	1071	1020	4.8
35-degree helical gear	961	924	4.1

has shown a good agreement with the AGMA calculations. Hence, verifying the FE procedure of analysis.

7.4 RESULTS AND DISCUSSION

The validated FEM technique for frictionless conditions was further extended for analyzing the gears individually for spur and helical gear pairs with the inclusion of friction. In this section the results were evaluated with FEA and RomaxDESIGNER model for different gears.

7.4.1 SPUR GEAR ANALYSIS

The static stress analysis was conducted on the spur gear models. The contact stress distributions of the spur gears for different friction coefficients are shown in Figures 7.4–7.7.

The maximum von Mises contact stress plot for different friction coefficients (0 to 0.3), along the contact path is shown in Figure 7.7. The contact stresses increased significantly with increasing frictional coefficients. The spur gear analysis revealed that the contact stresses increased by around 15% for a rise of coefficient of friction from 0 to 0.3 at the pitch point contact. The normalized stress plot was used for spur and for other helical gears. The importance of using the normalized stress plot is to get the index factor at any point along the line of action and this factor can be multiplied to other gear analysis to obtain the frictional contact stress rise. This normalized plot is a generalized form of gear contact stress.

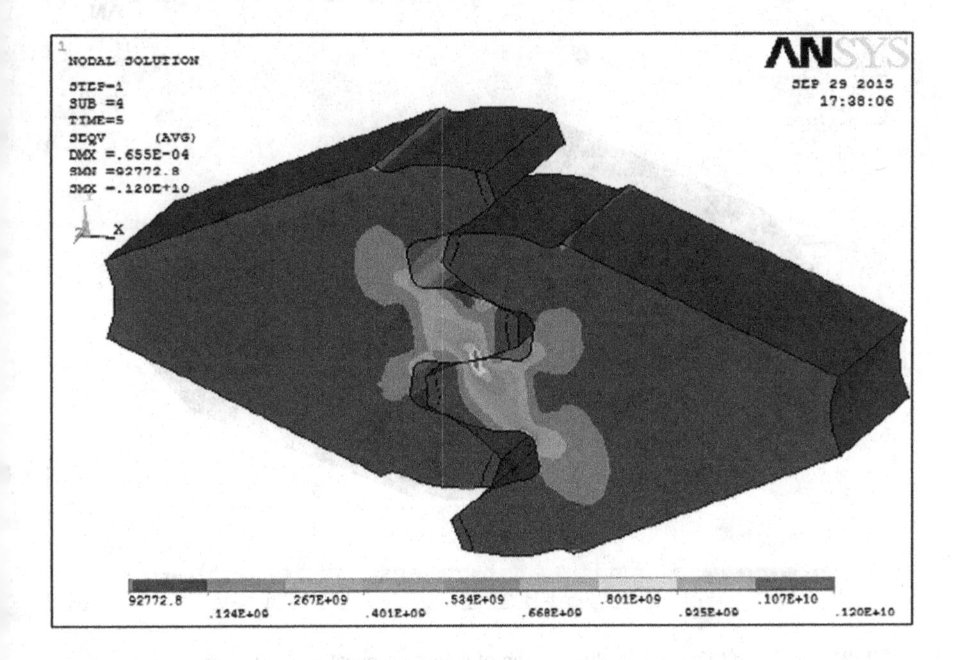

FIGURE 7.4 von Mises stress distribution of spur gear pair for $\mu = 0$ case.

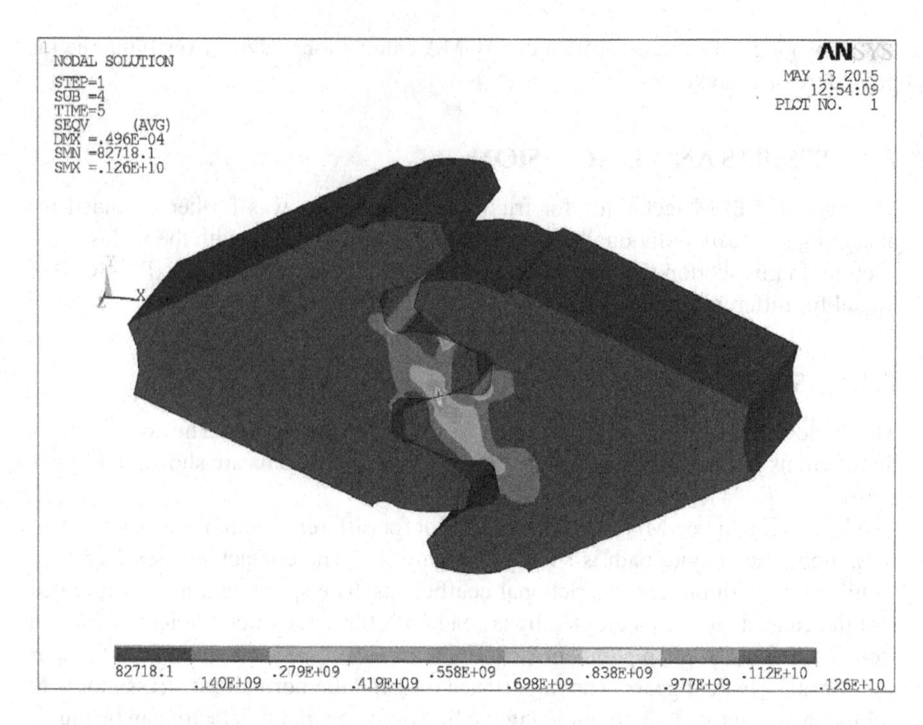

FIGURE 7.5 von Mises stress distribution of spur gear pair for $\mu = 0.1$ case.

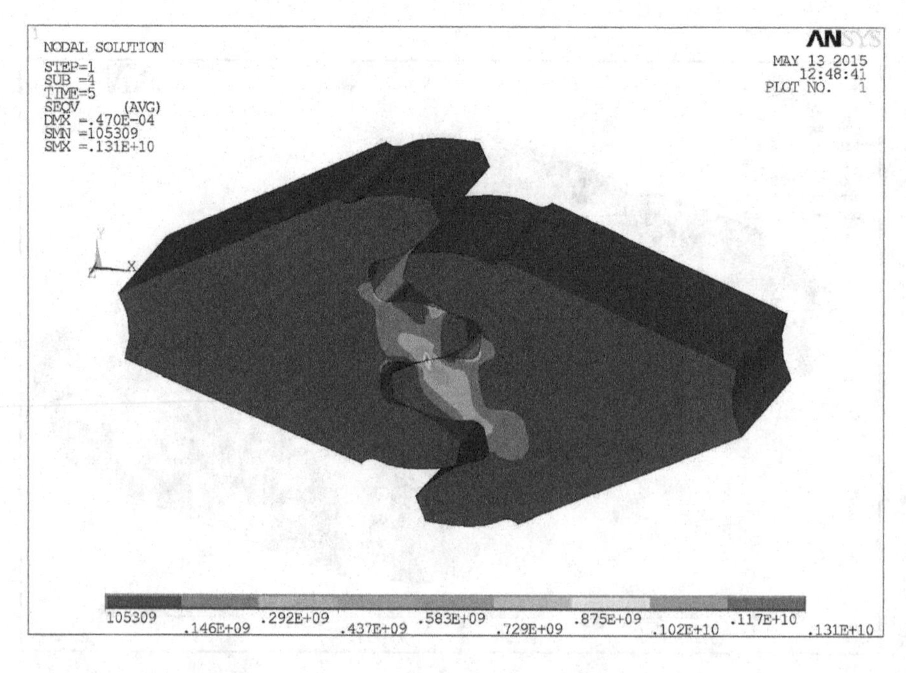

FIGURE 7.6 von Mises stress distribution of spur gear pair for $\mu = 0.2$ case.

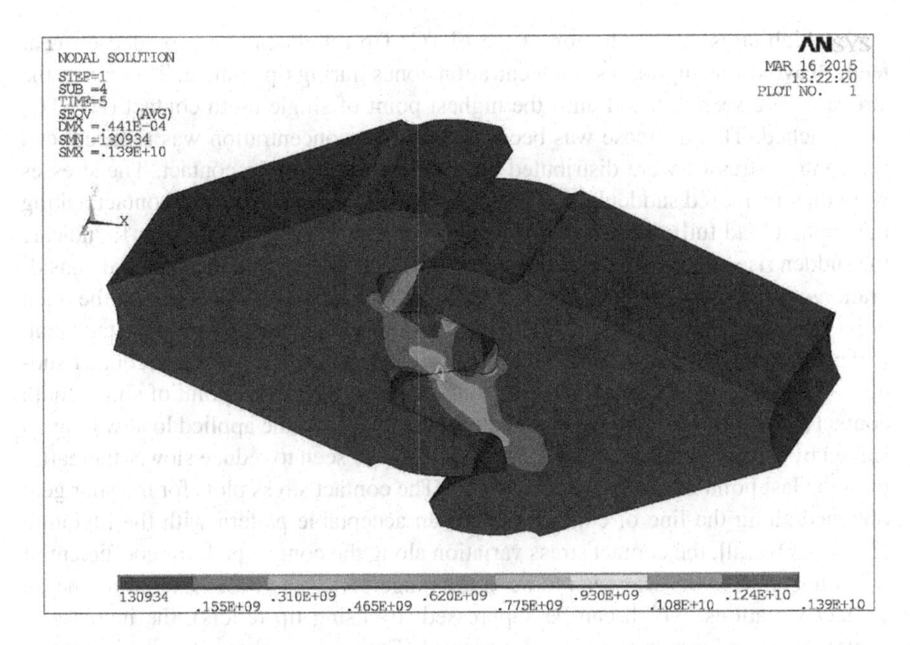

FIGURE 7.7 von Mises stress distribution of spur gear pair for $\mu = 0.3$ case.

Figure 7.8 shows that the plots for different coefficients of friction are similar in pattern and are rising with increasing friction coefficient values. At first point of tooth contact (FPTC) the stresses were at a maximum because of the high stress concentration at the contact tip. Furthermore, digging of tooth tip into the meshing tooth pair may take place. The increase in contact stresses due to friction was around 36%

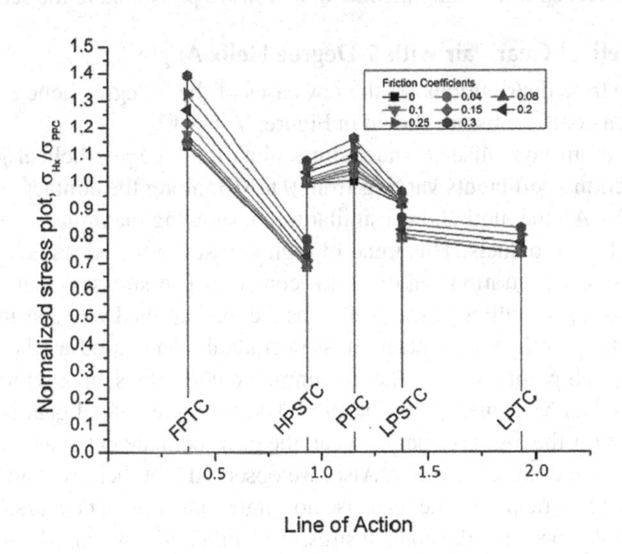

FIGURE 7.8 Spur gear normalized stress plot along the contact path for different coefficient of friction.

due to high stress concentration at the FPTC. Tip reliefs can be provided on gear teeth to avoid this high-stress concentration zones during tip contact. Thereafter, the stresses were seen reduced until the highest point of single tooth contact (HPSTC) was reached. This decrease was because the stress concentration was reduced, and the contact stresses were distributed evenly over two teeth in contact. The stresses were then increased suddenly at HPSTC, as a single tooth comes into contact, taking the applied load fully. The straight-line plot was considered for spur gear to indicate the sudden rise at HPSTC and sudden drop at LPSTC. However, the standard consideration of maximum contact stress between contacting spur gears occurs at the pitch point contact (PPC) [12, 23, 24]. The results were in agreement with this statement, if the tip contact stresses were suppressed. Furthermore, the stresses reduced suddenly when two teeth came back into contact, i.e. at the lowest point of single tooth contact (LPSTC). Here the stresses reduced drastically as the applied load was again shared by the two teeth in contact. The stresses were seen to reduce slowly thereafter until the last point of tooth contact (LPTC). The contact stress plots for the spur gear obtained along the line of contact showed an acceptable pattern with the literature [23–25]. Overall, the contact stress variation along the contact path for coefficient of friction from 0 to 0.3 was seen to be in the range of 12% to 36%. Excluding the tip contact variations, which can be suppressed (by using tip reliefs), the increase in contact stresses for spur gears can be derived. The analysis showed a rise in contact stresses of around 15%, which is a significant increase in stresses.

7.4.2 Helical Gear Analysis

Similar to the spur gear model above, Finite Element Models were generated and analysed for helical gears in order to compute the contact stresses along the contact path. The contact stresses along the line of action are presented in the sections below.

7.4.2.1 Helical Gear Pair with 5-Degree Helix Angle

The contact stress distributions for the few cases of the 5-degree helical gear for different friction coefficients are shown in Figures 7.9–7.12.

The maximum von Mises contact stress plots for 5-degree helical gear pair for different friction coefficients varying from 0 to 0.3, along the contact path is shown in Figure 7.13. All the plots show a similar trend, showing maximum contact stresses at the tip and root contacts. This trend of high stresses at the edges is again because of high stress concentration region at tip contacts. The stresses after the teeth tip contact reduced gradually by around 40% as the load applied was distributed among three contacting teeth. The contact stresses gradually increased as the contact was close to the pitch point contact. The maximum contact stress, apart from the tip and root contacts, can be seen close to the PPC. The 5-degree helical gear is similar to a spur gear, but for the 5-degree helical gear, the gear teeth are cut at an angle and thus single tooth contact may not occur. Also, we observed that there was no HPSTC and LPSTC characteristic points, i.e. because no single tooth contact occurs. The normalized stress plot shows that the contact stresses significantly increased with increasing frictional coefficients. The increase of contact stresses due to a rise in the coefficient of friction was in the range of 15 to 20% throughout the contact path, except for tip

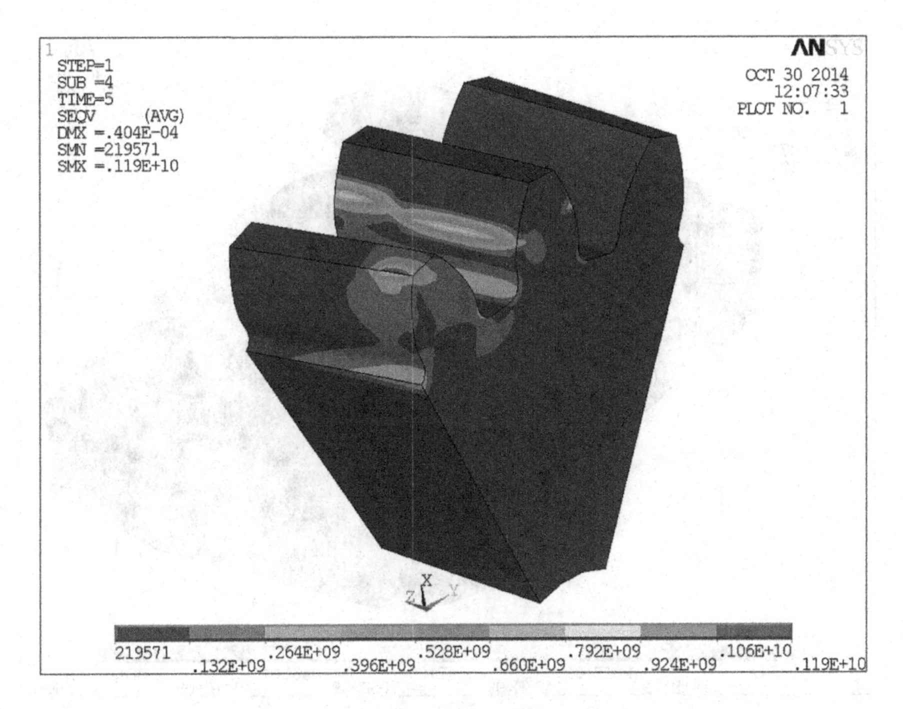

FIGURE 7.9 von Mises stress distribution of 5-degree helical gear pair for $\mu = 0$ case.

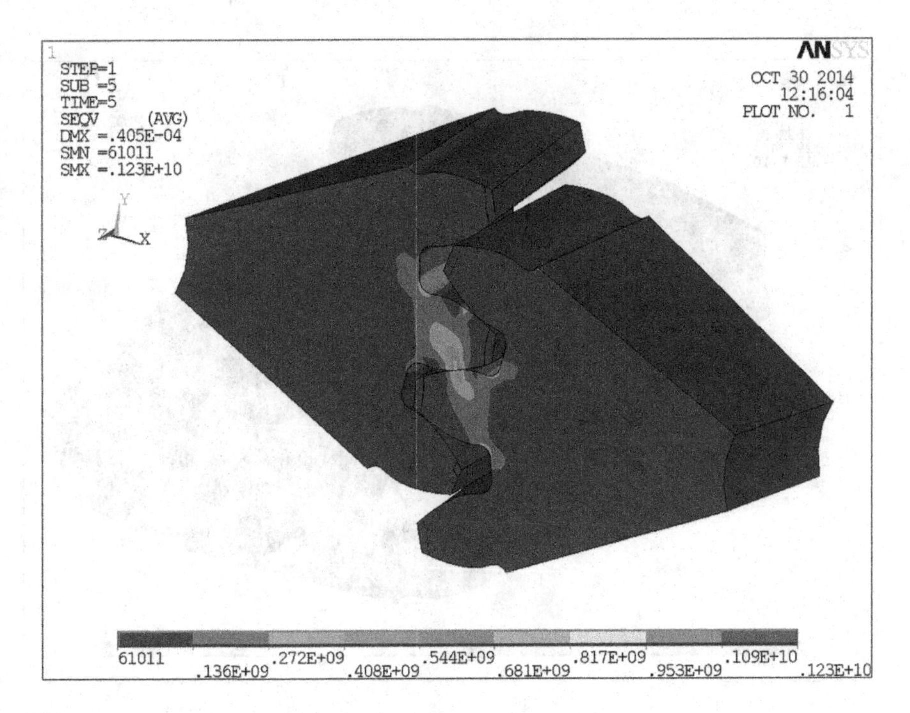

FIGURE 7.10 von Mises stress distribution of 5-degree helical gear pair for $\mu = 0.1$ case.

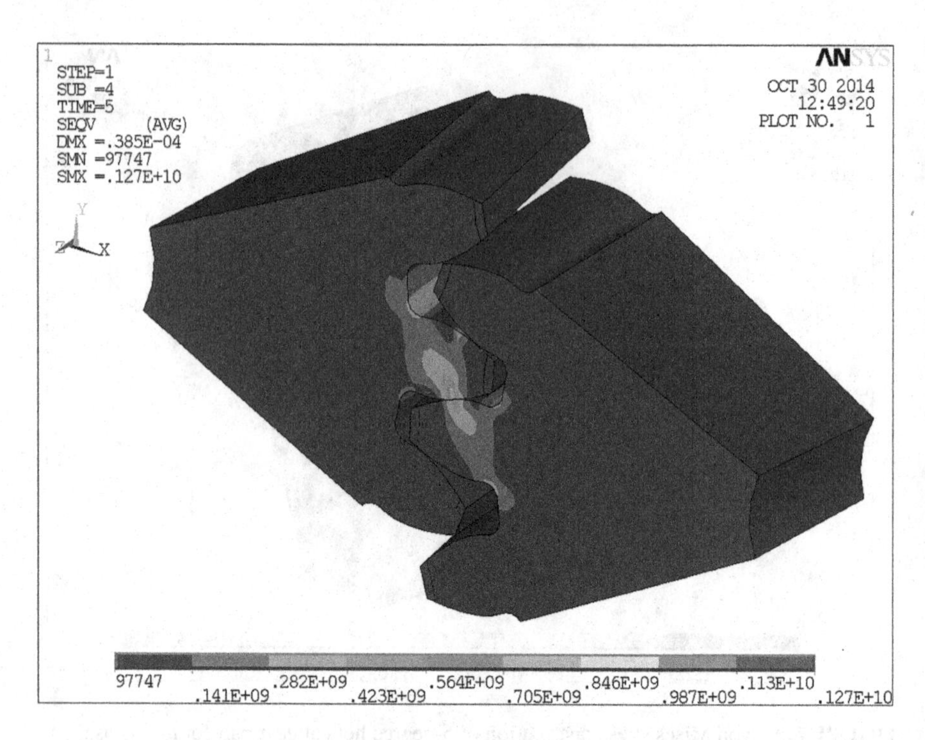

FIGURE 7.11 von Mises stress distribution of 5-degree helical gear pair for $\mu = 0.2$ case.

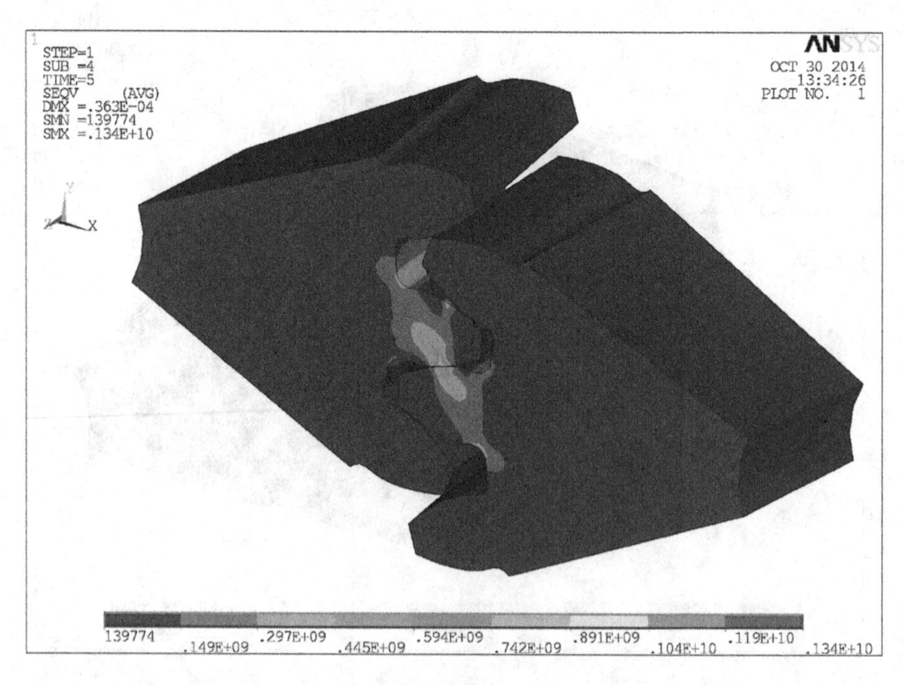

FIGURE 7.12 von Mises stress distribution of 5-degree helical gear pair for $\mu = 0.3$ case.

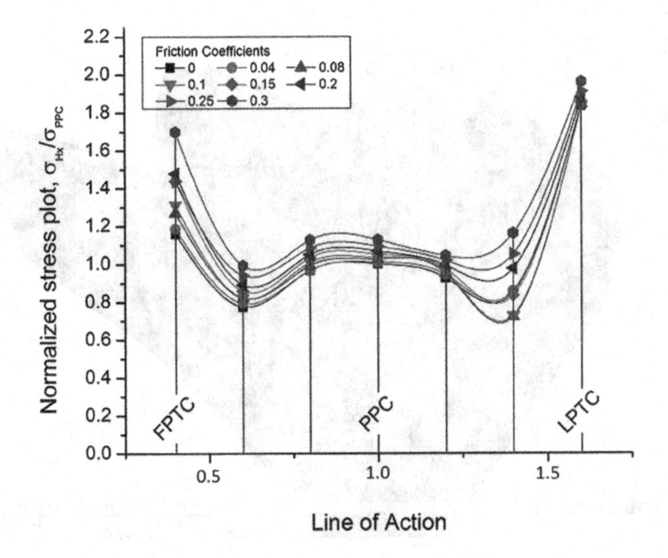

FIGURE 7.13 Helical Gear Analysis (5-degree helical gear pair).

contact. At FPTC (tip contact) the increase of stresses was around 46%, which was due to the frictional effects combined with high stress concentration reasons. The 5-degree helical gear analysis revealed that the contact stresses on increased, on average, by around 18% at PPC for a rise of coefficient of friction from 0 to 0.3. The contact stress increase is significant and cannot be considered as negligible. The stresses should decrease with an increasing helix angle, but since the frictional effects are more the overall stresses are seen increasing.

7.4.2.2 Helical Gear Pair with 15-Degree Helix Angle

The 15-degree helical gear model was generated, and static stress analysis followed. The contact stress distributions in the 15-degree helical gear for different friction coefficients are shown in Figures 7.14–7.17.

The large tip contact stresses were also seen in 15-degree helical gear contact. The normalized contact stress plots for friction coefficients varying from 0 to 0.3 along the contact path are shown in Figure 7.18. The contact stress increase rate for gear pair with 15-degree helix angle is higher, in the range of 18 to 25% throughout the contact path. This shows that for helix angle of 15 degrees, the contacting area is producing higher frictional variations. The surface contact is more compared to spur and 5-degree helix angled gear, and the increased contacting area has produced higher sliding friction at PPC. The higher frictional contact area has increased the frictional contact stresses. The plot trend along the contact path is different from that of the spur gear pair, as there is no single gear tooth contact between these gear pairs. There are either two or three teeth in contact at all times and the stress patterns vary, with maximum von Mises stress shifted slightly away from the PPC. This might have occurred due to the non-uniform load distribution on gear teeth in contact.

The 15-degree helical gear analysis revealed that the contact stresses increased by around 22% for a rise of coefficient of friction from 0 to 0.3. The rise of contact

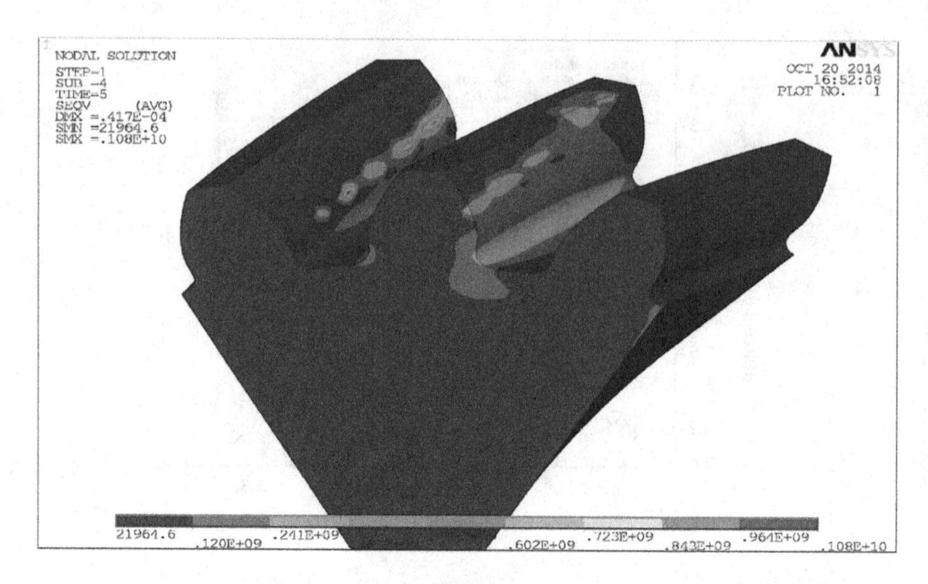

FIGURE 7.14 von Mises stress distribution of 15-degree helical gear pair for $\mu = 0$ case.

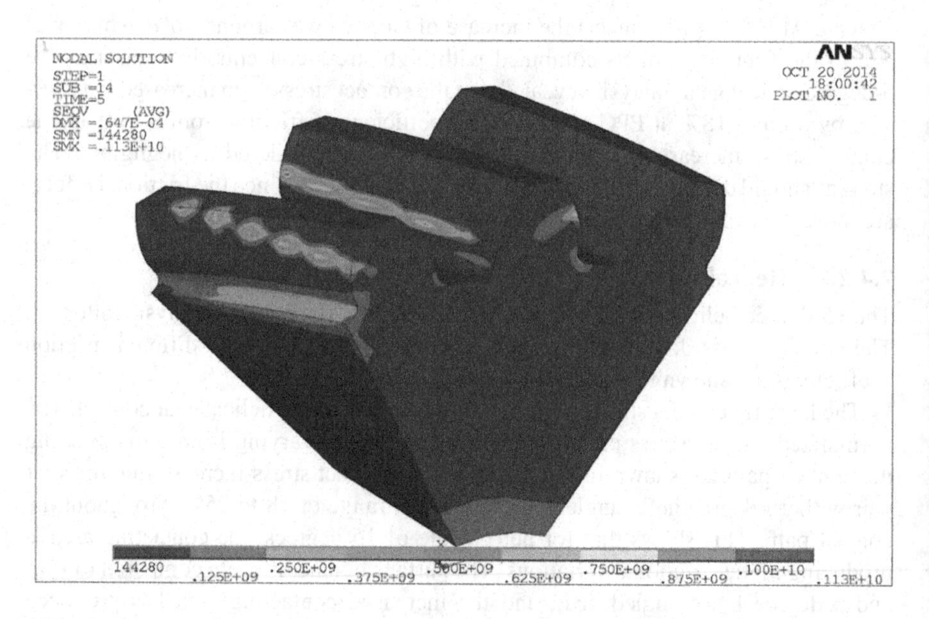

FIGURE 7.15 von Mises stress distribution of 15-degree helical gear pair for $\mu = 0.1$ case.

stresses by 22% is quite a substantial increase and needs to be incorporated during the contact stresses calculations (AGMA based calculations). Thus, a dimensionless factor needs to be developed, based on the effect of friction on contact gears. In turn, a function is also necessary to be developed, which can evaluate the percentage rise of stresses for a particular frictional coefficient acting on a pair of helical gear.

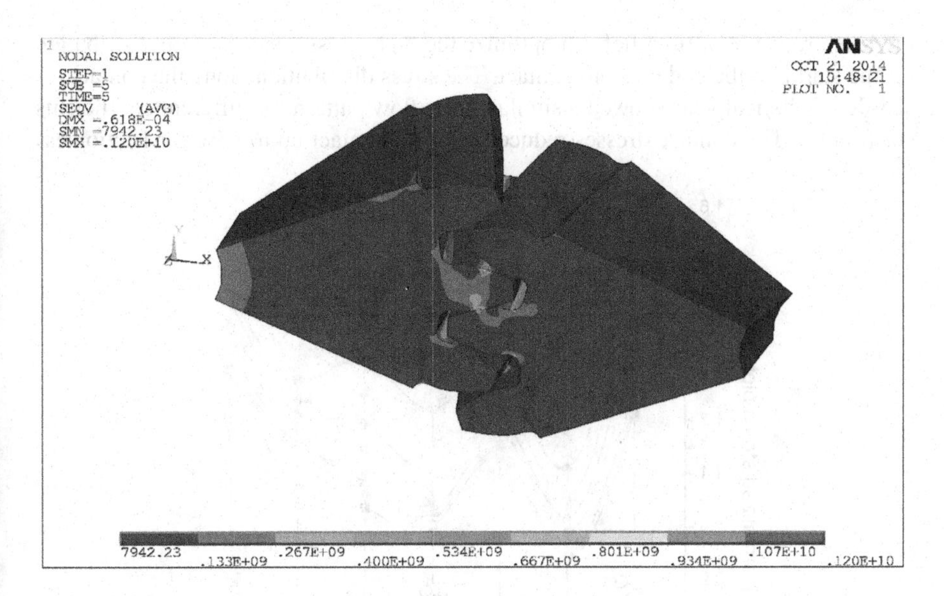

FIGURE 7.16 von Mises stress distribution of 15-degree helical gear pair for $\mu = 0.2$ case.

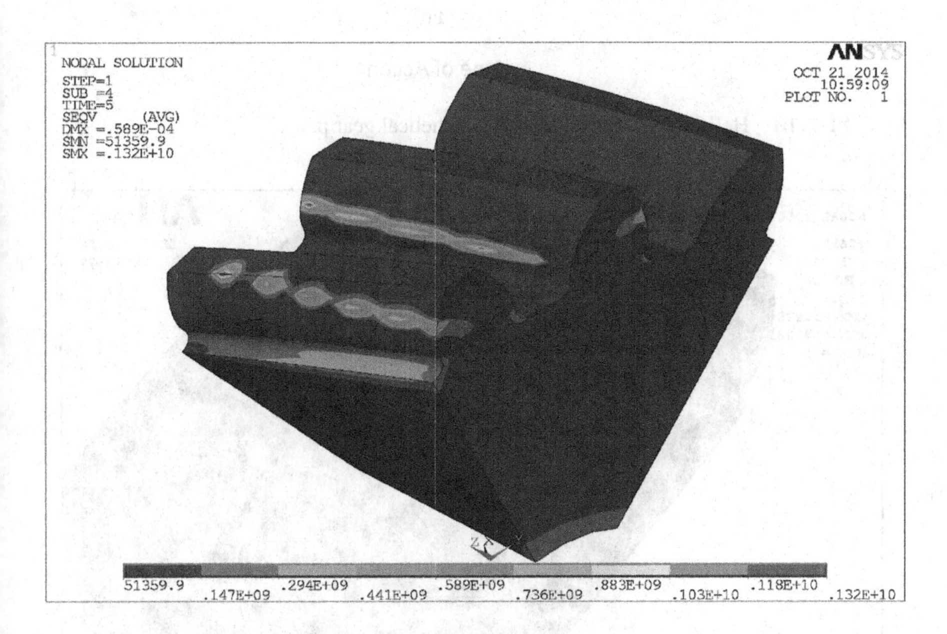

FIGURE 7.17 von Mises stress distribution of 15-degree helical gear pair for $\mu = 0.3$ case.

7.4.2.3 Helical Gear Pair with 25-Degree Helix Angle

The 25-degree helical gear pair was modelled similarly, and static stress analysis was conducted on the model. The contact stress distributions on the 25-degree helical gear for different friction coefficients are shown in Figures 7.19–7.22.

The provision of tip relief can optimize the high stress concentration distribution at the start and the end of tooth contact. The stress distribution along the contact for 25-degree helical gear showed a similar stress flow pattern for different coefficients of friction. The contact stresses reduced after tip contact up to 15% for frictionless

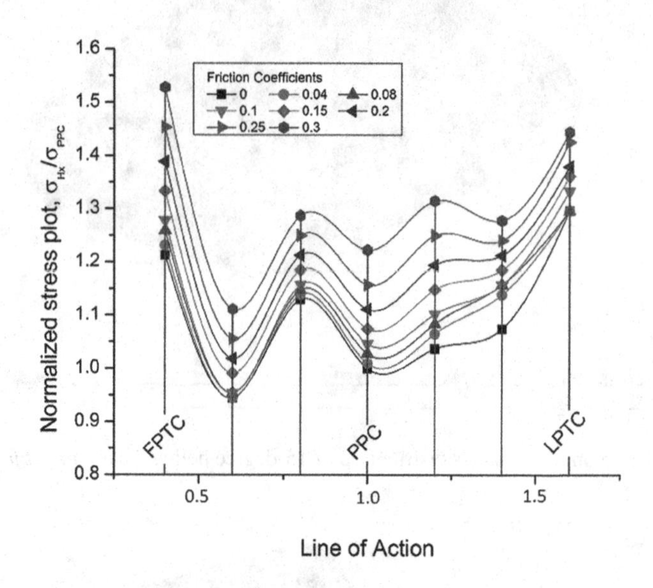

FIGURE 7.18 Helical gear analysis (15-degree helical gear pair).

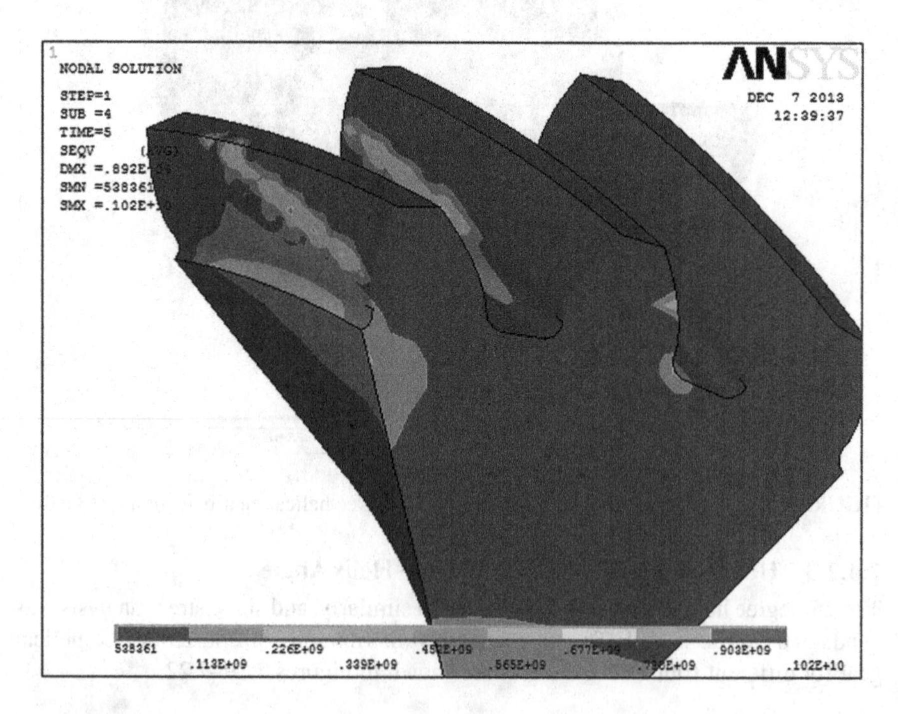

FIGURE 7.19 von Mises stress distribution of 25-degree helical gear pair for $\mu = 0$ case.

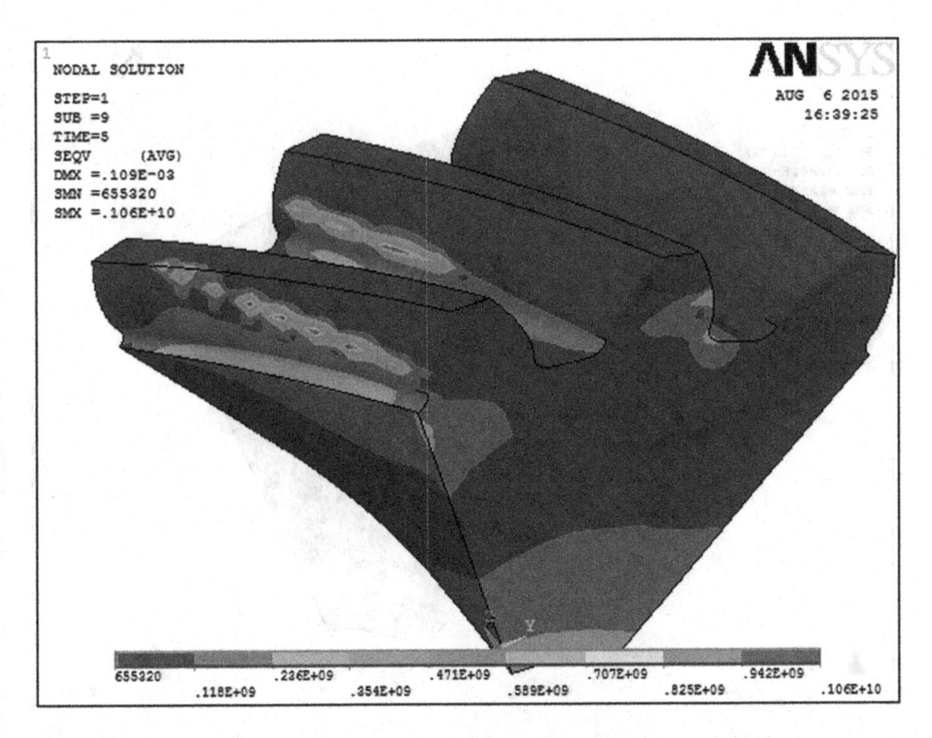

FIGURE 7.20 von Mises stress distribution of 25-degree helical gear pair for $\mu = 0.1$ case.

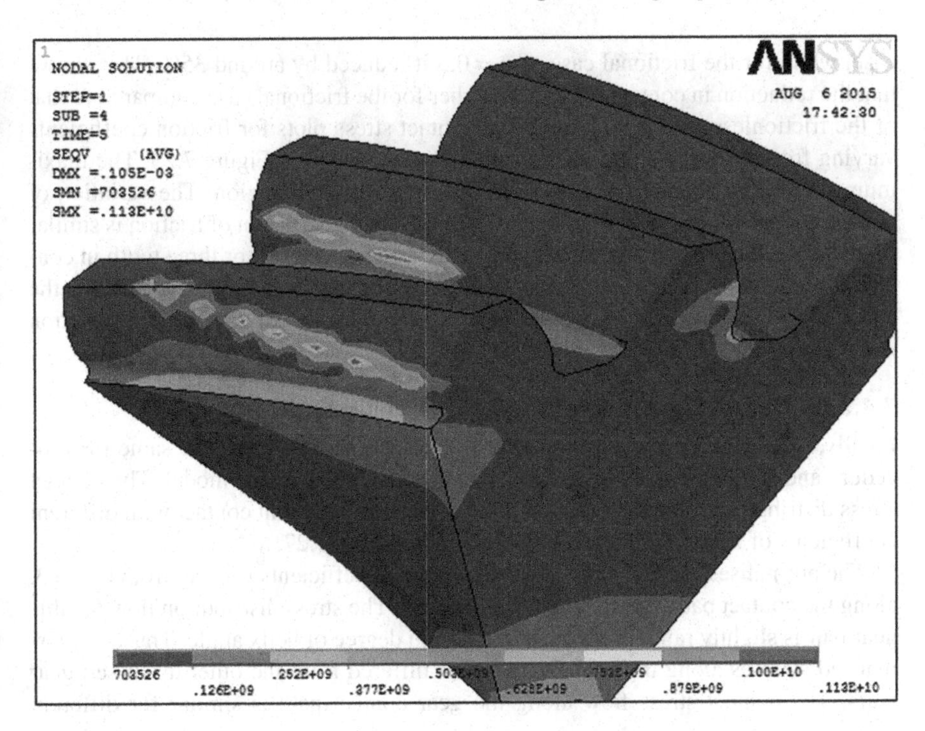

FIGURE 7.21 von Mises stress distribution of 25-degree helical gear pair for $\mu = 0.2$ case.

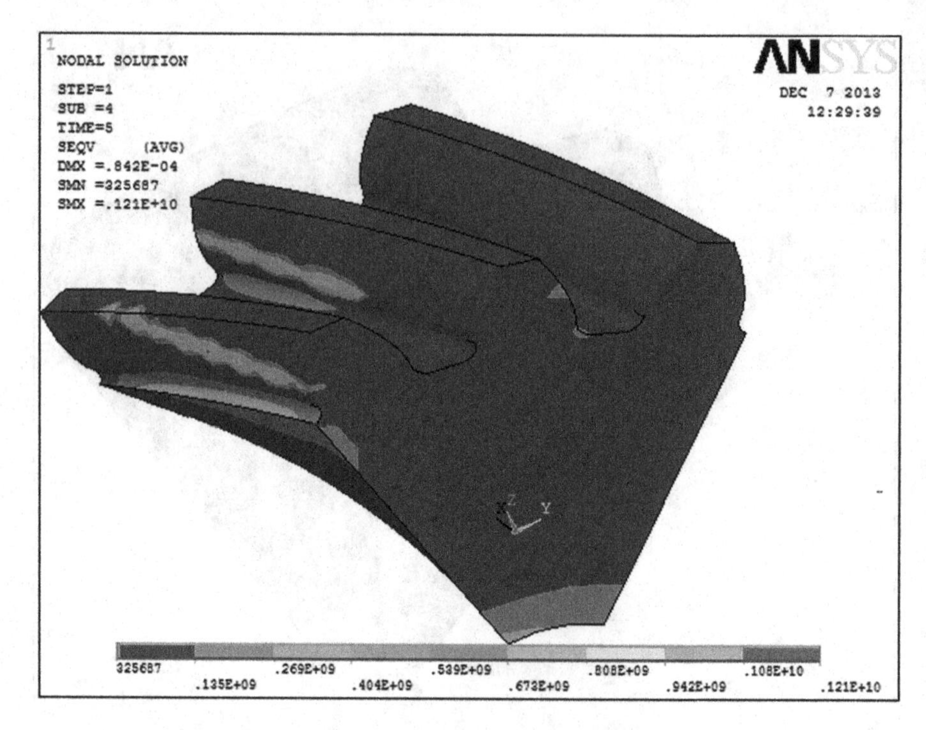

FIGURE 7.22 von Mises stress distribution of 25-degree helical gear pair for $\mu = 0.3$ case.

case, while for the frictional case of $\mu = 0.3$ it reduced by around 35%. This shows that the reduction in contact stresses is higher for the frictional case compared to that of the frictionless case. The normalised contact stress plots for friction coefficients varying from 0 to 0.3, along the contact path, are shown in Figure 7.23. The maximum von Mises contact stress was seen close to the PPC region. The variation of contact stresses of 25-degree helical gear for different coefficient of friction is similar to that of 15-degree helical gear and the gears experienced two or three tooth in contact throughout the contact path. The 25-degree helical gear analysis revealed that the contact stresses increased by around 18% at PPC for a rise of coefficient of friction from 0 to 0.3.

7.4.2.4 Helical Gear Pair with 35-Degree Helix Angle

Finally, a 35-degree helical gear model was also generated using the same FE procedure and static stress analysis was conducted on the gear pair model. The contact stress distributions on the 35-degree helical gear at pitch point contact with different coefficients of friction are represented in Figures 7.24–7.27.

The normalised contact stress plots for friction coefficients varying from 0 to 0.3, along the contact path, are shown in Figure 7.28. The stress distribution flow for this gear pair is slightly random because of the high degree of helix angle. Thus, the normalised stresses along the gear contact path differed from the other discussed gear pairs. The contact stress flow along the gear contact path is similar for different

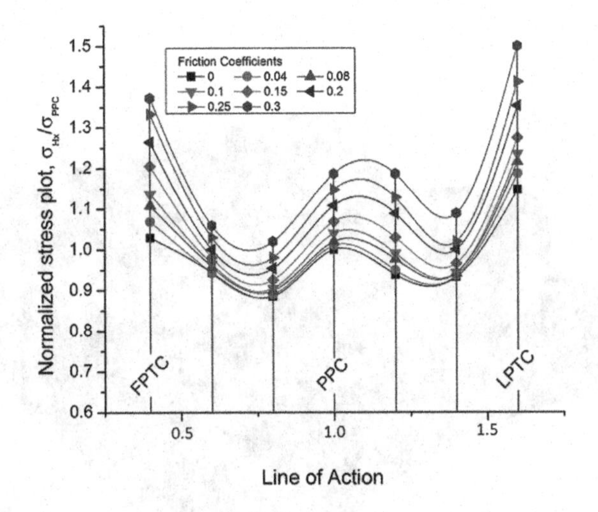

FIGURE 7.23 Helical gear analysis (25-degree helical gear pair).

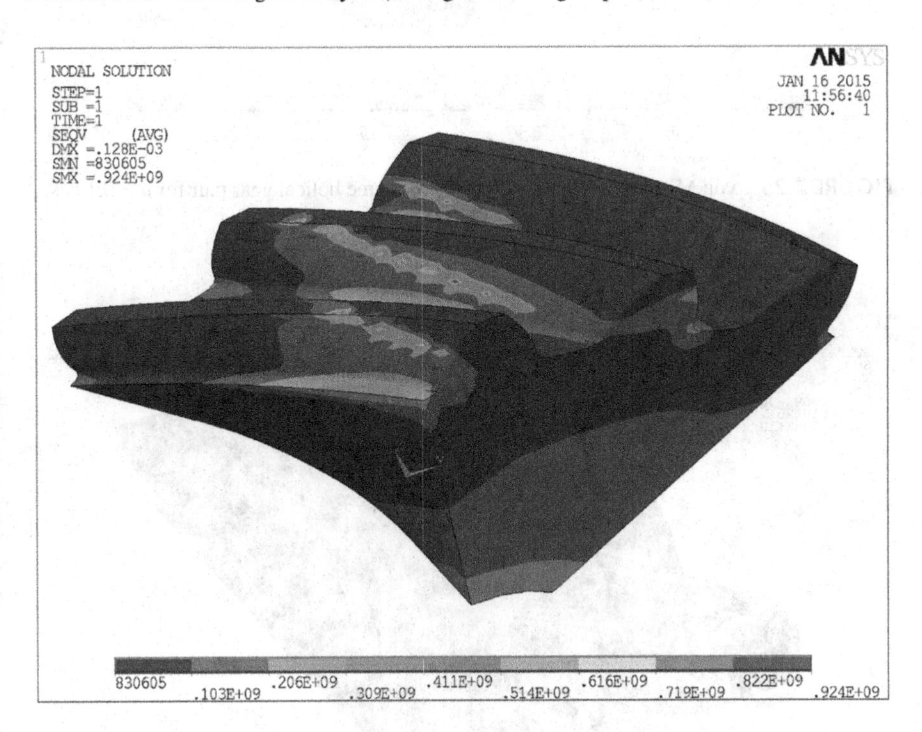

FIGURE 7.24 von Mises stress distribution of 35-degree helical gear pair for $\mu = 0$ case.

coefficients of friction and contact stresses significantly increased with the increasing frictional coefficients.

The 35-degree helical gear analysis revealed that the contact stresses on an average increased by around 18% at PPC as coefficient of friction increased from 0 value

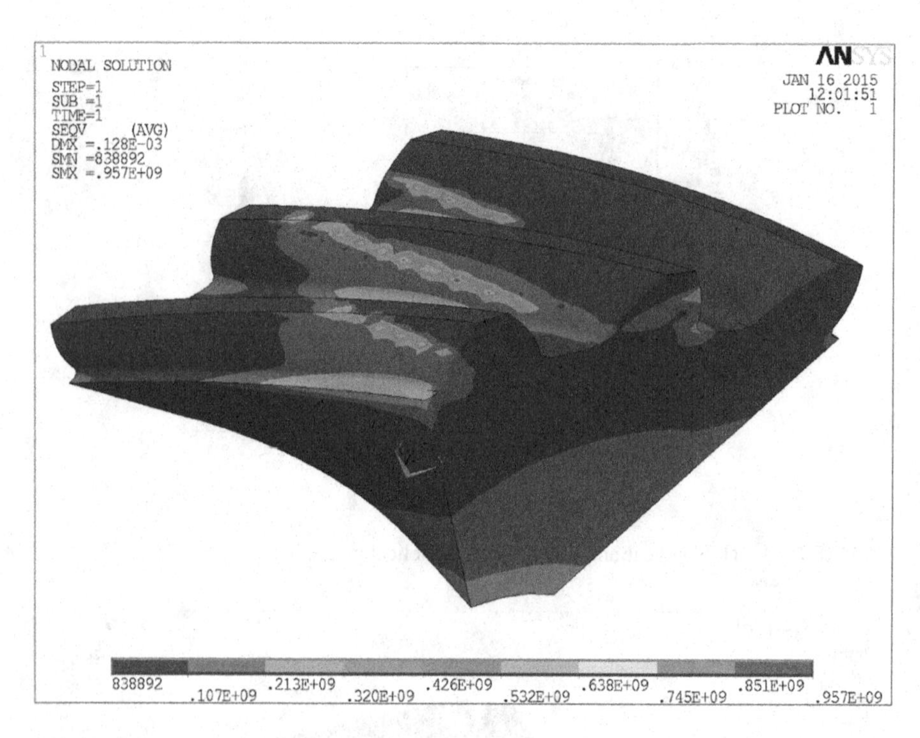

FIGURE 7.25 von Mises stress distribution of 35-degree helical gear pair for $\mu = 0.1$ case.

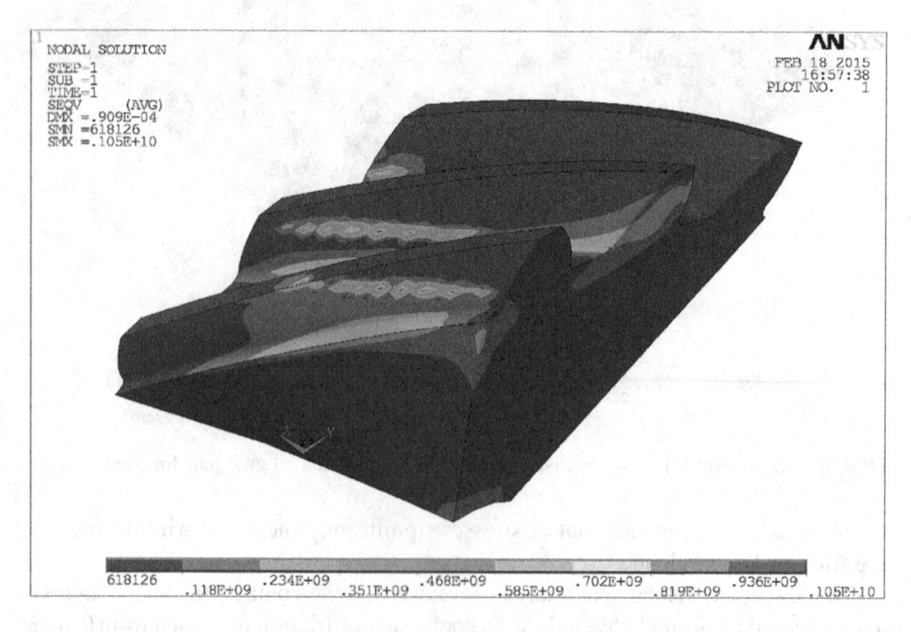

FIGURE 7.26 von Mises stress distribution of 35-degree helical gear pair for $\mu = 0.2$ case.

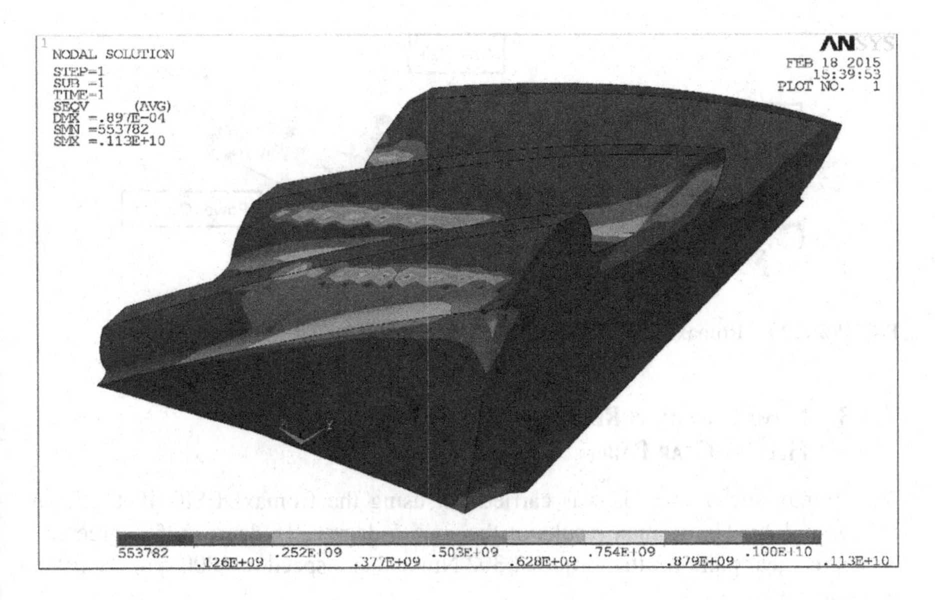

FIGURE 7.27 von Mises stress distribution of 35-degree helical gear pair for $\mu = 0.3$ case.

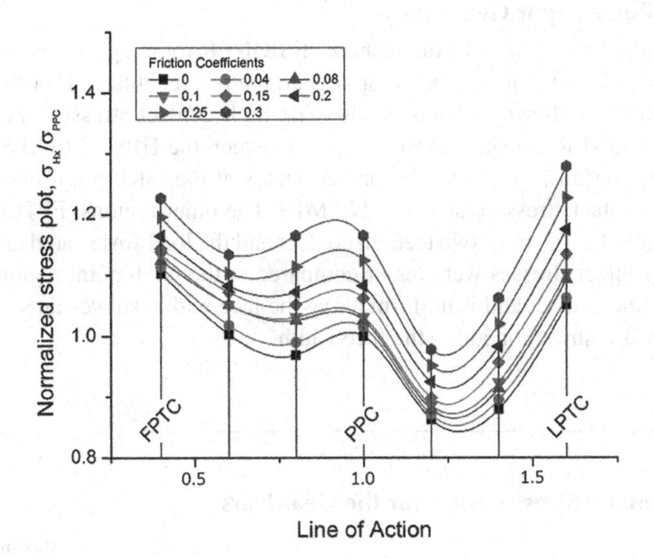

FIGURE 7.28 Helical gear analysis (35-degree helical gear pair).

to 0.3 value. The stress variation showed maximum contact stresses close to the pitch point region. The stress distribution along the contact path showed a different pattern when compared to the previously analysed gears. The contact stress increase rate in this set of gears was also more (approximately 18%) and there is a necessity to include the frictional effect into the contact stress calculations. As mentioned earlier, a dimensionless factor and a function to derive the factor need to be developed. The factor can then be incorporated in the contact stress calculations.

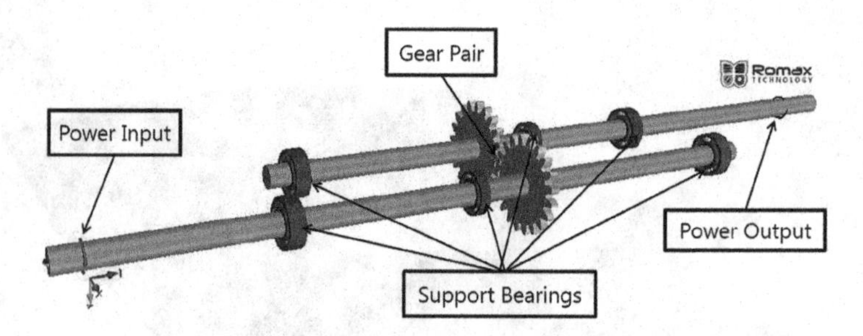

FIGURE 7.29 RomaxDESIGNER model based on the GDSTR gear setup.

7.4.3 Romax Analysis Results of the Spur and Helical Gear Pairs

The Romax stress analysis was carried out using the RomaxDESIGNER, shown in Figure 7.29. The Romax results of the spur, 5-degree, 15-degree, 25-degree and 35-degree gear pairs for the torque of 302 Nm and at a speed of 1000 rpm are listed in Table 7.3.

7.4.3.1 Romax Spur Gear Analysis

The contact pattern obtained from RomaxDESIGNER for the gears above are shown and discussed below. Firstly, for the spur gear, the results obtained for the load case of 302 Nm is as shown in Figure 7.30. The high contact stresses obtained from RomaxDESIGNER are shown in the region between the HPSTC to LPSTC, which was obvious because single tooth contact occurs at the pitch point region and the maximum contact stress value was 1279 MPa. The other regions, FPTC to HPSTC and LPSTC to LPTC, have two teeth in contact and the load was shared among them, hence the contact stresses were lesser compared to the pitch point region. The spur gear has a line contact and is uniform along the face width, so we can see a uniform contact stress distribution along the face width.

TABLE 7.3
Romax Contact Stress Results for the Gear Pairs

	Gear Type	Coefficient of Friction, μ (from Romax Result[a])	Maximum Contact Stress, MPa Romax Results
Analysis of Gears	Spur gear	0.094	1279
	5-degree helical gear	0.096	1260
	15-degree helical gear	0.095	1215
	25-degree helical gear	0.093	1092
	35-degree helical gear	0.090	1006

[a] Ref. [22]

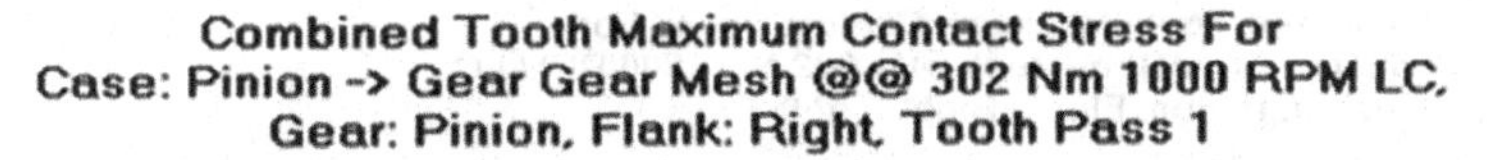

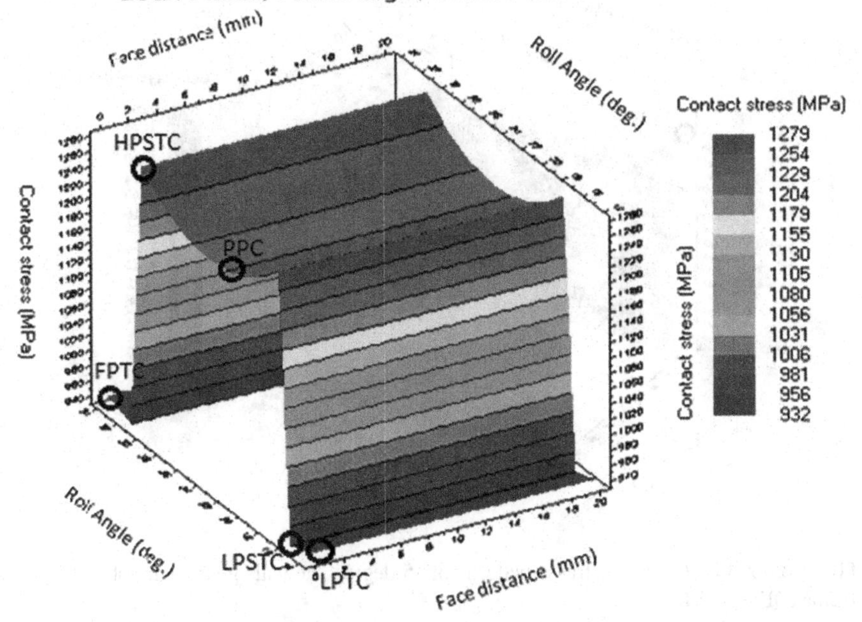

FIGURE 7.30 Contact stress pattern of spur gear pair obtained from RomaxDESIGNER.

7.4.3.2 Romax 5-Degree Helical Gear Analysis

The contact stress pattern of the 5-degree helical gear obtained from the RomaxDESIGNER is presented in Figure 7.31. The contact stress distribution showed a similar trend with that of the spur gear pair, but the stress variations were slightly inclined, due to the small helix angle. The maximum contact stress obtained for 302 Nm loading was 1260 MPa and was found at the pitch point contact region. There is no prominent single tooth contact in helical gears and hence we have two to three teeth contact pairs. Thus, the contact stresses are reduced in helical gears as the load getting shared among the contacting teeth. The contact stresses for the 5-degree helical gear pair were found to be in a higher range at the tip contacts, due to stress concentration and possible digging effect of tip into the corresponding gear flank. Optimal tip relief can be introduced to lower these unwanted contact stresses.

7.4.3.3 Romax 15-Degree Helical Gear Analysis

The contact stress contours obtained for the 15-degree helical gear pair from the RomaxDESIGNER are shown in Figure 7.32. The contact pattern showed that the highest contact stress obtained was 1215 MPa, for the 302 Nm load condition. There was no single tooth contact in 15-degree helical gears and hence we have two to three teeth contact pairs. Thus, the contact stresses were lower than the 5-degree helical gear pair as the load getting shared more among contacting teeth compared to the

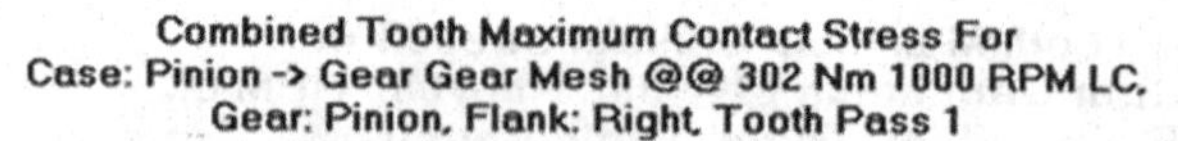

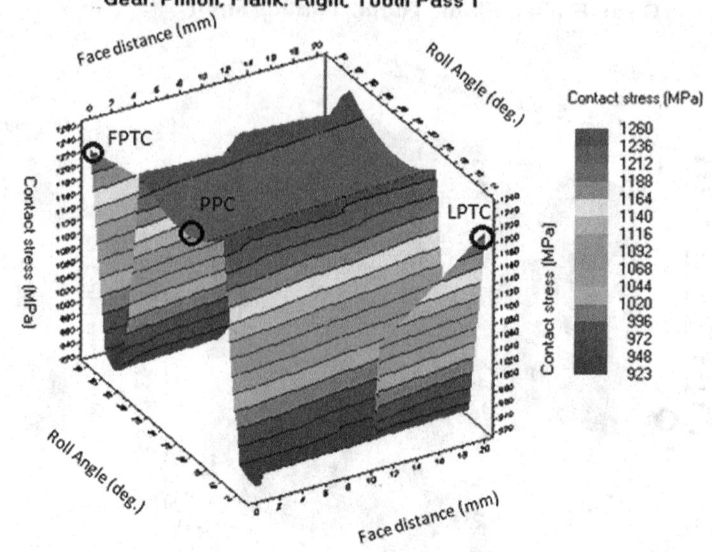

FIGURE 7.31 Contact stress pattern of 5-degree helical gear pair obtained from RomaxDESIGNER.

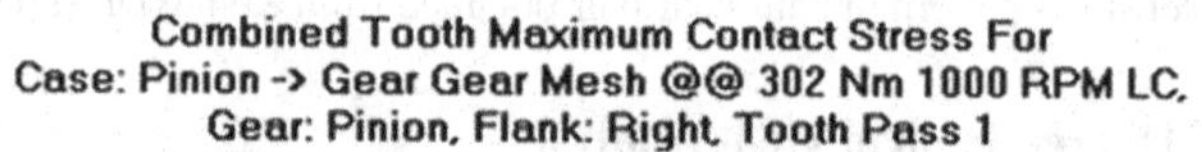

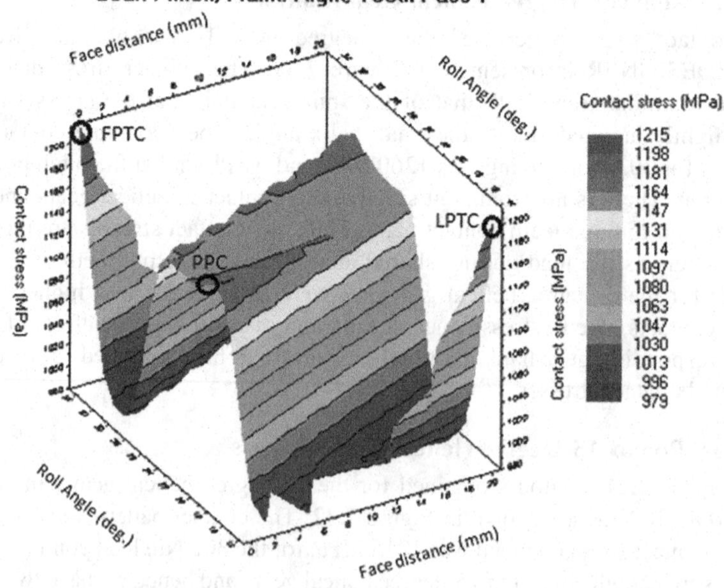

FIGURE 7.32 Contact stress pattern of 15-degree helical gear pair obtained from RomaxDESIGNER.

5-degree helical gear. The stress distribution showed a slight shift of the maximum contact stress from the pitch point region. This similar shift was also seen in the finite element contact 2D stress plot along the path of contact (Figure 7.18). Hence, results from RomaxDESIGNER showed a close trend with the finite element results for 15-degree helical gear pair. The higher tip stresses can be resolved with optimal tip relief.

7.4.3.4 Romax 25-Degree Helical Gear Analysis

The Romax contact stress pattern for the load case of 302 Nm and 1000 rpm, for 25-degree helical gear, is shown in Figure 7.33. The contact stress pattern showed that the maximum contact stress obtained was 1092 MPa and was found near the pitch point region. The contact stresses were less compared to previous helical gear pairs as the load was distributed more among contacting teeth compared to the 15-degree helical gear. The contact stress varied along the face width as the teeth gradually came in and went out of contact and was not uniform. The high tip contact stresses can be minimised by introducing a proper tip relief. The Romax contact stress trend for 25-degree helical gear pair along the contact point was close to the contact stress trend obtained from finite element method (Figure 7.23).

7.4.3.5 Romax 35-Degree Helical Gear Analysis

The contact stress pattern, obtained from Romax for 35-degree helical gear, is shown in Figure 7.34. The contact stress contour showed that the maximum contact stress

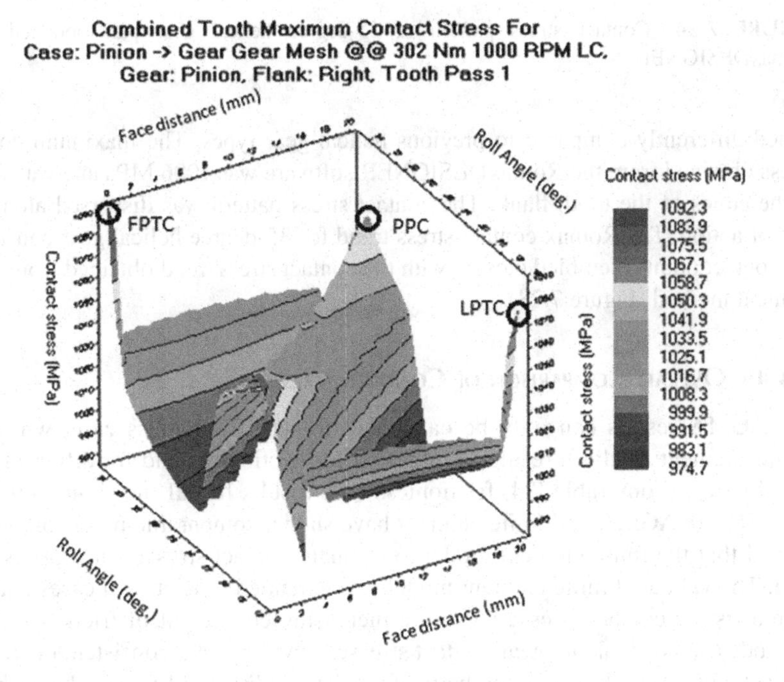

FIGURE 7.33 Contact stress pattern of 25-degree helical gear pair obtained from RomaxDESIGNER.

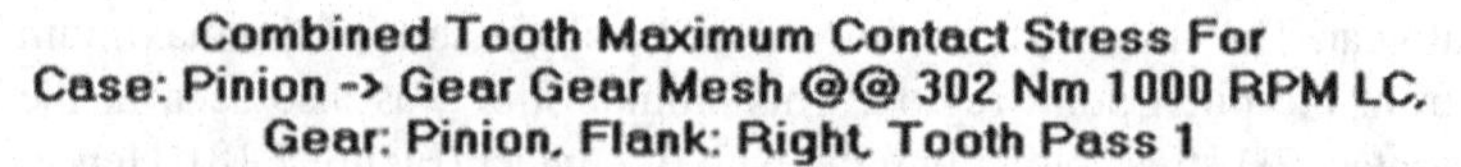

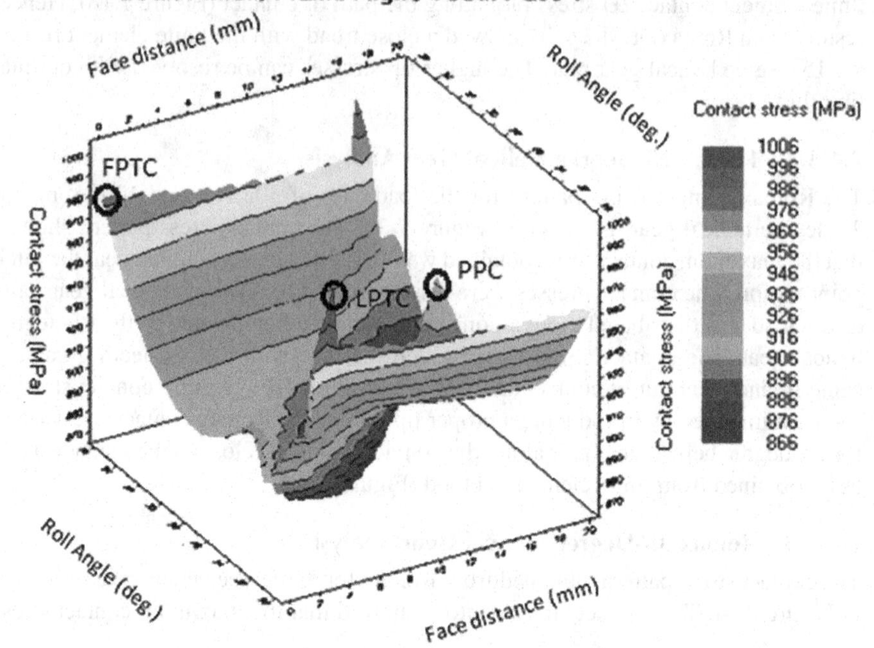

FIGURE 7.34 Contact stress pattern of 35-degree helical gear pair obtained from RomaxDESIGNER.

varied differently compared to previous helical gear types. The maximum contact stress obtained from the RomaxDESIGNER software was 1006 MPa and was found at the edges of the tooth flank. The contact stress pattern was dispersed along the line of action. The Romax contact stress trend for 35-degree helical gear pair along the contact point resembled closely with the contact stress trend obtained from finite element method (Figure 7.28).

7.4.4 Overall Comparison of Contact Stresses

The AGMA results can only be calculated for the frictionless case, while the Finite Element Analysis results were found for frictionless and for different frictional cases. From Table 7.4, frictionless case results for all the simulating gear pairs of both AGMA and finite element have shown comparable precision, which proved that the finite element model to evaluate contact stresses in gears is reliable. The validated finite element model was extended to frictional cases and has seen a rise in contact stresses with the increasing coefficient of friction. All the methods used to evaluate gear contact stresses have shown a consistent decline of contact stresses with increasing helix angle. The AGMA and Romax have shown results for particular cases, while FEM can be used to evaluate contact stresses at

TABLE 7.4
Contact Stress Overall Comparison of Gears

Gears		Contact Stress, MPa Torque 302 Nm, Speed 1000 rpm		
Gear Type	**Coefficient of Friction**	**Theoretical (AGMA)**	**Finite Element (ANSYS)**	**Romax**
Spur	0	1210	1200	1279
	0.1		1260	
	0.2		1310	
	0.3		1390	
5-degree	0	1208	1190	1260
	0.1		1230	
	0.2		1270	
	0.3		1340	
15-degree	0	1165	1090	1215
	0.1		1130	
	0.2		1210	
	0.3		1330	
25-degree	0	1071	1020	1092
	0.1		1060	
	0.2		1130	
	0.3		1210	
35-degree	0	961	924	1006
	0.1		957	
	0.2		1050	
	0.3		1130	

different frictional cases and have shown consistent results. The FEM evaluated contact stresses have increased in a range of 15–22% with the increase in friction coefficient from 0 to 0.3. Also, the Romax results for the realistic gear contact condition (with real time friction) showed a good match with the finite element model results.

The results obtained from the FE analysis for the same condition $\mu = 0.1$ was compared with that of Romax to check for the similarities. The stress distribution from the set of gears for $\mu = 0.1$ has been compared and listed in Table 7.4. The comparison of both the results shows that the percentage variations between them are below 10%. Hence the finite element results are again verified by the Romax software simulations.

7.5 CONCLUSIONS

The FE simulations have been successfully validated with the available frictionless gear calculations. The validated FE methodology (with a new APDL involute profile program) were further extended for the present parametric studies. The contribution of the simulations and validations carried out have been explained well. Further, gear contact stress analyses, including friction were performed on spur and different helical gears, to understand the effect of friction and helix angle. RomaxDESIGNER was also used to model the experimental simulation-based results and compared with the FE results.

Based on the study made on the different sets of gear pairs, the results obtained have been analysed and following conclusions are presented.

1. Spur gear individual analysis showed the contact stress variations along the line of action, with 5 characteristic points, LPTC, HPSTC, PPC, LPSTC and LPTC as discussed. The tip relief at the start (FPTC) was suggested to avoid high stress concentration region and stress increase was limited to 15%.
2. Helical gear individual analysis showed contact stress variations along the line of action, where HPSTC and LPSTC points are not visible due to the obvious reasons of 2–3 teeth are in contact always. However, tip relief was recommended to reduce the stress concentration and thereby the stresses by around 10% to 20%.
3. For spur and different angled helical gear ranging between 0 to 35-degree, the individual analysis of different set of gears for increased frictional coefficients showed 15% to 22% rise in gear contact stresses.
4. RomaxDESIGNER is a great tool to simulate the actual experimental gear pair and evaluate the contact stresses along the line of action and across the face width.

REFERENCES

1. H. Hertz, "On the contact of elastic bodies," *J. Pure Appl. Math.*, vol. 92, pp. 156–171, 1881.
2. J. F. Archard, "Contact of rubbing flat surfaces," *J. Appl. Phys.*, vol. 24, no. 8, pp. 981–988, 1953.
3. K. L. Johnson, *Contact Mechanics*, 9th Prinit. Cambridge, England: Cambridge University Press, 2003.
4. R. C. Juvinall and K. M. Marshek, *Fundamentals of Machine Components Design*, 5th ed. John Wiley & Sons, Inc., 2012.
5. V. Ramamurti and M. Ananda Rao, "Dynamic analysis of spur gear teeth," *Comput. Struct.*, vol. 29, no. 5, pp. 831–843, 1988.
6. S. Sundarajan and B. G. Young, "Finite-element analysis of large spur and helical gear systems," *J. Propuls. Power*, vol. 6, no. 4, pp. 451–454, 1990.
7. C. R. M. Rao and G. Muthuveerappan, "Finite element modelling and stress analysis of helical gear teeth," *Comput. Struct.*, vol. 49, no. 6. pp. 1095–1106, 1993.
8. Y.-C. Chen and C.-B. Tsay, "Stress analysis of a helical gear set with localized bearing contact," *Finite Elem. Anal. Des.*, vol. 38, no. 8, pp. 707–723, June 2002.
9. C. Gosselin, "A review of the current contact stress and deformation formulations compared to finite element analysis," in *International Gearing Conference, UK*, 1994.
10. S. M. Vijayakar, H. R. Busby, and D. R. Houser, "Linearization of multibody frictional contact problems," *Comput. Struct.*, vol. 29, no. 4, pp. 569–576, January 1988.
11. J. S. Chen, F. L. Litvin, and A. A. Shabana, "Computerized simulation of meshing and contact of loaded gear drives," in *International Gearing Conference, UK*, 1994.
12. S. Barone, L. Borgianni, P. Forte, and I. Meccanica, "CAD/FEM procedures for stress analysis in unconventional gear applications," *Int. J. Comput. Appl. Technol.*, vol. 15, no. 1, pp. 1–11, 2001.

13. R. D. Chacón, L. J. Andueza, M. A. Díaz, and J. A. Alvarado, "Analysis of stress due to contact between spur gears," in *9th WSEAS International Conference on Computational Intelligence, Man-Machince Systems and Cybernetics (CIMMACS '10)*, 2010, pp. 216–220.
14. R. G. Parker, S. M. Vijayakar, and T. Imajo, "Non-linear dynamic response of a spur gear pair: Modelling and experimental comparisons," *J. Sound Vib.*, vol. 237, pp. 435–455, 2000.
15. I. Atanasovska, "Influence of load distribution on the load capacity of cylindrical involute gears," PhD Thesis, University of Kragujevac, Serbia (In Serbian), 2004.
16. S. S. Patil, S. Karuppanan, and I. Atanasovska, "A short review on frictional contact stress distribution in involute gears tribology in industry," *Tirbology Ind.*, vol. 41, no. 2, pp. 254–266, 2019.
17. S. S. Patil, S. Karuppanan, and I. Atanasovska, "Contact stress evaluation of Involute Gear Pairs, including the effects of friction and helix angle," *J. Tribol.*, vol. 137, no. 4, p. 44501, 2015.
18. S. Patil, S. Karuppanan, and A. A. Wahab, "Contact pressure evaluation of a gear pair along the line of action using finite element analysis," *Appl. Mech. Mater.*, vol. 393, pp. 403–408, 2013.
19. S. Patil, S. Karuppanan, I. Atanasovska, A. A. Wahab, and M. R. Lias, "Contact stress analysis for gears of different helix angle using finite element method," in *MATEC Web of Conferences*, vol. 13, 2014.
20. C.-M. Hsieh, R. L. Huston, and F. B. Oswald, "Contact stresses in meshing spur gear teeth: Use of an incremental finite element procedure," NASA Technical Report 90, pp. 1–22, 1992.
21. I. Atanasovska, "3D spur gear FEM model for the numerical calculation of face load factor," *Facta Univ.-Ser.: Mech., Autom. Control Robot.*, vol. 6, no. 45, pp. 131–143, 2007.
22. S. S. Patil, *Evaluation of Frictional Contact Stresses in Spur and Helical Gears*, Universiti Teknologi PETRONAS, 2017.
23. P. Somprakit, R. L. Huston, and F. B. Oswald, "Contact stresses in gear teeth – A new method of analysis," in *27th Joint Propulsion Conference*, Cleveland, OH, 1991.
24. S.-C. Hwang, J.-H. Lee, D.-H. Lee, S.-H. Han, and K.-H. Lee, "Contact stress analysis for a pair of mating gears," *Math. Comput. Model.*, vol. 57, no. 1–2, pp. 40–49, Jan. 2013.
25. A. R. Hassan, "Contact stress analysis of spur gear teeth pair," *World Acad. Sci. Eng. Technol.*, vol. 58, pp. 611–616, 2009.

13. [illegible] L. [illegible] [illegible] Analysis of [illegible] [illegible]

14. [illegible] Nonlinear [illegible] gear pair [illegible]

15. [illegible] on the load [illegible] of cylindrical [illegible]

16. [illegible]

17. [illegible]

18. [illegible] [illegible] Technical Report [illegible]

19. [illegible]

20. [illegible] 2016.

21. [illegible]

22. [illegible]

23. [illegible]

8 Squeaking in Total Hip Arthroplasty

A Scoping Review on a Biotribological Issue

Alessandro Ruggiero and Marco De Stefano
University of Salerno, Fisciano, Italy

T.V.V.L.N. Rao
The Assam Kaziranga University, Jorhat, India

CONTENTS

8.1 INTRODUCTION

The hip joint is the articulation of the pelvis with the femur [1] and it is one of the most particular and essential joints in the human body [2] for its support function since the loads that hip joint suffers are very high [3]. It is no coincidence that it is subject, especially as it ages, to clinical diseases [4]. In fact, it is estimated that more than 1 million people per year in the worldwide undergo hip surgery [5] caused by pathologies, such as hip osteoarthritis [6, 7] (Figure 8.1).

Total Hip Arthroplasty (THA) currently represents one of the best solutions to a great part of joint disorders; the term arthroplasty means the orthopedic surgical procedure in which a certain joint is replaced [8] or rebuilt or even eliminated [9] to improve the standard of living of patients over the next 15–20 years [10]. This limit, whose reason originates mainly from implant integrity loosening and wear [11], is now object of studying and improving thanks to a better understanding and experience of this field [12]. The goal is to avoid a future surgery replacement because it is risky and expensive [13].

DOI: 10.1201/9781003243205-8

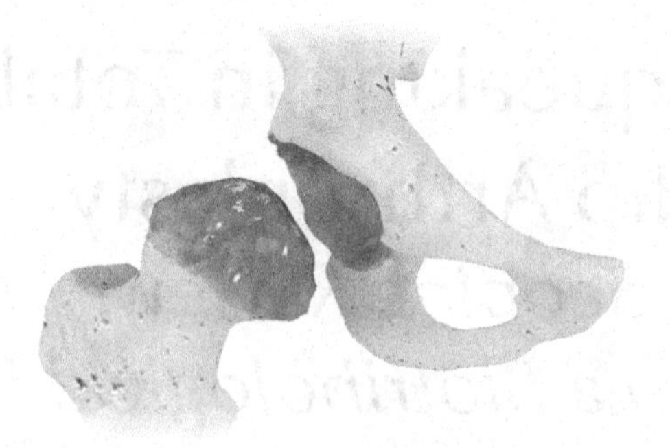

FIGURE 8.1 Hip osteoarthritis.

The main elements of a total hip replacements $(THR)_n$ can be summarized in Figure 8.2.

The stem is fixed in the femur canal and a cup (liner), embedded in the acetabular cavity of the pelvis [14] and covered with a shield (shell) in contact with the femoral head, connected to the superior part of the stem, i.e., the neck of the prosthesis.

Regarding materials, or rather biomaterials, i.e. [15], the first stems were made of stainless steel, which is no longer in use because of its poor biocompatibility. Titanium and its alloys (commercially, $\alpha+\beta$ titanium alloys, such as titanium-6Al-4 V) are the most commonly used alloys for stem. Titanium alloys are also chosen due to *their low wear resistance* and their high mechanical strength [16]. The choice of the head/cup materials depends on several factors [17, 18], including high biocompatibility, chemical (phase and microstructure composition), mechanical (yield strength, hardness, resistance to wear) and topographic (surface texture, roughness) properties and market demands (availability, price, reliability). Currently, several options are available to the surgeon when choosing the bearing surface in THA (i.e., ceramic-on-ceramic (CoC), ceramic-on-polyethylene (CoPE), metal-on-polyethylene (MoPE)), each with advantages and drawbacks. The most common material for acetabular liners is polyethylene (PE), either ultra-high molecular weight PE (UHMWPE) (the so-called 'standard' or 'conventional' PE) or cross-linked UHMWPE (XLPE), or ceramics or metal. The choice of polyethylene as cup material comes from its high wear resistance [16]. In fact, the last can also determine the known osteolysis, which causes the destruction of bone tissue [19]. Although combinations such as MoM (metal on metal) or CoC (ceramic on ceramic) are very frequent because of their high hardness properties, wear resistance [20] and low friction coefficients [21], the current THA implants mainly use cross-linked Polyethylene plus Vitamin E which performs very well in the clinical practice. Wear and osteolysis are described as occurring mainly with conventional PE bearings associated with metal or ceramic heads (MoPE or CoPE). XLPE has been reported with less wear but also with a decrease in mechanical proprieties; ceramic-on-ceramic (CoC) is related to much less wear and the highest bio-tolerability but carries the risk of breakage and noise from the implant following arthroplasty.

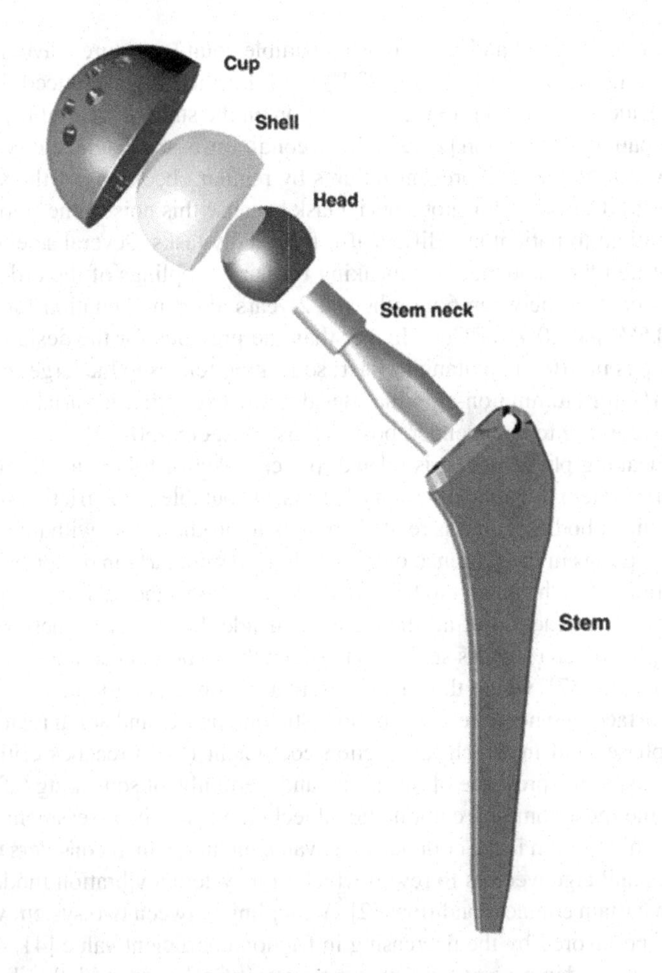

FIGURE 8.2 Elements of a total hip replacement.

Therefore, it can be seen that the choice of THR material is a difficult task, so that the investigation is aimed at new materials and design [22] such as polymeric composite material made of a biopolymeric matrix reinforced with fiberglass [23, 24].

8.2 THE SQUEAKING

The potential complications that can be observed by using ceramic bearings is the possibility of some defects due to crack formation, although, with the last generation of ceramic materials [13], ceramic failure is hardly ever reported. Rather, some clinical studies report on the presence of audible noise in some ceramic-on-ceramic THA patients. The so-called "clicks" or "grinds" or "squeak" can be produced on THAs during the gait [25].

Although the exact etiology of squeaking is still debated in the scientific literature, it is common opinion that it is probably related to a complex tribomechanical phenomenon, also connected to the implant design. In this perspective, the phenomenon of squeaking is inserted and defined as an audible sound pressure wave, if their frequencies are in the range of 20–20,000 Hz [26]. Further, it was noticed that the frequency of squeaking could provide information on the state of wear of the prosthesis and on the patient's condition [27, 28]. So, a condition-based diagnostic system could be conceived to detect and prevent failures by regular check-ups of the state of the prothesis [29]. Certainly, it is not a trivial task because this noise generation is different from patient to patient and differs also from daily tasks. Several scientific works have quantified the incidence of squeaking for hard couplings of the order of 1–5% at 5 years and 10% between 6 months and 2 years after implantation for MoM and between 0.5% and 20% for CoC, higher than the previous for the design where the ceramic cup is inserted in a titanium liner: squeaking relates to the large rim. Instead, is about 11% for alumina-on-alumina. Besides, the cycle-life, is variable from 10 to 25 years, according to age, gender, possible diseases, etc. [30–33].

The squeaking phenomenon is related to a complex of tribodynamic phenomena effecting the system dynamical stability. This is attributable to the friction interactions between contact bodies, and it is related not only to prothesis, but with many tribomechanical couplings in the presence of friction [34], particularly in dry or limit lubrication conditions, which conduct to friction-induced self-excited oscillations (vibrations induced by the interaction of all the elements inside the system). These oscillations happen in phenomena such as stick-slip between the femoral head and the acetabular cup surfaces [35–37], where the friction oscillates from static to kinetic respectively when the surfaces are not in relative motion (sticking phase) and are in relative motion (slipping phase) and in which the friction coefficient (CoF) reaches critical values [38–40], causing the presence of instability and, certainly, of squeaking (effect).

One of the most common computational techniques for the assessment of the stability of the hip motion is the complex eigenvalue method which considers the system eigenvalues and eigenvectors to reveal which of a system's vibration modes became unstable in certain contact conditions [2]. A coupling between two system vibrational modes can be favored by the decreasing in friction coefficient value [41, 42], which affects the system damping and stiffness matrices; if CoF varies with the sliding speed it may also affect the incidence of squeaking (negative friction-slope instability).

It means that two vibrational modes, according to the system properties, becoming closer and closer in frequency, get coupled [43].

Assuming the ith eigenvalue of the system made of a real part and the imaginary one, the imaginary part of the eigenvalue furnishes the eigenfrequency of the mode while the real part represents the modal damping. The eigenvalues with positive real parts (negative modal damping) are identified as an unstable mode. For certain values of CoF, two system eigenvalues collide to a single value and then become with positive real parts (negative modal damping), while the imaginary part of the collided eigenvalue splits in two, conducting to the unstable behavior of the system (coupling mode) [43].

This instability, also reported by Kang [44], depends not only on the friction coefficient, but also on the system's operative conditions. Moreover, the contact area can cause instability [45], but this happens when the friction–velocity curve has a negative slope, while in the opposite case, squeaking is not guaranteed [46, 47]. In any case, a relation between friction and noise can be found [48, 49].

Many variables influence the described phenomenon, such as the contact forces, and thus the contact area, the type of materials, the presence of wear and lubrication, and the contact surface topography. Generally, the larger the contact area, the higher the noise generation (if any) as well as the greater the surface roughness. These general literature results should be transferred to "squeak" phenomena in hip joints, where other factors can determine squeaking [50], even if aspects, such as the patients' characteristics and prostheses orientations, should also be considered.

In particular, the first factor is at a clinical level, where variables such as obesity and body mass index [51, 52], lifestyles (e.g., intense physical activity) and body structure [53–56] could influence the phenomenon, for example ligamentous laxity that is related to rim impingement [57]. In any case, proofs are currently being made of this and more investigations are required. Similarly, prosthesis orientation (acetabular position, tilt and anteversion angles) could have an impact on squeaking [58, 59].

In particular, the anteversion and the inclination of the acetabular part might be linked with squeaking because it could determine impingement and edge-loading. Hence, the position of the prothesis should follow this rule for head diameter more than 32 mm: cup inclination of 45° and the sum of the cup anteversion plus 0.7 times the stem ante torsion equals 42° [60]. In addition, both the diameter of the neck and the hip center medialization are also key variables [2].

The term "might" underlines, instead, that other studies stated the absence of correlation between these variables, but, in any case, every consideration depends on the patient and his conditions.

The design impacts [61, 62] on the natural prosthesis' vibrational frequencies: the larger prosthesis (in terms of head diameter) present lower frequencies and they more easily generate squeaking [63]; secondly, radial clearance between the inner surface of the acetabular cup and the outer surface of the femoral head plays a key role on pressure and deformation values during walking [64]: the higher the clearance, the higher the pressure, because of the modification of the contact distribution. The choice of materials is as crucial as implant design with regard to the generation of squeaking, as demonstrated in the investigation by Ecker et al. [65]. About that, the optimal design, as in many engineering applications, is almost impossible to achieve. Firstly, the word optimal should be referred to the single patient rather than just in general: for instance, the long neck allows a higher degree of motion, but they are more sensitive to squeak. In addition, problems like laxity or other clinical aspects can favor the use of ceramic insert as protection with its disadvantages (the diffusion of metal debris). Consequently, each choice comes from a specific and accurate trade-off.

Moreover, several studies confirmed that the three main mechanisms of prosthetic wear, defined as the loss of material from prosthesis due to the contact between surfaces [66], are adhesive, abrasive and fatigue cycle. Unfortunately, they are not exclusive, but they can take place at the same time [67]. Many variables contribute to the development of wear such as head size, third body debris [68], the kind of material, but, also, age, activity level [69], prosthesis design [70], type of synovial lubrication [71–73]) (boundary, mixed hydrodynamic/elastohydrodynamic). Moreover, the latter could favor, in some kinematic/dynamic joint condition, the synovial film rupture with the increasing of friction forces and then the possibility of squeaking phenomena. In any case, the theme of wear, characterized by several particles of varying shape, size and composition, is crucial because it could be one of the causes of the failure [74, 75] and diseases like osteolysis [76, 77] since this debris stimulates a foreign-body response, resulting in bone loss [78]; definitively, it has been accepted that wear is associated with squeaking [79–82].

Of course, wear is strictly connected to the lubrication mode. It is therefore essential that the hip replacement is lubricated both for a long-term reliability [83] and for decreasing the friction coefficients of the tribosystem to avoid squeaking phenomena. This is guaranteed by the presence of synovial fluid, a viscous liquid which is produced by the synovial membrane. Its main limitation is protein degradation, being the fluid composed of glycoproteins, so it has been suggested to investigate alternative bio lubricants [84].

In the end, all these factors (clinical level, prothesis design, materials, patient conditions, friction, lubrication) could represent, whether simultaneously or not, the causes of problems during a hip joint's life cycle. The potential effects might be not only the squeaking, but also wear, failure of the prothesis, diseases such as osteolysis, etc.

8.3 SQUEAKING INVESTIGATIONS METHODS

Previously, it has been noted that the natural frequencies of the system could possibly correlate to squeaking. The four parts of the hip replacement have different eigenfrequencies: shells range between 4.3 kHz and 9.2 kHz, liners above 16 kHz, the femoral heads above 20 kHz while the stem is between 2 and 20 kHz. So, the shell, and in particular the stem, have frequencies low enough to generate sound pressure waves perceptible by the human auditory system. Obviously, the value reached depends on the system and on the conditions explained before. In literature, different approaches are recognized to the study of the behavior of prostheses and its squeaking: *in-vivo* [63, 85, 86] by using specific sensors, *in-vitro* techniques by using hydraulic or electromechanical machines, known as Hip Simulation Machines [87–90] and computational methods such as Finite Element Analysis (FEA) [62, 91, 92].

For this reason, it could be useful to run a modal analysis of the hip implant, to study those vibration frequencies that could be linked to squeaking. This analysis could be performed on multi-physics software where a modal analysis could be realized with a finite element model, modelling the contact forces of the considered hip replacement prothesis and eventually a fluid simulation of the synovial interface, since the fluid stresses are involved in the problem. The above approach needs the knowledge of the prostheses' kinematics and load conditions during the gait as investigated by Kang [93], who analyzed the stability of the hip joint system in full lubrication regime (constant CoF) and in mixed lubrication conditions (negative slope CoF). Kang found that in the case of full lubrication the system becomes dynamically only when the hip extended, making tilting angle between the axis of reaction force and the hip joint angle a key parameter for the onset of hip squeak, while in the case of mixed lubrication mode, due to the friction coefficient increasing in decreasing the sliding speed, the system was dynamically unstable over the entire gait cycle and the unstable bending frequency changes. Hence, kinematics parameters variation during the gait [94] joined to lubrication phenomena for the hip squeaking generation.

8.3.1 Finite Element Analysis

Today, Finite Element Analysis (FEA) is very common in the tribology field and in prosthesis calculations [95, 96]. The study of the eigenvalues by conducting a numerical modal analysis allows to determine the eigenfrequencies for each mode as criteria for stability and squeaking.

The finite element method operates by decomposing the complex and continuous domain of the system into a set of numerous subdomains of variable geometric nature (finite elements) and into a set of points (nodes), solving them and reassembling the entire prostheses domain solutions.

The resulting system is no longer continuous; it is now discrete with finite degrees of freedom constituted by finite elements (triangles, squares) first, and then to nodes (points) [97]. The relationship between the finite element and nodes is expressed by shape functions which link the displacement of a generic point inside the finite element as a function of nodal displacements.

In particular, the shape function can be evaluated by different methods such as polynomial series (the largest used), Lagrange, and Hermite series [98].

The choice of the shape function (one-, two- or three-dimensional) depends on the kind of finite element considered for the specific problem, geometry, and properties.

After the evaluation of the shape function, it is possible to define the general motion equations that will be considered in squeaking analysis in the form:

$$\left[M_t\right]\ddot{u}+\left[C_t\right]\dot{u}+\left[K_t\right]u=F_t \tag{8.1}$$

Where $[M]$, $[C]$ e $[K]$ are the matrices of mass, damping, and stiffness, with the subscript t to underline the whole of the finite elements.

The expansion of the $[M]$, $[C]$, $[K]$ matrices and the F vector, which contains the concentrated and distributed loads components, deriving also from head/cup contact conditions (frictional), from the local dimension, relating only to the degrees of freedom of the finite element, to the global one, connected to all the degrees of freedom of the prostheses, is required. An assembly operation is needed: all the terms corresponded to the single node d.o.f are summed.

It is worthwhile emphasising that due to friction within the contact area the stiffness matrix has specific properties connected both to the structural stiffness matrix and to the asymmetrical friction induced stiffness matrix [2]. In this way the head/cup CoF variation could induce a modification of the stiffness and damping matrices and then drive the dynamical instability phenomena.

Equation (8.1) is valid until the linear regime, which can be hypothesized but not always guaranteed in prosthesis dynamical analysis.

In general, the non-linearities could derive from:

1. Material behavior (plastic deformation)
2. No small-displacements analysis (theory of small deformations not satisfied)
3. Constraints conditions (backlashes).

In these cases, the computational complexity increases, and numerical algorithms are required to solve structural and dynamical problems.

From an operating point of view, Finite Element Analysis involves five main steps [99] that will be described with reference to the topic:

- Materials definition
  Prostheses are usually analyzed as homogeneous bodies, isotropic and linear-elastic behavior. A further variable is the temperature that is set at 25°C (environmental) or 37°C (body). The second is preferred because it is closer to the behavior of the body.
- Geometry definition
  It is the phase of translation of the body into dimensional sizes [100]. The geometric model is a set of equations in which some variables are defined. Nowadays, all simulation systems require CAD modeling, and they are increasingly advanced in features (Figure 8.3). In the case of the prosthesis, for convergence problems, stem and cup are often simplified as cylinders [2].
- Meshing
  This regards the type of mesh to adopt. The choice of mesh is always crucial for a finite element approach as its quality influences the final solution. It can be evaluated mainly by [101, 102]:
  a. Skewness = measure of the deviation of a cell from the ideal one (optimal value ≅ 0).
  b. Smoothness = size variation between adjacent cells (should be gradual).
  c. Aspect ratio = ratio of longest and shortest edge (optimal value ≅ 1).

  In the case of prostheses, an 8-noded hexahedral element mesh [2] is preferred, since the analyses made with these elements are more precise, especially regarding deformation tests.

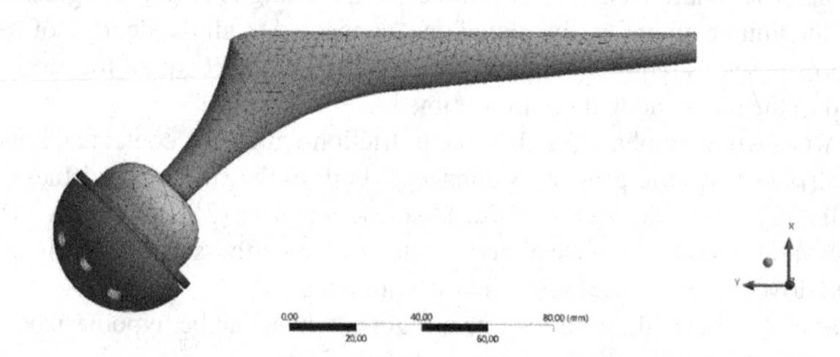

FIGURE 8.3 Example of a 3D prosthesis finite elements model.

- Boundary Conditions
  Boundary conditions, like forces, pressures, or displacements, fixed or variable over time, and any translation or rotation constraints in the three directions x, y, z have to be imposed. In general, all variables that can affect the body and its interactions with other objects or with the external environment are inserted. For prostheses, the force components are introduced on the stem and rotations on the head, usually considering the gait cycle [35, 72, 73, 94, 103, 104] as represented in Figure 8.4, and the total reaction force which determines the contact pressure values and friction stress matrix. Additional constraints, such as contact conditions, can be added to the model.

The required calculation time depends mainly on the number of elements and nodes of the mesh, and on the complexity of the problem in terms of the number of the

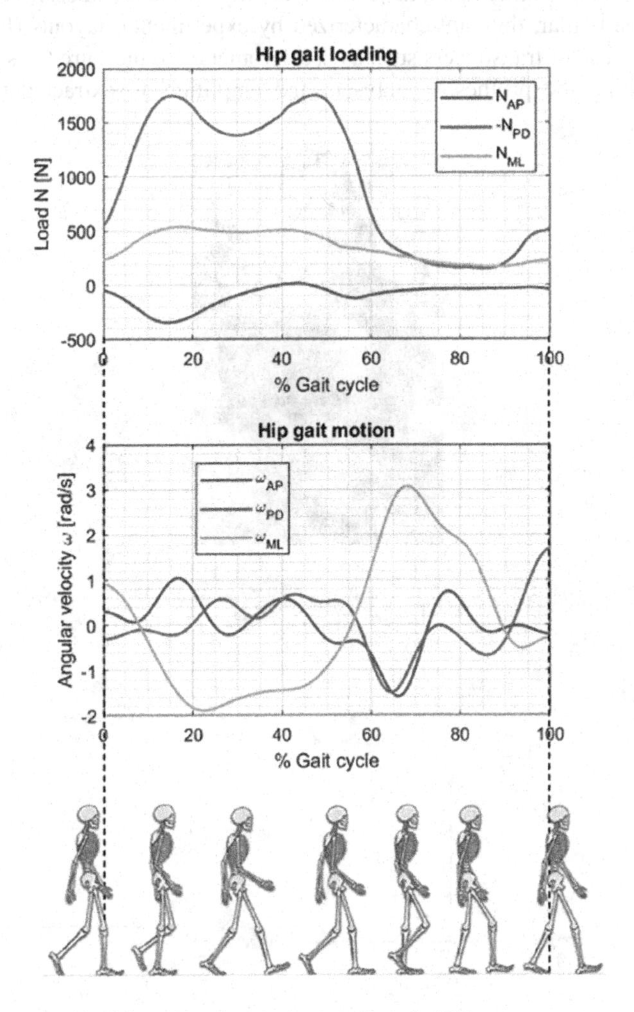

FIGURE 8.4 Hip loading and motion during a gait cycle [73].

desired vibration modes. About that, standard EN 1998-1 states that this number must be such that the sum of the modal masses is at least 90% of the total mass of the object.

The main output are definitely the modes of vibration of the THR, identifying the unstable ones.

8.3.2 *In-vivo* and *In-vitro* Experiments

By these approaches, analyses are done directly on the real prosthesis. *In-vivo* techniques act directly on the patient (clinical approach) to which very precise sensors of acoustic emissions are connected [105]. They have the goal to capture the vibrations of the prosthesis during daily movements such as getting up, sitting. walking etc.

The signal is then processed and filtered before being acquired. *In-vitro* techniques [106], on the other hand, have an effect only on the prosthesis and not on the patient. In particular, they are characterized by experimental layouts (Figure 8.5a) involving the use of transducers such as accelerometers to measure the signal, shakers for exciting the prothesis, analysers and amplifiers to extract data from the

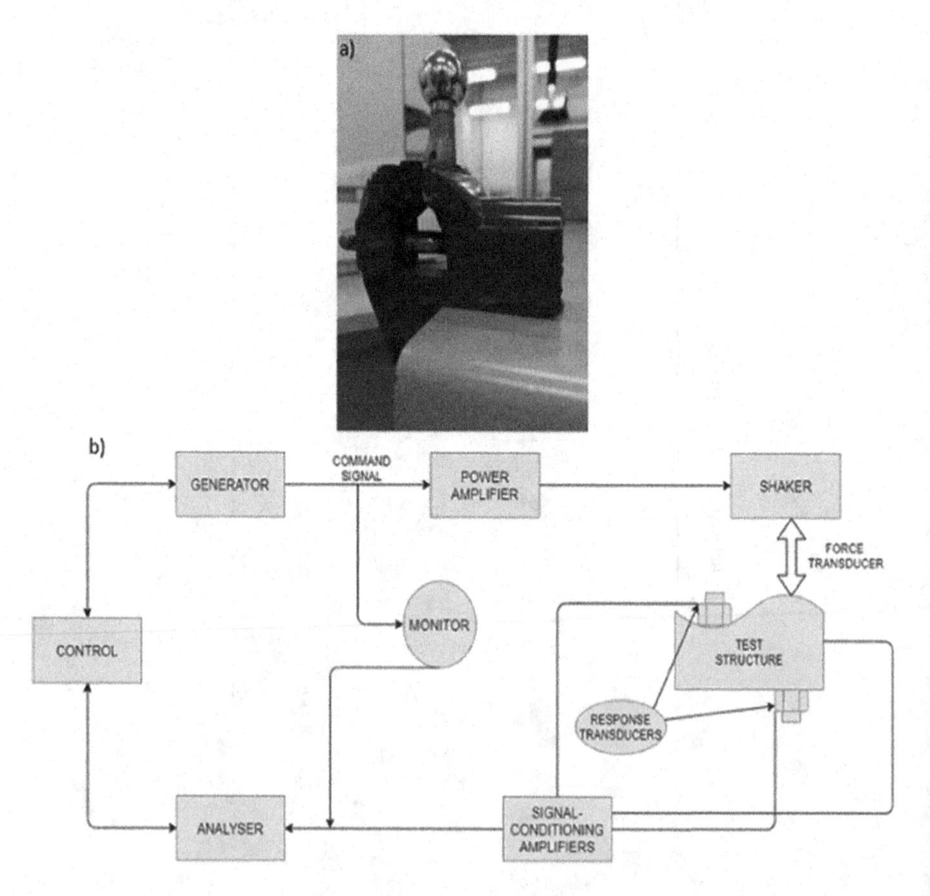

FIGURE 8.5 (a) Experimental modal analysis setup (b) Example of an experimental layout.

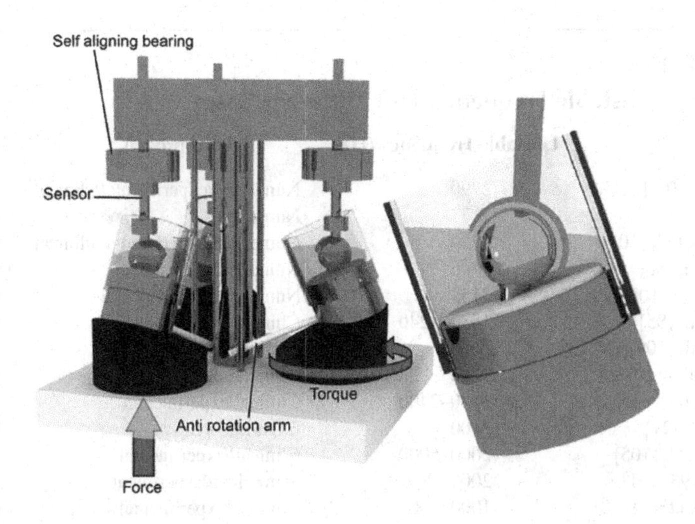

FIGURE 8.6 A schematization of a station of a typical hip joint simulator.

obtained signal and a control to monitor all the operations (Figure 8.5b), or by hip joint simulators (Figure 8.6).

The analyzed THR configuration can be free or constrained, whereas excitation can be obtained by contact (mechanical), or no-contact shakers (electrical) and it can be periodic, random, or impulsive. In any case, it is important to control the experiment, especially considering other external factors that could distort the acquired signals.

8.4 LITERATURE OVERVIEW ON SQUEAKING

Several researches provided the values of squeaking frequencies of investigated hip prosthesis. Obviously, each result depends on the hypotheses used by the researchers, but the differences are quite low. In Table 8.1, various outcomes are summarized in terms of prostheses type and unstable frequencies found. They were listed in order of time to emphasize, ahead in time, the use of more powerful and complex techniques. Indeed, the models or the algorithm proposed, year by year, involved more and more aspects and complexity in order to move closer to reality.

More in detail:
Walter et al. (2008) [88], after a review about this field, confirmed that factors such as age, weight, height, and position of the prothesis are potential causes of squeaking (as discussed in chapter 1), evaluated by a boxplot. Also, by a FEM analysis, they discovered an audible squeak for a mismatch between shell and liner. According to them, the origin of squeaking derives from high friction values and happens only for metallic parts and not for ceramics due mainly to edge loading. In conclusion other influences exist for example the control of surgeon and the surgery.

Sariali et al. (2009) [109] analyzed, in a lubrication regime, by the use of a hip simulator, the friction coefficient in particular conditions of load and motion.

TABLE 8.1
Squeaking Unstable Frequencies for Different Cases

Author	Unstable Frequency (Hz)	Approach	Material
Piriou et al. [61]	2790	Numerical/experimental/clinical	CoC
Fan et al. [103]	3177	Numerical	CoC
Ouenzerfi et al. [107]	1759	Numerical/experimental/clinical	CoC
Walter et al. [88]	1546	Numerical	CoC
Hothan et al. [108]	3159	Numerical/experimental	CoC
Sariali et al. [85]	1450–2240	Clinical/experimental	CoC
Sariali et al. [109]	2600	Clinical/Experimental	CoC
Askari et al. [110]	1700	Numerical	CoC
Currier et al. [86]	1540–2400	Clinical/experimental	CoC
Weiss et al. [21]	3400	Experimental	CoC
Rodgers et al. [105]	2000–5000	Clinical/Experimental	CoC-MoM
Kang [47, 93, 111]	2000	Numerical/Experimental	MoC
Fitzpatrick et al. [112]	1000–4000	Clinical/Experimental	CoC-MoM
Sariali et al. [113]	2200–2600	Clinical/ Experimental	CoC
Askari et al. [114]	2600–3000	Numerical	CoC

Successively, they run tests positioning a third body made of alumina between head and cup, noticing that this had increased about 26 times friction and had caused squeaking. This underlines the importance of lubrication for preventing the development of wear, regarding not only squeaking problem but in general for the life cycle of a hip joint. Other remarkable results are linked to a high abduction that generates high wear.

Currier et al. (2010) [86] observed, by a clinical and experimental approach, that squeaking happens over a range of frequencies, far lower than natural ones. This entails that this phenomenon is caused not by natural frequencies but by other factors such as the rolling/sliding mechanism. This is a different point of view where the attention is now focused on the motion of the hip prothesis and not on its mechanical and geometrical characteristics. With regard to this, they stated the relation with hip rotation speed and friction in the range of 0.06–0.18. Moreover, the latter is independent of the force since squeaking happens in both light and heavy patients. This statement is, actually, strongly discussed because of other opposite results. Finally, this study did not consider a numerical simulation whose results conflict with it.

Hothan et al. (2011) [108] investigated the influence of component design on squeaking and its frequencies through the adoption of a numerical and experimental approach. Their idea was to analyze the prothesis geometrically by varying the dimensions of the cup and the stem as well as the position. They found that these factors play a key role for squeaking (friction rising due to a metal transfer or wear), whereas the head does not influence observed squeaking and so can be considered to be rigid bodies. With regard to the contact, a high load influences squeaking in contrast to the bearing clearance. Finally, they stated that squeaking happens only when the friction reaches a critical value because in this way the energy given to the stem is enough to produce audible vibrations. Hence, the aim is to control friction coefficient under precise levels.

Weiss et al. (2012) [21], studied squeaking, reconsidering the kinematics of the coupling. As discussed before, during the gait cycle, the load and the rotation vectors change. Secondly, they found, experimentally, that the rotational component had determined the presence of squeaking. In particular, the friction induced whirl vibration derived from a motion resulting elliptical. This statement is a novelty in the squeaking field because vibrations were previously associated with a one-dimensional character. Weiss et al. instead identified two stem bending modes, leading to a new view of THA. Consequently, the mathematical description of the model is a fundamental aspect in order to discover the exact behavior of the prothesis.

Fan et al. (2012) [103], who confirmed Hothan et al.'s [108] statement about the critical friction coefficient limit value ($\mu = 0.15$ in this study) by a FEM model. In addition, they proposed a valid alternative to contrast squeaking: the first by adding damping material (UHMWPE), the second by imposing a reasonable stiffness on acetabular components. The scope is to increase friction coefficient so that, also the stability of the hip joint increases.

In the same year, again Sariali et al. (2012) [85], by comparing *in-vivo* and *in-vitro* experiments, discovered an intermittent squeaking (rising from a bent position) with a lower fundamental frequency (1450 Hz) compared to the walking squeakers (2240 Hz) that suggested the presence of two kind of squeaking. The interesting aspect of their work was the consideration of lubrication in both tests: in this way the experiments simulate a behavior more likely close to reality in contrast to other investigations made in dry regime (for instance numerical ones).

Askari et al. (2014) [110] proposed a multibody system (that is, a valid alternative to FEM approach for force modelling) combined with a FFT analysis (for audible sounds analysis) to investigate hip squeaking. Variables such as hip size and implantation and friction were discussed. In particular, they stated the relation between size and squeaking frequencies: the lower the size, the higher the frequency (as stated earlier). Head angular speed, stick-slip phenomena, and force changes caused vibration in the femoral head and thus squeaking, in contrast to the initial position that did not have a great impact.

In the same year, again Askari et al. [114] developed a detailed spatial multibody dynamic model, obtained by a combination of Hertzian theory and friction-velocity law, analyzing the gait cycle in three-dimensional motion. Several common causes of squeaking were deducted, including implant size, friction-induced vibration, and so on. The interesting aspect is the comparison between vibration amplitude in three axes: it was observed that along the internal-external axis rotation is very small in respect to the others. Hence vibrations should be investigated in all possible directions.

Rodgers et al. (2014) [105] compared experimental (by automated robot) and clinical (by a set of four ultrasonic sensors) testing on total hip replacement implants to classify any differences. Precisely, the *in-vivo* measurements are made before the revision surgery while the *in-vitro* measurements after (stem and shell were not replaced). The results showed a strong similarity between the two approaches about the characteristic frequencies, indicating that tissue attenuation decreased signal magnitude, but not the period shifting of signals. Anyway, the investigation was related to a single patient and, therefore, not statistically significant. Finally, the article underlines that the occurrence of squeaking, although it could not lead to dangerous consequences, represents a condition of discomfort to patients causing difficulties in the daily life. This factor must be considered as well as clinical problems.

Ouenzerfi et al. (2015) [107] confirmed the previous statements by numerical, *in-vivo* and experimental approaches. In addition, they investigated the cause of squeaking and stated that it could be generated by the coupling of two modes of vibration under frictional contact. In addition, the friction coefficient has a relevant impact on stability (high values generate squeaking). As Fan [103], Ouenzerfi et al. also faced the problem of reducing squeaking, but they proposed, apart from the design and the materials, a clinical solution by minimizing factors or situations that could generate squeaking (for instance, edge-loading). Finally, a mismatching between simulation and *in-vivo* experiments was found due to anatomy of the human body not considered during simulation.

Piriou et al. (2016) [61] provided a FEM model and compared it with experiments and *in-vivo* to find any differences. The material used was ceramic for the head and cup, titanium for the stem, and no lubrication was considered (this is not an unrealistic scenario because it may simulate the destruction of synovial fluid). Because of the low discrepancy between the results, the model results proved reliable. Finally, an unstable vibration of the stem, due to an increasing friction force, caused squeaking with a mean frequency of 2790 Hz. The authors underlined one of the most important advantages of FEM simulation (apart from its reliability): the possibility to modify in real time the boundary conditions or the material properties in order to find the best configuration for the case study.

Kang (2017) [111] realized a finite element algorithm to investigate hip squeak re-modelling the contact kinematics (Hertz theory and Coulomb law) analytically, together with a mode discretization, and compared the results obtained from a non-linear (as told before linearity is not always guaranteed) transient analysis with those obtained experimentally. He found that tilt angles increase the positive real part and so determines squeaking (according to his study the design and implantation are considerable). The same effect is determined by rotating the direction of the stem and, of course, by the friction coefficient (stick-slip phenomenon). The comparison with experiments confirms that FE algorithm is reliable, and it can be used for other tests in the future.

Fitzpatrick et al. (2017) [112] developed an acoustic emission prototype and used it to test 90 patients with different age and implant conditions. In particular, the study found that the instrument can detect any relevant sound, squeaking included. Anyway, in order to validate the outcomes, they also used an experimental approach. The natural frequencies were very similar, so that the authors have proposed the tool as a future diagnostic device. In more detail, they proposed a diagnostic procedure to find any problems in hip joints. This is an interesting look in the future where more and more precise machines will be in action.

Finally, Sariali et al. (2018) [113] reported an *in-vitro* lubricated test and compared it with *in-vivo* experiments. For each approach, the impact of edge loading was irrelevant for squeaking, which was different from the addition of an alumina ceramic third-body particle in the contact head-cup. In any case, during *in-vitro* tests, squeaking was intermittent. On the other hand, in *in-vivo* tests, the squeak frequencies were lower, because of the potential dampening effect of soft tissue. Hence the anatomy of our body can facilitate or not (as in this case) the occurrence of squeaking.

Overall, during these 15 years, the procedure (numerical as FEM, clinic directly applied to the patient and experimental by appropriate instruments) has changed from study to study, and, accordingly, the squeaking frequencies have varied, mainly, in the range 2–4 Hz. The differences are very low, demonstrating that today the high-tech software, as well as the instruments adopted, are valid choices for investigating squeaking or other phenomena. However, any discrepancies are present and derived from the procedure applied, the characteristics of the materials and other variables such as the coefficient of friction, orientation, the position of the prosthesis and the clinical condition of the patient. About materials the works always provide a hard coupling, metal, or ceramic, and not plastic. This confirms the tendency to prefer hard materials than soft ones. Moreover, many of the projects tried to understand which variables have an impact on squeaking, and to what extent they do so. Most of the works confirmed that friction is the main cause of squeaking, and this is an essential statement for future directions. The other factors discussed before such as implantation or material are common reasons of squeaking, but not in every study. Furthermore, sometimes two or three methods, have been compared, to secure more answers about instable frequencies. The comparison showed a good agreement, and it is necessary to validate the model. Lastly, the possibility to contrast squeaking is as important as its investigation. Several options are suggested as well, as indicated by Fan et al. [103] or by Ouenzerfi et al. [107].

In particular, FEM provided results quite similar to the experiments as confirmed by Piriou [61] and Walter [88], justifying its common use among scientists. Probably its main limitation is the impossibility to simulate synovial lubrication regimes and to consider human body anatomy.

8.5 CONCLUSIONS AND RESEARCH DIRECTIONS

The aim of this scoping review on the THR squeaking was to examine how research is conducted on this topic, highlighting the theoretical key concepts and the approaches used in squeaking analysis, so as to identify knowledge gaps and research directions.

The analysis of the scientific framework allows us to conclude that the phenomenon of squeaking in hip prostheses is certainly influenced by the vibrating modes of the prostheses. It has been demonstrated that squeaking is linked to the articulation of the femoral head and cup and that high friction could induce vibrations. Each of the four parts of the hip replacement have different eigenfrequencies: shells range between 4.3 kHz and 9.2 kHz; liners above 16 kHz; the femoral heads above 20 kHz; while the stem is between 2 and 20 kHz. So, the shell, and in particular the stem, has frequencies low enough to generate sound pressures audible by humans, and for this reason could be a major cause in noise generation. For this reason, it could be useful to run a modal analysis of the hip implant, in order to study those vibration frequencies that could be linked to squeaking.

Finite Element Models provide a reliable tool for squeaking analysis allowing the modal analysis of the system, which permits us to understand its vibrational properties in the frequency domain. Finite Element Analysis allows the evaluation of the eigenfrequencies of the system and its mode shapes, only requiring the evaluation of the mass and stiffness matrix of the investigated prosthesis.

The calculated frequencies are all in the order of kilohertz and are lightly influenced by the adopted materials; thus, similar behavior can be expected for similar prostheses.

However, future developments are required to consider more a more detailed model considering the non-linear hip contact forces and the effect of the synovial lubricating interface, to better investigate the relation between friction phenomena and squeaking in more approximate models.

Finally, although many results and relations between key variables have been already found, more research is required to validate the proposed theories and studies; for instance, the impact of wear on squeaking, or the possibility to use more suitable materials in terms of stiffness and structural dampening, or the development of innovative numerical models considering the effect of the synovial lubrication effects during daily activities. Moreover, the investigation can be driven to study contact models, from a microscopical point of view, considering the complex topography of the contact surfaces. The impact of other variables such as temperature, and its connection to friction, could be interesting to explore. Finally, innovative models, based on the energy analysis of friction-induced vibrations, could be adopted for a more and more approximate THR dynamical models.

FUNDING

This research was funded by MIUR PRIN 2017 BIONIC.

ACKNOWLEDGMENTS

The authors thank Luigi Lena (IRCCS Instituto Ortopedico Rizzoli) for the original pictures.

REFERENCES

1. S. Kishner, "Hip joint anatomy," *Medscape*, 2017.
2. E. Askari, et al. "A review of squeaking in ceramic total hip prostheses," *Tribology International*, 93: 239–256, 2016.
3. L. Ambrosio, *Biomedical Composites*, 2nd edition Woodhead Publishing Series in Biomaterials, 2017.
4. C.P.W. Chong, J.A. Savige, W.K. Lim, "Medical problems in hip fracture patients," *Archives of Orthopaedic and Trauma Surgery*, 130(11): 1355–1361, 2010.
5. T.P. Sculco, "The economic impact of infected total joint arthroplasty," *Orthopedics*, 18(9): 871, 1995.
6. National Institutes of Health, *Osteoporosis overview*, 2019.
7. V. Bottai, G. Dell'Osso, F. Celli, G. Bugelli, N. Cazzella, E. Cei, G. Guido, S. Giannotti, "Total hip replacement in osteoarthritis: The role of bone metabolism and its complications," *Clinical Cases in Mineral and Bone Metabolism*, 12(3): 247–250, 2015.
8. G.D. Baura, *Medical Device Technologies*, Academic Press, 2012, pp. 381–404.

9. H.E. Jergesen, J.S. Siopack, "Total hip arthroplasty," *The Western Journal of Medicine*, 162(3): 243–249, 1995.
10. T. Karachalios, G. Komnos, A. Koutalos, "Total hip arthroplasty," *EFORT Open*, 3(5): 232–239, 2018.
11. K. Colic, A. Sedmak, "The current approach to research and design of the artificial hip prosthesis: A review," *Rheumatology and Orthopedic Medicine*, 1: 1–17, 2016.
12. S.G. Ghalme, A. Mankar, Y. Bhalerao, "Biomaterials in hip joint replacement," *International Journal of Materials Science and Engineering*, 4(2): 113–125, 2016.
13. S. Affatato, A. Ruggiero, M. Merola. "Advanced biomaterials in hip joint arthroplasty. A review on polymer and ceramics composites as alternative bearings," *Composites Part B: Engineering* 83 (2015): 276–283.
14. Standing, Anatomy of Grey, *Edra*, 41st edition, 2017.
15. A. Aherwar, A. Sing, A. Patnaik, "Current and future biocompatibility aspects of biomaterials for hip prosthesis," *AIMS Journal*, 3(1): 23–43, 2016.
16. M. Merola, S. Affatato, "Materials for hip prostheses: A review of wear and loading considerations," *Materials*, 12: 495, 2019.
17. J.J. Callaghan, A.G. Rosenberg, H.E. Rubash, *The Adult Hip*, Lippincott Williams and Wilkins, Philadelphia, USA, 2007.
18. M. Choroszyński, M.R. Choroszyński, S.J. Skrzypek, "Biomaterials for hip implants – Important considerations relating to the choice of materials," *Bio-Algorithms and Med-Systems*, 13(3): 133–145, 2017.
19. Adam et al., *Monesi's Histology*, Piccini, 7th edition, 2018.
20. I. Hutchings, *Tribology*, Edward Arnold, 1992.
21. C. Weiss, A. Hothan, G. Huber, M.M. Morlock, N.P. Hoffmann, "Friction-induced whirl vibration," *Journal of Biomechanics*, 45(2): 297–303, 2012.
22. C.Y. Yoon, T.R. Hu, "Recent updates for biomaterials used in total hip arthroplasty," *Biomaterials Research*, 22: 1–12, 2018.
23. R.R.C. da Costa, F.R.B. de Almeida, A.A.X. da Silva, S.M. Domiciano, A.F.C. Vieira, "Design of a polymeric composite material femoral stem for hip joint implant," *Polímeros*, vol. 29(4): 2019.
24. H. Yildiz, F.-K. Chang, S. Goodman, "Composite hip prosthesis design. II. Simulation," *Journal of Biomedical Materials Research*, 39(1): 102–119, 1998.
25. S. Affatato, F. Traina, C. Mazzega-Fabbro, V. Sergo, M. Viceconti, "Is ceramic-on-ceramic squeaking phenomenon reproducible in vitro? A long-term simulator study under severe conditions," *Journal of Biomedical Materials Research Part B: Applied Biomaterials*, 91B: 264–271, 2009
26. H. Amstutz, *Evolution of Hip Arthroplasty*, Amstutz HC, 1991.
27. C.L. Brockett, S. Williams, Z. Jin, G.H. Isaac, J. Fisher, "Squeaking hip arthroplasties: A tribological phenomenon," *The Journal of Arthroplasty*, 28(1): 2013.
28. J. Cluett, *Squeaking Hip Replacements: Is There a Problem with Your Artificial Hip?*, Verywellhealth, 2020.
29. A. Koons-Stapf, "Condition based maintenance: Theory, methodology, & application," in *Conference: Reliability and Maintainability Symposium*, Tarpon Springs, 2015.
30. A. Cogan, R. Nizard, L. Sedel, "Occurrence of noise in alumina-on-alumina total hip arthroplasty. A survey on 284 consecutive hips," *Orthopaedics & Traumatology, Surgery & Research*, 97(2): 206–210, 2011.
31. D.J. Berry, W.S. Harmsen, M.E. Cabanela, B.F. Morrey, "Twenty-five-year survivorship of two thousand consecutive primary Charnley total hip replacements," *Journal of Bone and Joint Surgery*, 84-A(2): 2002.

32. C. Delaunay, I. Petit, I.D. Learmonth, P. Oger, P.A. Vendittoli, "Metal-on-metal bearings total hip arthroplasty: The cobalt and chromium ions release concern," *Orthopaedics & Traumatology, Surgery & Research*, 96(8): 894–904, 2010.
33. M.T. Manley, K. Sutton, "Bearings of the future for total hip arthroplasty," *Journal of Arthroplasty*, 23(7): 47–50, 2008.
34. D. Guida, M.C. De Simone, Z.B. Riviera, "Finite element analysis on squeal-noise in railway applications," *FME Transactions*, 46(93–100): 93, 2016.
35. C. Weiss, A. Hothan, M.M. Morlock, N. Hoffmann, "Friction-induced vibration of artificial hip joints," *GAMM-Mitteilungen*, 32(2): 193–204, 2009.
36. C. Weiss, P. Gdaniec, N.P. Hoffmann, A. Hothan, G. Huber, M.M. Morlock, "Squeak in hip endoprosthesis systems: An experimental study and a numerical technique to analyze design variants," *Medical Engineering & Physics*, 32(6): 604–609, 2010.
37. A. Ruggiero, R. D'Amato, F.B. Haro, "A finite element model for the analysis of squeaking hips," in *Proceedings of the Seventh International Conference on Technological Ecosystems for Enhancing Multiculturality*, 2019.
38. C. Liguori, A. Ruggiero, D. Russo, P. Sommella, "A statistical approach for improving the accuracy of dry friction coefficient measurement," *IEEE Transactions on Instrumentation and Measurement*, 68(5): 1412–1423, 2019.
39. J.-J. Sinou, L. Jezequel, "On the stabilizing and destabilizing effects of damping in a non-conservative pin-disc system," *Acta Mechanica*, 199: 43–52, 2008.
40. A.J. McMillan, "A non-linear friction model for self-excited vibrations," *Journal of Sound and Vibration*, 205(3), 323–335, 1997.
41. D. Guida, M.C. De Simone, "Modal coupling in presence of dry friction," *Machines*, 6(1): 8, 2018.
42. A. Ruggiero, G. Di Leo, C. Liguori, D. Russo, P. Sommella, "Accurate measurement of reciprocating kinetic friction coefficient through automatic detection of the running-in," *IEEE Transactions on Instrumentation and Measurement*, 69(5): 2398–2407.
43. T.D. Rossing, N.H. Fletcher, "Coupled Vibrating Systems," in *Principles of Vibration and Sound*, Springer, New York, 2004.
44. J. Kang, "Theoretical model for friction-induced vibration of ball joint system under mode-coupling instability," *Tribology Transactions*, 58: 807–814, 2015.
45. J. Kang, "The onset condition of friction noise in ball joint under concentric loading," *Applied Acoustics* 89: 57–61, 2015.
46. J. Kang, "Dynamic instability of a spherical joint under various contact areas," *Proceedings of the Institution of Mechanical Engineers, Part C: Journal of Mechanical Engineering Science*, 229(1): 54–58, 2015.
47. J. Kang, "Theoretical model of ball joint squeak," *Journal of Sound and Vibration*, 330(22): 5490–5499, 2011.
48. K. Lontin, M. Khan, "Interdependence of friction, wear, and noise: A review," *Friction*, 2021.
49. Q.S. Nguyen, "Instability and friction," *Comptes Rendus Mécanique*, 331(1): 99–112, 2003.
50. E. Askari, et al., "A review of squeaking in ceramic total hip prostheses," *Tribology International*, 93: 239–256, 2016.
51. S.J.C. Stanat, J.D. Capozzi, "Squeaking in third- and fourth-generation ceramic-on-ceramic total hip arthroplasty," *The Journal of Arthroplasty*, 27(3): 445–453, 2012.
52. S.-C. Ki, B.-H. Kim, J.-H. Ryu, D.-H. Yoon, Y.-Y. Chung, "Squeaking sound in total hip arthroplasty using ceramic-on-ceramic bearing surfaces," *Journal of Orthopaedic Science*, 16(1): 21–25, 2011.

53. T. Kiyama, T.L. Kinsey, O.M. Mahoney, "Can squeaking with ceramic-on-ceramic hip articulations in total hip arthroplasty be avoided?" *Journal of Arthroplasty*, 28(6): 1015–1020, 2013.
54. W.L. Walter, G.C. O'Toole, W.K. Walter, A. Ellis, B.A. Zicat, "Squeaking in ceramic-on-ceramic hips: The importance of acetabular component orientation," *Journal of Arthroplasty*, 22(4): 496–503, 2007.
55. S.A. Sexton, E. Yeung, M.P. Jackson, S. Rajaratnam, J.M. Martell, W.L. Walter, B.A. Zicat, W.K. Walter, "The role of patient factors and implant position in squeaking of ceramic-on-ceramic total hip replacements," *Journal of Bone and Joint Surgery*, 93: 439–442, 2011.
56. W.L. Walter, E. Yeung, C. Esposito, "A review of squeaking hips," *JAAOS-Journal of the American Academy of Orthopaedic Surgeons*, 18(6): 319–326, 2010.
57. J.A. Rodriquez, "The squeaking hip is a multifactorial concern," *Orthopedics Today*, 28, 92, 2008.
58. T.V. Swanson, D.J. Peterson, R. Seethala, R.L. Bliss, C.A. Spellmon, "Influence of prosthetic design on squeaking after ceramic-on-ceramic total hip arthroplasty," *The Journal of Arthroplasty*, 25(6): 36–42, 2010.
59. W.L. Walter, A. Watson, B. Zicat, *Stripe Wear and Squeaking in Ceramic Total Hip Bearings*, Elsevier, vol. 17, pp. 190–195, 2006.
60. T. Hisatome, H. Doi, "Theoretically optimum position of the prosthesis in total hip arthroplasty to fulfill the severe range of motion criteria due to neck impingement," *Journal of Orthopaedic Science*, 16(2): 229–237, 2011. doi: 10.1007/s00776-011-0039-1. Epub 2011 February 26. PMID: 21359509.
61. P. Piriou, G. Ouenzerfi, H. Migaud, E. Renault, F. Massi, M. Serrault, "A numerical model to reproduce squeaking of ceramic-on-ceramic total hip arthroplasty. Influence of design and material," *Orthopaedics & Traumatology, Surgery & Research*, 102(4): S229–S234, 2016.
62. S. Affatato, M. Merola, A. Ruggiero, Development of a novel in silico model to investigate the influence of radial clearance on the acetabular cup contact pressure in hip implants," *Materials*, 11(8): 1282, 2018.
63. F. Baruffaldi, R. Mecca, S. Stea, A. Beraudi, B. Bordini, M. Amabile, A. Sudanese, A. Toni, "Squeaking and other noises in patients with ceramic-on-ceramic total hip arthroplasty," *Hip International*, 30(4): 438–445, 2020.
64. A. Ruggiero, M. Merola, S. Affatato. "Finite element simulations of hard-on-soft hip joint prosthesis accounting for dynamic loads calculated from a musculoskeletal model during walking," *Materials*, 11(4): 574, 2018.
65. M. Ecker, C. Robbins, G. van Flandern, D. Patch, S.D. Steppacher, B. Bierbaum, S.B. Murphy, "Squeaking in total hip replacement: No cause for concern," *Orthopedics*, 31(9): 875–884, 2008.
66. A.J. Sampagar, "Total Hip Arthroplasty," in *Seminar*, Davangere, 2011.
67. J.-P. Hung, J.S.S.-S. Wu, "A comparative study on wear behavior of hip prosthesis by finite element simulation," *Biomedical Engineering: Applications, Basis and Communications*, 14: 139–148, 2002.
68. J.J. Callaghan, D.R. Pedersen, R.C. Johnston, T.D. Brown, "Clinical biomechanics of wear in total hip arthroplasty," *The Iowa Orthopaedic Journal*, 23, 1–12, 2003.
69. P.A. Devane, J.G. Horne, "Assessment of polyethylene wear in total hip replacement," *Clinical Orthopaedics and Related Research*, 369: 59–72, 1999.
70. G. Matsoukas, Y. Kim, "Design optimization of a total hip prosthesis for wear reduction," *Journal of Biomechanical Engineering*, 131(5): 2009.

71. A. Ruggiero, A. Sicilia, "Lubrication modeling and wear calculation in artificial hip joint during the gait", *Tribology International*, 142: 2020.
72. A. Ruggiero, A. Sicilia, S. Affatato, "In silico total hip replacement wear testing in the framework of ISO 14242-3 accounting for mixed elasto-hydrodynamic lubrication effects," *Wear*, 460: 2020.
73. A. Ruggiero, A. Sicilia," A mixed elasto-hydrodynamic lubrication model for wear calculation in artificial hip joints," *Lubricants*, 8(7): 72, 2020.
74. L. Grillini, S. Affatato, "How to measure wear following total hip arthroplasty," *The Journal of Clinical and Experimental Research on Hip Pathology and Therapy*, 3(23): 233–242, 2013.
75. B. Mjoberg, "Theories of wear and loosening in hip prostheses: Wear-induced loosening vs loosening-induced wear-a review," *Acta Orthopaedica Scandinavica*, 65(3): 361–371, 1994.
76. Y.H. Zhu, K.Y. Chiu, W.M. Tang, "Polyethylene wear and osteolysis in total hip arthroplasty," *Journal of Orthopaedic Surgery*, 9(1): 91–99, 2001.
77. C.J. Grobbelaar, T.A. Du Plessis, F.E. Steffens, M.J. van der Linde, "Longterm evaluation of polyethylene wear in total hip replacement - a statistical analysis of the association between the degree of wear versus pain, interface change, osteolysis and implant failure," *SA Orthopaedic Journal*, 10(1): 2011.
78. K.J. Bozic, M.D. Ries, "Wear and osteolysis in total hip arthroplasty," *JAAOS-Journal of the American Academy of Orthopaedic Surgeons*, 16: 142–152, 2005.
79. E. Askari, P. Flores, D. Dabirrahmani, R. Appleyard, "Dynamic modeling and analysis of wear in spatial hard-on-hard couple hip replacements using multibody systems methodologies," *Nonlinear Dynamics*, 2015.
80. E. Askari, P. Flores, D. Dabirrahmani, R. Appleyard, "Wear Prediction of Ceramic-on-Ceramic Artificial Hip Joints, in New Trends," in *Mechanism and Machine Science, From Fundamentals to Industrial Applications*, Springer, 2015.
81. N. Jaehyun, K. Jaeyoung, "Squeak noise of ceramic-on-ceramic hip joint using FEM," *The Korean society for Noise and Vibration Engineering*, 23(12): 1090–1095, 2013.
82. J.V. Bono, L. Sanford, J.T. Toussaint, "Severe polyethylene wear in total hip arthroplasty: Observations from retrieved AML Plus hip implants with an ACS polyethylene liner," *The Journal of Arthroplasty*, 9(2): 119–125, 1994.
83. K. Tanaka, D. Uchijima, K. Hasegawa, T. Katori, R. Sakai, K. Mabuchi, "Limitation of the lubricating ability of total hip prostheses with hard-on-hard sliding material," *Tribology Online*, 8(5): 272–277, 2013.
84. A.P. Harsha, T.J. Joyce, "Challenges associated with using bovine serum in wear testing orthopaedic biopolymers," *Journal of Engineering in Medicine*, 225(10): 948–958, 2011.
85. E. Sariali, Z. Jin, T. Stewart, J. Fisher, "Spectral characterization of squeaking in ceramic-on-ceramic total hip arthroplasty: Comparison of in vitro and in vivo values," *Journal of Orthopaedic Research*, 30(2): 185–189., 2012.
86. J.H. Currier, D.E. Anderson, D.W. Van Citters, "A proposed mechanism for squeaking of ceramic-on-ceramic hips," *Wear*, 269(11–12): 782–789, 2010.
87. C. Restrepo, J. Parvizi, S.M. Kurtz, P.F. Sharkey, W.J. Hozack, R.H. Rothman, "The noisy ceramic hip: Is component malpositioning the cause?" *Journal of Arthroplasty*, 23(5): 634–639, 2008.
88. W.L. Walter, T.S. Waters, M. Gillies, S. Donohoo, S.M. Kurtz, A.S. Ranawat, W.J. Hozack, M.A. Tuke, "Squeaking hips," *Journal of Bone and Joint Surgery*, 90(4): 102–111, 2008.

89. C. Restrepo, W.Y. Matar, J. Parvizi, R.H. Rothman, W.J. Hozack, "Natural history of squeaking after total hip arthroplasty," *Clinical Orthopaedics and Related Research*, 468(9): 2340–2345, 2010.
90. S. Affatato, A. Ruggiero, M. Merola, S. Logozzo, "Does metal transfer differ on retrieved Biolox® Delta composites femoral heads? Surface investigation on three Biolox® generations from a biotribological point of view," *Composites Part B: Engineering*, 113: 164–173, 2017.
91. J. Nam, C. Hoil, J. Kang, "Finite element analysis for friction noise of simplified hip joint and its experimental validation," *Journal of Mechanical Science and Technology*, 30(8): 3453–3460, 2016.
92. R.M. Gillies, S.M. Donohoo, W.L. Walter, "Finite element analysis of squeaking hips," *Orthopaedic Proceedings*, 91-B(SUPP_I): 89–89, 2009.
93. J. Kang, Numerical calculation of hip squeak over the normal gait cycle, *International Journal of Precision Engineering and Manufacturing*, 20: 2205–2214, 2019.
94. D. Levine, J. Richards, M.W. Whittle, *Whittle's Gait Analysis-E-Book*, Elsevier Health Sciences, 2012.
95. R. Brighenti, *Analisi numerica dei solidi e delle strutture*, Società editrice Esculapio, 2014.
96. G. Diana, F. Cheli, *Dinamica dei sistemi meccanici*, Polipress, 2011.
97. A. Ruggiero, R. D'Amato, S. Affatato, "Comparison of meshing strategies in THR finite element modelling," *Materials*, 12(14): 2332, 2019.
98. J. Fish, T. Belytschko, *A First Course in Finite Elements*, John Wiley & Sons, 2007.
99. H.-H. Lee, *Finite Element Simulation with Ansys Workbench 14*, SDC Publications, 2012.
100. S. Affatato, A. Ruggiero, S. Logozzo, "Metal transfer evaluation on ceramic biocomponents: A protocol based on 3D scanners," *Measurement*, 173, 108574, 2021.
101. Ansys Inc. Proprietary, "Introduction to ANSYS Mechanical," in *Meshing in Mechanical*, 2010.
102. T. Ghisu, *Introduction to Meshing*, 2015.
103. N. Fan, G.X. Chen, L.M. Qian, "Analysis of squeaking on ceramic hip endoprosthesis using the complex eigenvalue method," *Wear*, 271(9–10): 2305–2312, 2011.
104. N. Fan, G.X. Chen, "Numerical study of squeaking suppresses for ceramic-on-ceramic hip endoprosthesis," *Tribology International*, 48: 172–181, 2012.
105. G.W. Rodgers, J.L. Young, A.V. Fields, R.Z. Shearer, T.B.F. Woodfield, G.J. Hooper, J.G. Chase, "Acoustic emission monitoring of total hip arthroplasty implants," in *Proceedings of the 19th World Congress the International Federation of Automatic Control*, 2014.
106. C. Lee, L. Zhang, D. Morris, K.Y. Cheng, R.A. Ramachandran, M. Barba, D. Bijukuma, D. Ozevin, M.T. Mathew, "Non-invasive early detection of failure modes in total hip replacements (THR) via acoustic emission (AE)," *Journal of the Mechanical Behavior of Biomedical Materials*, 118: 2021.
107. G. Ouenzerfi, F. Massi, E. Renault, Y. Berthier, " Squeaking friction phenomena in ceramic hip endoprosthesis: Modeling and experimental validation," *Mechanical Systems and Signal Processing*, 58–59: 87–100, 2015.
108. A. Hothan, G. Huber, C. Weiss, N. Hoffmann, M. Morlock, "The influence of component design, bearing clearance and axial load on the squeaking characteristics of ceramic hip articulations," *Journal of Biomechanics*, 44: 837–841, 2011.
109. E. Sariali, T. Stewart, Z. Jin, J. Fisher, *In Vitro Investigation of Friction under Edge Loading Conditions for Ceramic-on-Ceramic Total Hip Prosthesis*, Wiley InterScience, 2009.

110. E. Askari, P. Flores, D. Dabirrahmani, R. Appleyard, “A computational analysis of squeaking hip prostheses,” *Journal of Computational and Nonlinear Dynamics*, 10: 024502, 2015.
111. J. Kang, “Finite element algorithm reproducing hip squeak measured in experiment,” *Journal of Sound and Vibration*, 393: 374–387, 2017.
112. A.J. Fitzpatrick, G.W. Rodgers, G.J. Hooper, T.B.F. Woodfield, “Development and validation of an acoustic emission device to measure wear in total hip replacements in-vitro and in-vivo,” *Biomedical Signal Processing and Control*, 33: 281–288, 2017.
113. A.E. Sariali, “In vitro and in vivo analysis of squeaking frequencies in ceramic-on-ceramic total hip arthroplasty,” *Orthopaedic Proceedings*, 94-B(SUPP_XXV): 2018.
114. E. Askari, P. Flores, D. Dabirrahmani, R. Appleyard, “Nonlinear vibration and dynamics of ceramic-on-ceramic artificial hip joints: A spatial multibody modelling“, *Nonlinear Dynamics*, 76: 1365–1377, 2014.

9 Synovial Lubrication Modeling of Total Hip Replacements Using Musculoskeletal Multibody Dynamics

Alessandro Ruggiero and Alessandro Sicilia
University of Salerno, Fisciano, Italy

CONTENTS

9.1 INTRODUCTION: THE BIOTRIBOLOGY OF THE HUMAN ARTICULATIONS

The biotribology is a branch of the biomechanics which applies the notions of the tribology to biological environments. In the case of human articulations, the biotribological properties of the joints belonging to a musculoskeletal system play a key role in terms of functionality, stability and mobility of the whole system, since an articulation linking two moving bones has to guarantee load capacity and wide range of motion.

The human articulations are essentially composed by the cartilage, which is a deformable, porous connective tissue. The presence of cells called chondrocytes within the cartilage is responsible for the production of collagen fibers, which are

DOI: 10.1201/9781003243205-9

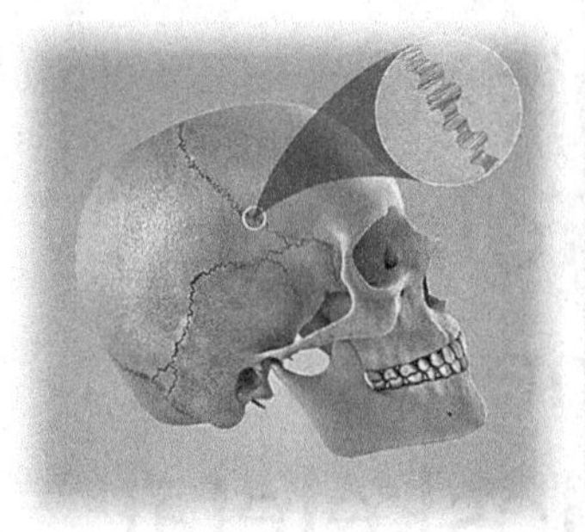
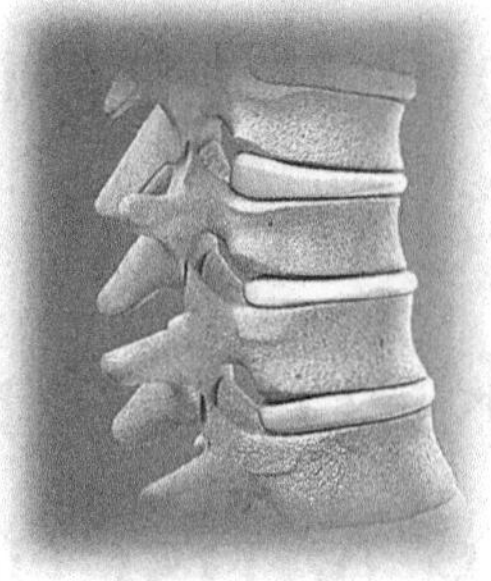
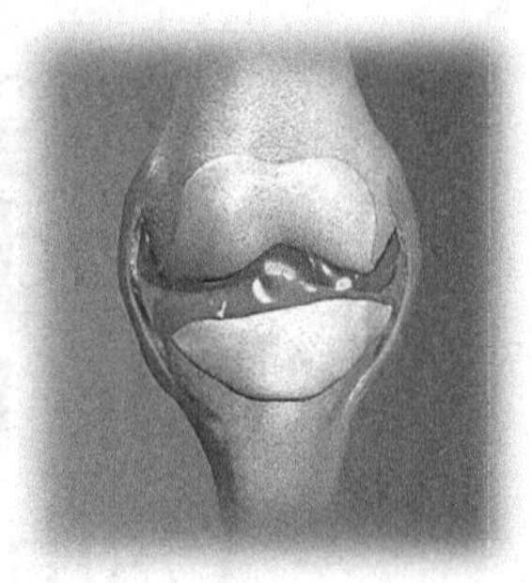

FIGURE 9.1 The human articulations classes.

filamentous proteins characterized by high resistance to tensile stresses along the fibrils.

Three classes of human articulations (Figure 9.1) are distinguished in dependence on the density of the collagen fibers and on their orientation with respect to the surfaces of the linked bones:

- the *fibrous joints*, that is the fixed ones, since they present a very high density of collagen fibers oriented orthogonally with respect to the linked bony surfaces, in order to prevent the relative motion between the coupled bones (e.g., the sutures between the skull's bones);
- the *cartilaginous joints*, which allow a limited range of motion, are characterized by an intermediate configuration of the collagen fibers between the fibrous joint and the next one (e.g., the intervertebral discs);
- the *synovial joints*, which permit the widest range of motion thanks to the parallel orientation of the collagen fibers with respect to the articulated surfaces and to the presence of the synovial cavity separating the bones which are joined by ligaments (e.g., the shoulder, the elbow, the wrist, the hip, the knee, the ankle, etc.).

The synovial joint is the most interesting from a tribological point of view among the human articulations because of the synovial cavity [1]. Initially, the cartilage covering the extremities of the bones linked by a synovial joint presents a layered structure (Figure 9.2): in the superficial zone, towards the cavity, the collagen fibers are arranged in parallel direction with respect to the surface, in order to absorb the shear stresses generated by the surfaces' relative sliding motion; then, going down along the cartilage thickness towards the bone, the collagen fibers' density increases and their orientation tilts to become orthogonal with respect to the close bony surface (in this deep zone the calcification occurs, locating the transition zone from the cartilage to the bone).

The synovial cavity is surrounded by a synovial membrane which supplies the natural lubricant of the articulation, called the synovial fluid, and provides to the ejection of undesired products in the cavity. The synovial fluid is substantially made of blood plasma, hyaluronic acid macromolecules and lubricin proteins: it is the main

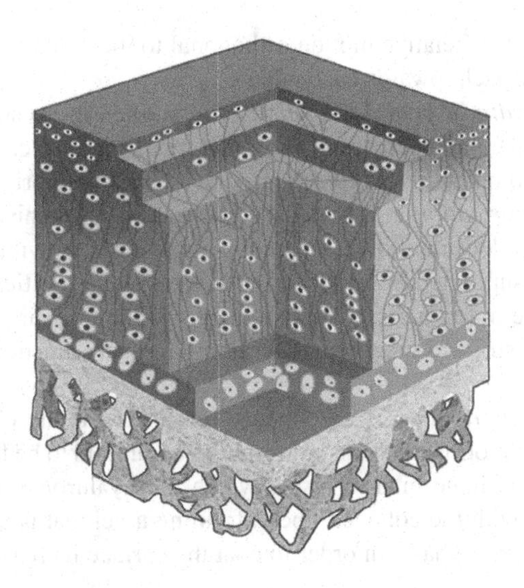

FIGURE 9.2 Layered structure of the cartilage.

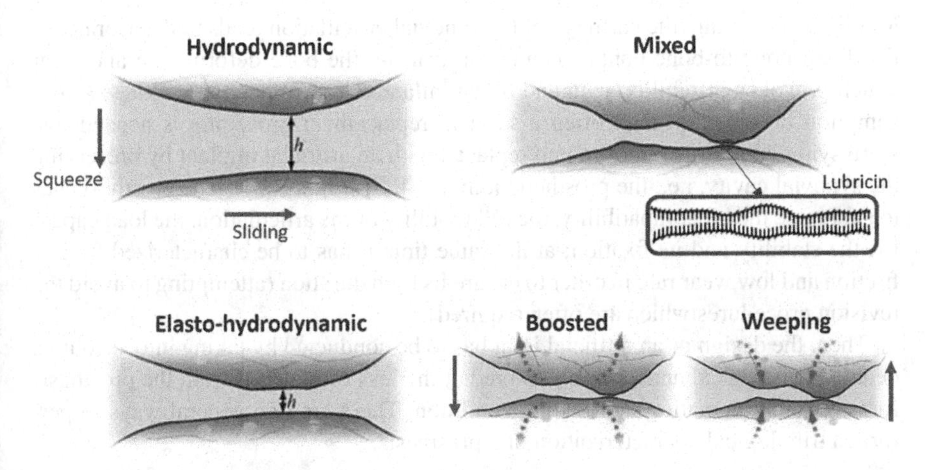

FIGURE 9.3 Synovial lubrication modes.

component responsible of the tribologically optimal behaviour of the synovial articulation, which allows to perform a wide motion with a very low friction coefficient.

The synovial joint configuration constitutes a very effective tribo-system in which several complex lubrication modes combine in space and time [1, 2] (Figure 9.3). In particular, the classical full film lubrication and the boundary one can be distinguished:

- the *hydrodynamic lubrication* belongs to the full film category and it occurs when the bearing load is caused by the relative motion of the articulated surfaces, which can be the sliding (relative motion lying in the contact plane, when the surfaces constitute a converging duct along the motion direction)

and the squeeze (relative motion orthogonal to the contact plane, when the surfaces approach towards each other);
- the *elasto-hydrodynamic lubrication* occurs when the synovial fluid pressure within the gap becomes so high that it deforms the cartilage surfaces, in order to completely separate them with a very thin lubricant film;
- the *mixed lubrication* is a boundary lubrication mechanism in which the too high load doesn't allow a separation of the surfaces with a full lubricant film, so it is supported by both lubricated areas and asperities in contact (the asperities are not in a direct contact thanks to the lubricin, which sticks to the cartilage surfaces creating an ultrathin layer that drastically reduces the friction);
- the *boosted lubrication* is a boundary lubrication mode characterized by high squeeze motion that constrains the synovial fluid to be filtrated through the porous cartilage matrix, inside which the hyaluronic acid macromolecules react with the collagen fibers, forming a gel that is exudated during the next *weeping* phase, in order to coat the surface increasing the lubrication effect.

Mainly due to aging, the cartilage of the synovial articulation tends to deteriorate, so the direct bone-to-bone contact can occur, causing the bone deformation and wear which provokes instability, pain and tissue inflammation. When the cartilage's consumption becomes critical, often a surgical replacement procedure is needed: the worn synovial joint is removed and replaced with an artificial implant by preserving the synovial cavity, i.e., the prosthetic joint [3, 4, 5] (Figure 9.4). The prosthesis has to guarantee the biocompatibility, the full mobility of the articulation, the load capacity, the stability and its fixation; at the same time it has to be characterized by low friction and low wear rate in order to ensure its high duration (attempting to avoid the revision procedures which are often required).

Then, the design of an artificial joint has to be conducted by taking into account a detailed tribological analysis able to predict the loss of material from the prosthesis for a certain kinematics and loading condition. There are two general ways to perform a tribological characterization of a prosthesis:

- the *in vitro* approach, which takes advantage of sophisticated experimental instrumentation to simulate the functioning of a tribo-system, producing results very close to the reality but excessively time/money-consuming;
- the *in silico* approach, which aims to simulate the same phenomena in a virtual computational environment (calculator), so it is cheaper than the *in vitro* one but it needs a detailed mathematical and physical description of the problem in order to obtain accurate outcomes.

The prosthesis can be viewed as a mechanical tribo-pair, so the tribology of an artificial joint is simpler to model than the natural synovial one [5, 4] and its lubrication/contact state, for a given lubricant, is generally determined by the intensity of the relative velocity between the surfaces (kinematics) and by the magnitude of

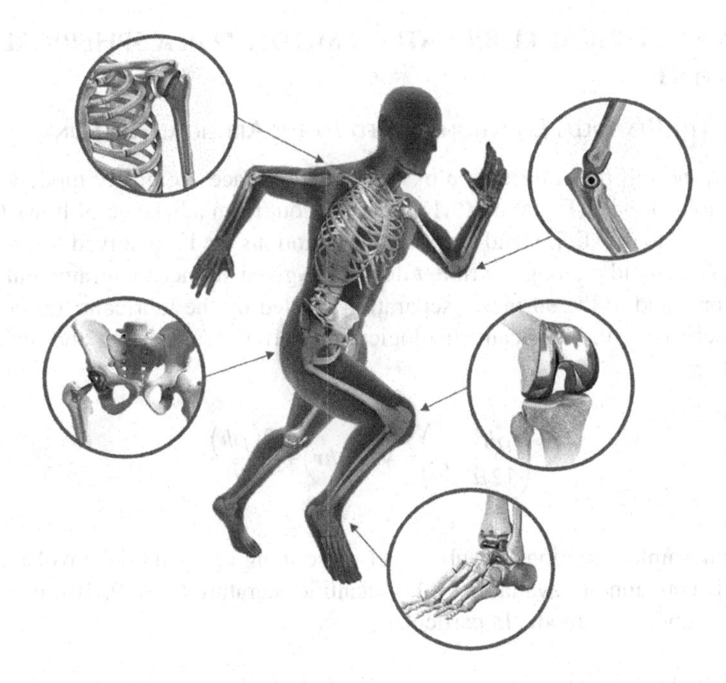

FIGURE 9.4 Typical joint replacements.

the load that pushes the surfaces against each other (dynamics). Furthermore, it is well known that the loading condition acting on a joint is a consequence of the joint reactions, which are generated by the kinematics and the dynamics of the whole system that includes the analysed joint. Then, the aim of this chapter is to describe two numerical models already developed in MatLab environment by the authors regarding:

- the mixed elasto-hydrodynamic lubrication of the Total Hip Replacement (THR), which simulates its tribological behaviour in terms of eccentricity, fluid/contact pressure, surfaces' separation and wear depth, starting from the time evolution of the load and relative angular velocity vectors, taking into account the non-Newtonian behaviour of the synovial fluid and the progressive gap geometry modification due to the wear [6];
- the mechanics of the multibody musculoskeletal system associated with the lower limb, which solves the inverse dynamics in order to elaborate the muscles' activation and the joint reactions, starting from a certain kinematics of the bones and considering the muscles' wrapping around them [7].

These two stand-alone models are merged in order to assemble a pipeline that starts from the musculoskeletal kinematics and leads to the tribological states associated to a synovial joint (the hip in this chapter).

9.2 A NUMERICAL LUBRICATION MODEL FOR A SPHERICAL JOINT

9.2.1 The Reynolds Equation Applied to the Artificial Hip Joint

Actually, the full film lubrication of two paired surfaces is mostly modelled with the Reynolds model (Equation (9.1)): it comes out from a balance of flows (due to the pressure, to the sliding and to the squeeze motions) and it is solved for the lubricant pressure field p along the time t due to the given surfaces' entrainment velocity vector v and to the surfaces' separation h filled by the lubricant; furthermore, it is governed by the lubricant rheological properties of density ρ and dynamical viscosity μ.

$$\nabla\cdot\left(\frac{\rho h^3}{12\mu}\nabla p\right)=\nabla\cdot(\rho h v)+\frac{\partial(\rho h)}{\partial t} \tag{9.1}$$

Several implementations of sub-models describing the quantities involved in the Reynolds equation are available in the scientific literature [8, 3, 9, 10], in order to detail the lubrication mode. In particular:

- if the hydrodynamic regime is considered, the lubricant rheological properties ρ and μ are considered as constants, while the lubricant thickness h depends only on the relative approach between the surfaces, so that the numerical solution of the Reynolds equation is given by the inversion of a linear system;
- moving from the hydrodynamic mode to the elasto-hydrodynamic one, the rheological properties ρ and μ need of relationships describing their variability with the high pressure reached within the gap, while the surfaces' separation h is composed also by the surfaces' elastic deformation contribution δ, which depends on the pressure p through a deformation model of the analysed geometry [11, 12, 13, 14], leading to a highly non-linear problem solvable through iterative techniques;
- if the possibility of a mixed lubrication regime is investigated, then the analysed domain Ω has to be distinguished in lubricated areas Ω_l, governed by the Reynolds equation, and contact areas Ω_c, in which a further surface deformation contribution due to the occurring contact δ_c has to be considered in order to prevent the interpenetration between the surfaces thanks to the rising of the contact pressure p_c [15, 16, 17, 18];
- the progressive geometry modification of the gap due to the wear can be taken into account considering an additional term equal to the wear depth u_w to the surfaces' gap h, which is updated every time step;
- the non-Newtonian behaviour of the lubricant (as in the case of the synovial fluid) can be introduced implementing a relationship describing the dependence of the viscosity μ on the fluid shear rate, where the latter can be approximated by the ratio between the sliding velocity and the surfaces' separation [19].

Regarding the model proposed by the authors, the lubricant density ρ dependence on the pressure p is governed by the classical Dowson-Higginson relationship [20] in the Equation (9.2), which increases the nominal density ρ_0 with the pressure according to the parameters a_ρ and b_ρ.

$$\rho(p) = \rho_0 \frac{a_\rho + b_\rho p}{a_\rho + p} \tag{9.2}$$

The viscosity μ varies both with the pressure through the well-known Barus relationship and with the shear rate following the Cross non-Newtonian model. In particular, it is written in Equation (9.3) as a product between:

the Cross term [19], which simulates the synovial fluid shear-thinning effect (Figure 9.5), that is the viscosity's increase for a decreasing shear rate in a finite range delimited by μ_0 and μ_∞ according to the parameters k_μ and n_μ, due to the aggregation of the hyaluronic acid macromolecules within the synovial fluid for low sliding velocity v_{sl};

the Barus term, characterized by the classical exponential dependence governed by the parameter α_μ.

$$\mu\left(p, \frac{v_{sl}}{h}\right) = \left[\mu_\infty + \frac{\mu_0 - \mu_\infty}{1 + k_\mu \left(\frac{v_{sl}}{h}\right)^{n_\mu}}\right] e^{\alpha_\mu p} \tag{9.3}$$

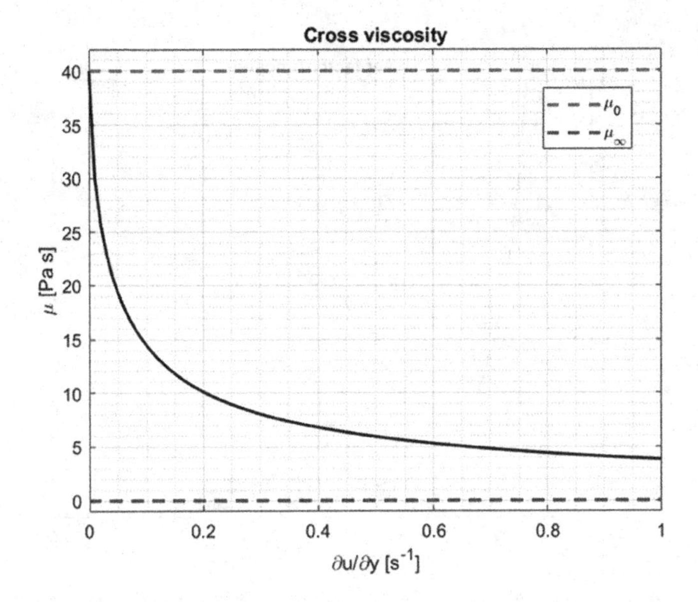

FIGURE 9.5 Shear thinning cross viscosity curve.

Since a Hard-on-Soft hip prosthesis will be analysed, as discussed in the following paragraphs, the chosen elastic deformation model is the column one [13, 21] reported in the Equation (9.4): this assumes that only the softer material is able to deform because of its very low Young modulus with respect to the harder counterpart and that the local deformation δ is proportional only to the local pressure p according to the gain coefficient k_d which depends on the mechanical and geometrical properties of the softer body (Young modulus E, Poisson ratio ν, radius R and thickness H).

$$k_d = \frac{R\left[\left(1+\frac{H}{R}\right)^3 - 1\right]}{E\left[\frac{1}{1-2\nu} + \frac{2}{1+\nu}\left(1+\frac{H}{R}\right)^3\right]} \rightarrow \delta(p) = k_d p \tag{9.4}$$

The wear depth u_w affecting the surfaces' separation h is elaborated with the classical Archard model [22, 23] in Equation (9.5), which assumes that the linear wear rate τ_w is proportional to the product between the pressure (the sum of the fluid and the contact terms, respectively p and p_c, in case of mixed lubrication mode) and the sliding velocity v_{sl} through the wear factor k_w. The wear factor is assumed to increase when the surfaces' separation h becomes comparable to the arithmetic roughness R_a and to extinguish over the full lubricated areas, according to the nominal wear factor k_{w_0} and the parameter α_w (Figure 9.6).

$$k_w\left(\lambda = \frac{h}{R_a}\right) = k_{w0}\lambda^{-\alpha_w} \rightarrow \tau_w = k_w\left(p + p_c\right)v_{sl} \tag{9.5}$$

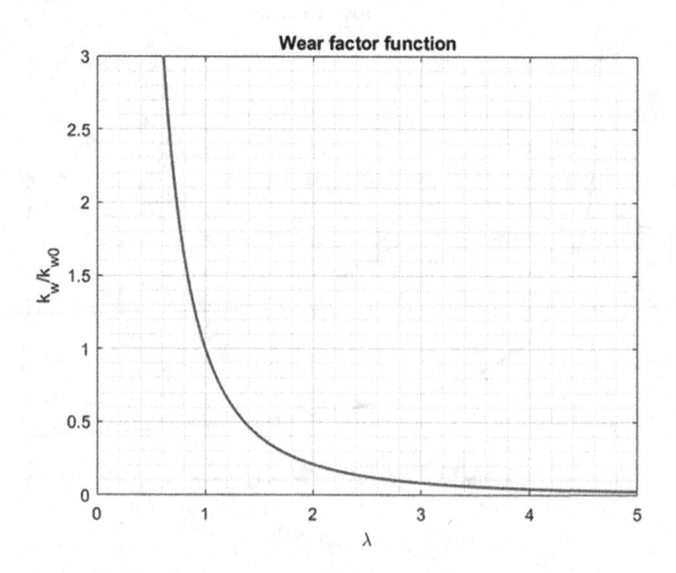

FIGURE 9.6 Wear factor curve.

Once the linear wear rate field is calculated at each time step, its integration over the time t leads to the wear depth u_w, in Equation (9.6), cumulated until that instant.

$$u_w = \int_0^t \tau_w dt \tag{9.6}$$

By knowing the geometrical separation h_g, which depends on the particular analysed geometry of the two bodies, the contact deformation δ_c is evaluated in (9.7) as the amount of elastic deformation needed to avoid the interpenetration [6] (assuming a boundary layer thickness Δ_b in order to have no numerical indeterminacy during the solving of the Reynolds equation).

$$\delta_c = \begin{cases} 0 & \text{if } h_g + \delta + u_w > \Delta_b \\ \Delta_b - \left(h_g + \delta + u_w\right) & \text{if } h_g + \delta + u_w \le \Delta_b \end{cases} \tag{9.7}$$

Then the total surfaces' separation h is obtained in the Equation (9.8).

$$h = h_g + \delta + \delta_c + u_w \tag{9.8}$$

The entrainment velocity vector $\boldsymbol{v}$ depends on the particular moving geometry of the tribo-system, so the final problem to be solved in order to evaluate the fluid pressure p and the contact one p_c is reported in Equation (9.9), coupling it with the boundary conditions on the pressure assumed as constant (p_0) in correspondence of the domain's boundaries $\partial\Omega$.

$$\begin{cases} \nabla \cdot \left(\dfrac{\rho h^3}{12\mu} \nabla p \right) = \nabla \cdot \left(\rho h \boldsymbol{v} \right) + \dfrac{\partial(\rho h)}{\partial t} & \text{in } \Omega_l \\ p_c = \dfrac{\delta_c}{k_d} & \text{in } \Omega_c \\ p = p_0 & \text{on } \partial\Omega \end{cases} \tag{9.9}$$

The problem in (9.9) has to be turned in spherical coordinates, in order to adapt it to the spherical joint case, which is the hip. The hip (Figure 9.7) is the human synovial articulation linking the femur bone to the pelvis bone, pairing the femoral head to the acetabulum; it allows three relative rotations around three successive rotation axes:

- the Flexion/Extension rotation θ_{FE} around the Medial/Lateral axis z_{ML};
- the Adduction/Abduction rotation θ_{AA} around the Anterior/Posterior axis x_{AP};
- the Internal/External Rotation θ_{IER} around the Proximal/Distal axis y_{PD}.

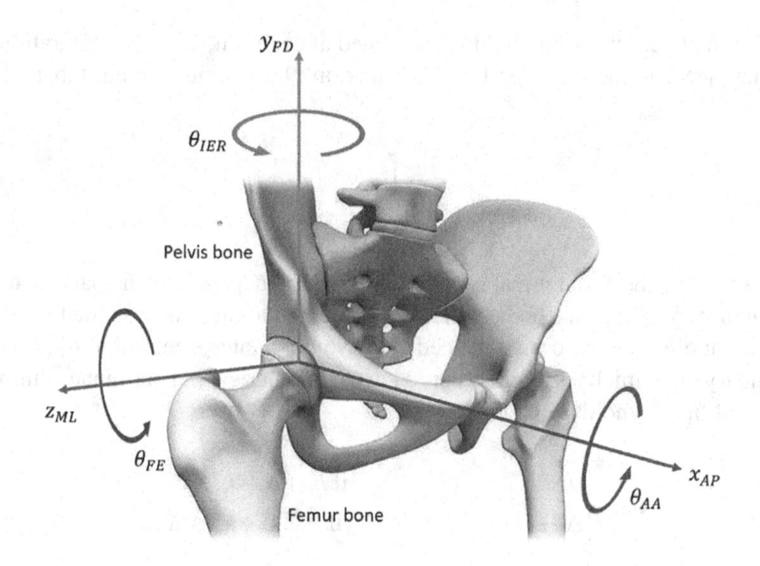

FIGURE 9.7 Hip joint scheme.

The hip prosthesis (Figure 9.8) is made up of the spherical femoral head, fixed to the femur bone through a metallic stem, and the hemispherical acetabular cup, fixed to the pelvis bone through a metallic back. The biomaterials actually used in the framework of the hip arthroplasty are metallics, polymerics and ceramics [24, 25, 26] in order to form Hard-on-Hard implants (Ceramic-on-Ceramic, Metal-on-Metal)

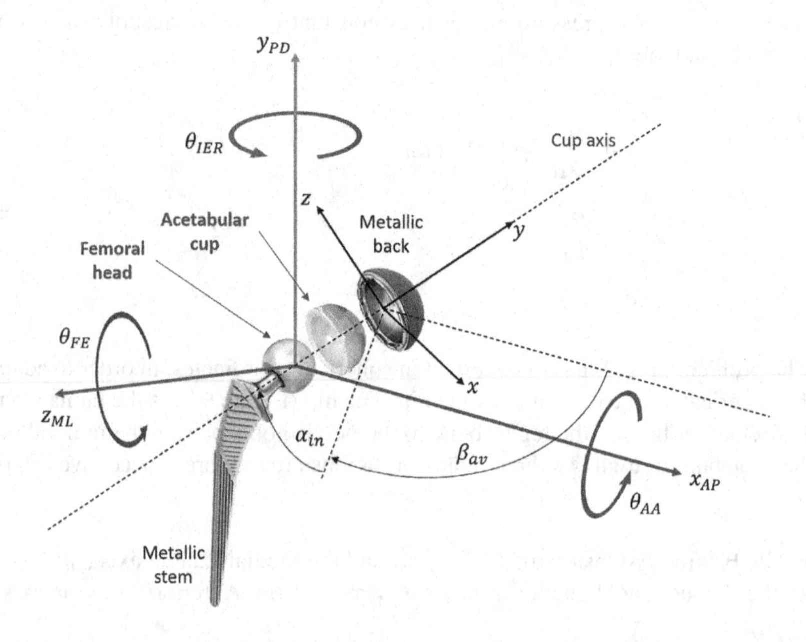

FIGURE 9.8 Hip joint prosthesis reference frames.

or Hard-on-Soft ones (Ceramic-on-Plastic, Metal-on-Plastic). The prosthesis more requested are of the type Hard-on-Soft [27], typically made of Ceramic or Metallic femoral head against Polymer acetabular cup (usually UHMWPE).

With reference to Figure 9.8, the Reynolds equation reference frame xyz is fixed to the acetabular cup, it has the origin in correspondence of the cup centre and it is composed by the y axis being the acetabular cup axis pointing inside it and by the x and z axes lying on the cup edges. The cup axis forms the inclination angle α_{in} with respect to the sagittal plane (containing x_{AP} and y_{PD}) and its projection on the latter is inclined by the anteversion angle β_{av} with respect to the longitudinal axis x_{AP}. Then the load vector $\boldsymbol{N}_h$ and the relative angular velocity $\boldsymbol{\omega}_h$ acting in the hip reference frame $x_{AP}y_{PD}z_{ML}$ have to be rotated, Equation (9.10), through the cup rotation matrix $\boldsymbol{R}_c$ (depending on the inclination and the anteversion angles α_{in} and β_{av}) in order to obtain the load and angular velocity vectors $\boldsymbol{N}$ and $\boldsymbol{\omega}$ defined in the xyz cup reference frame.

$$\boldsymbol{R}_c = \begin{bmatrix} \cos\left(\frac{\pi}{2} - \beta_{av}\right) & -\sin\left(\frac{\pi}{2} - \beta_{av}\right) & 0 \\ \sin\left(\frac{\pi}{2} - \beta_{av}\right) & \cos\left(\frac{\pi}{2} - \beta_{av}\right) & 0 \\ 0 & 0 & 1 \end{bmatrix} \begin{bmatrix} 1 & 0 & 0 \\ 0 & \cos(-\alpha_{in}) & -\sin(-\alpha_{in}) \\ 0 & \sin(-\alpha_{in}) & \cos(-\alpha_{in}) \end{bmatrix}$$

$$\left\{ \begin{aligned} \boldsymbol{N}_h &= \begin{bmatrix} N_{AP} \\ N_{PD} \\ N_{ML} \end{bmatrix} \\ \boldsymbol{\omega}_h &= \begin{bmatrix} \omega_{AP} \\ \omega_{PD} \\ \omega_{ML} \end{bmatrix} \end{aligned} \right. \rightarrow \left\{ \begin{aligned} \boldsymbol{N} &= \boldsymbol{R}_c^T \boldsymbol{N}_h \\ \boldsymbol{\omega} &= \boldsymbol{R}_c^T \boldsymbol{\omega}_h \end{aligned} \right. \tag{9.10}$$

Passing from the cartesian coordinates xyz to the spherical ones, the position $\boldsymbol{x}$ of any point P on the acetabular cup can be located by the spherical angles θ and φ, given the cup radius R, as shown in Figure 9.9 and written in the Equation (9.11). The transformation is executed through the radial unit vector $\hat{\boldsymbol{r}}$, which is useful to elaborate the geometrical approach h_g and the entrainment velocity vector $\boldsymbol{v}$ of the Reynolds equation in the case of a spherical joint.

$$\hat{\boldsymbol{r}} = \begin{bmatrix} \sin\theta\cos\varphi \\ \sin\theta\sin\varphi \\ \cos\theta \end{bmatrix} \rightarrow \boldsymbol{x} = R\hat{\boldsymbol{r}} \tag{9.11}$$

Given the femoral head radius r and the eccentricity vector of its centre with respect to the acetabular cup one $\boldsymbol{e}$, the geometrical gap h_g is calculated in Equation (9.12) by knowing the radial clearance c and the dimensionless eccentricity vector $\boldsymbol{n}$.

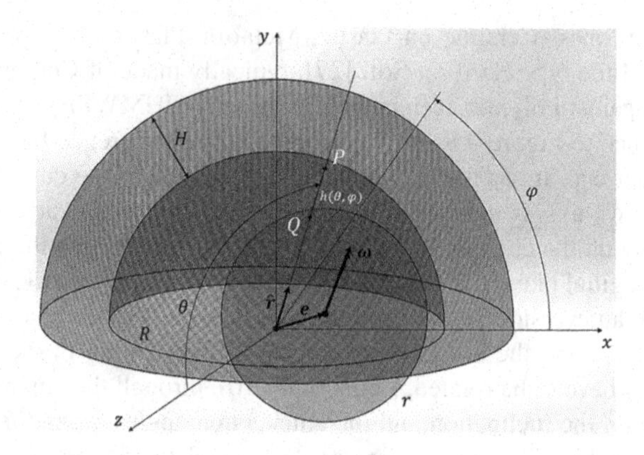

FIGURE 9.9 Spherical joint tribo-system reference frame.

$$\begin{cases} c = R - r \\ \boldsymbol{n} = \dfrac{\boldsymbol{e}}{c} \end{cases} \rightarrow h_g = c\left(1 - \boldsymbol{n}^T \hat{\boldsymbol{r}}\right) \tag{9.12}$$

The entrainment velocity vector $\boldsymbol{v}$ is given by the arithmetic mean between the velocity of a point P on the cup (which is null in the xyz reference frame) and the one referred to the associated point Q on the head in correspondence of the analysed position located by the spherical angles θ and φ. The velocity of the point Q is composed by a translational term related to the variation over time of the eccentricity $\boldsymbol{e}$ and by the rotational term generated by the relative angular velocity vector of the head with respect to the acetabular cup $\boldsymbol{\omega}$; therefore, the entrainment velocity vector $\boldsymbol{v}$ in spherical coordinates is obtained in Equation (9.13) by taking into account the spherical rotation matrix $\boldsymbol{R}_s$ depending on the angles θ and φ. It is worth noting that the only components of $\boldsymbol{v}$ needed in the Reynolds equation are the ones referred to the contact plane, i.e., U_θ and U_φ.

$$\begin{gathered} \boldsymbol{R}_s = \begin{bmatrix} \cos\varphi & -\sin\varphi & 0 \\ \sin\varphi & \cos\varphi & 0 \\ 0 & 0 & 1 \end{bmatrix} \begin{bmatrix} \cos\theta & 0 & \sin\theta \\ 0 & 1 & 0 \\ -\sin\theta & 0 & \cos\theta \end{bmatrix} \\ \rightarrow \boldsymbol{v} = \frac{1}{2}\boldsymbol{R}_s^T \left\{\dot{\boldsymbol{e}} + \boldsymbol{\omega} \times \left[(R - h)\hat{\boldsymbol{r}} - \boldsymbol{e}\right]\right\} = \begin{bmatrix} U_\theta \\ U_\varphi \\ U_\rho \end{bmatrix} \end{gathered} \tag{9.13}$$

At this point, the problem is completely adapted to the spherical joint in terms of geometry and motion (geometrical approach h_g and entrainment velocity vector $\boldsymbol{v}$) and coordinates (the spherical ones). Then, the Reynolds equation is reported in (9.14) by applying the spherical coordinate transformation to the differential operators.

$$\begin{cases} \sin\theta \dfrac{\partial}{\partial\theta}\left(\sin\theta \dfrac{\rho h^3}{12\mu}\dfrac{\partial p}{\partial\theta}\right) + \dfrac{\partial}{\partial\varphi}\left(\dfrac{\rho h^3}{12\mu}\dfrac{\partial p}{\partial\varphi}\right) \\ = R\sin\theta\left[\dfrac{\partial}{\partial\theta}(\sin\theta\rho h U_\theta) + \dfrac{\partial}{\partial\varphi}(\rho h U_\varphi) + R\sin\theta\dfrac{\partial}{\partial t}(\rho h)\right] \\ p(0,\varphi) = p(\pi,\varphi) = p_0 \\ p(\theta,0) = p(\theta,\pi) = p_0 \end{cases} \tag{9.14}$$

Once the mixed lubrication problem is numerically solved, the load vector $\boldsymbol{N}$ can be found by integrating the total pressure oriented with the radial unit vector $\hat{\boldsymbol{r}}$ over the domain's surface and the wear volume V_w is obtained by integrating in the same way the penetration depth u_w (Equation (9.15)).

$$\begin{cases} \boldsymbol{N} = \displaystyle\iint_{00}^{\pi\pi} (p + p_c)\hat{\boldsymbol{r}} R^2 \sin\theta d\theta d\varphi \\ V_w = \displaystyle\iint_{00}^{\pi\pi} u_w R^2 \sin\theta d\theta d\varphi \end{cases} \tag{9.15}$$

9.2.2 Numerical Algorithm

The Reynolds Equation (9.14) is written in every discrete point of the domains located by the angles θ_i and φ_j through the finite differences associated to the 5-points cross stencil [6]. The stencil form $\boldsymbol{u}_s$ of a quantity u is given in the Equation (9.16).

$$\boldsymbol{u}_s = \begin{bmatrix} u_{i-1,j} & u_{i,j-1} & u_{ij} & u_{i,j+1} & u_{i+1,j} \end{bmatrix}^T \tag{9.16}$$

With the stencil form (9.16) in mind, the Equation (9.14) is written in (9.17), in which some vectors and a matrix related to the finite differences are introduced and some product quantities are grouped in a single element.

$$\begin{cases} f = \dfrac{\rho h^3}{12\mu} \\ u_\theta = \rho h U_\theta \\ u_\varphi = \rho h U_\varphi \\ u_t = \rho h \end{cases} \rightarrow \boldsymbol{f}_s^T \boldsymbol{D}_{2_s} \boldsymbol{p}_s = \boldsymbol{D}_{\theta_s}^T \boldsymbol{u}_{\theta_s} + \boldsymbol{D}_{\varphi_s}^T \boldsymbol{u}_{\varphi_s} + \boldsymbol{D}_{t_s}^T \boldsymbol{u}_{t_s} \tag{9.17}$$

As stated above, the problem is highly non-linear: therefore, since the analytical form of all the involved quantities is known, the Newton iterative method is a suitable technique to use. So, the discrete Reynolds equation associated to the point I (typical 2-D grid ordering) R_I and its derivative with respect to the stencil pressure vector $\boldsymbol{J}_{R_I}$ are given in (18).

$$
\begin{aligned}
R_I(p_s) &= f_s^T D_{2_s} p_s - \left(D_{\theta_s}^T u_{\theta_s} + D_{\varphi_s}^T u_{\varphi_s} + D_{t_s}^T u_{t_s}\right) = 0 \\
J_{R_I}(p_s) &= \frac{\partial R_I}{\partial p_s} = f_s^T D_{2_s} + p_s^T D_{2_s}^T \frac{\partial f_s}{\partial p_s} - \left(D_{\theta_s}^T \frac{\partial u_{\theta_s}}{\partial p_s} + D_{\varphi_s}^T \frac{\partial u_{\varphi_s}}{\partial p_s} + D_{t_s}^T \frac{\partial u_{t_s}}{\partial p_s}\right)
\end{aligned}
\tag{9.18}
$$

The writing of Equation (9.18) for each point I belonging to the grid domain allows the assembly of the non-linear system of equation $\boldsymbol{R}$, in the unknown pressure vector $\boldsymbol{p}$, and its Jacobian $\boldsymbol{J_R}$ [6, 28]. The associated Newton update iteration scheme in (9.19) is applied. It is specified that in correspondence of each iteration the pressure cannot be negative, so all the negative elements of the pressure vector are set to zero.

$$
\begin{aligned}
&R(p) = 0 \\
&J_R(p) = \frac{\partial R}{\partial p} \rightarrow p^{(k+1)} = p^{(k)} - J_R^{-1}\left(p^{(k)}\right) R\left(p^{(k)}\right)
\end{aligned}
\tag{9.19}
$$

The resolution technique (9.19) is applied in correspondence of each time step in order to evaluate the time evolution of the pressure field $p(\theta, \varphi, t)$ and it requires the knowledge of the eccentricity $\boldsymbol{n}(t)$ and angular velocity $\boldsymbol{\omega}(t)$ vectors patterns: generally, the eccentricity is not a known input data, so the knowledge of the load vector time evolution is useful to build another equation in the eccentricity unknown $\boldsymbol{F}(\boldsymbol{n})$ which sets to zero the difference between the reference load $\boldsymbol{N}_{ref}$ and the ones calculated by integration of the pressure. The above procedure is described in the Equation (9.20), in which the Newton method is used again, but this time the Jacobian of $\boldsymbol{F}$, $\boldsymbol{J_F}$, is obtained in the numerical way through the finite differences defined with a small constant ε [6].

$$
\begin{aligned}
&F(n) = N(n) - N_{ref} = 0 \\
&J_F(n) = \frac{\partial F}{\partial n}, \left[\frac{\partial F}{\partial n}\right]_{ij} = \frac{F_i\left(n_j(1+\varepsilon)\right) - F_i\left(n_j\right)}{\varepsilon n_j} \\
&\rightarrow n^{(h+1)} = n^{(h)} - J_F^{-1}\left(n^{(h)}\right) F\left(n^{(h)}\right)
\end{aligned}
\tag{9.20}
$$

The resolution scheme in (9.20) constitutes an external iterative cycle which computes the internal one (9.19): the whole algorithm is able to elaborate the eccentricity vector time evolution $\boldsymbol{n}(t)$ of the artificial hip joint due to known reference load $\boldsymbol{N}_{ref}(t)$ and relative angular velocity $\boldsymbol{\omega}(t)$ vectors. Figure 9.10 gives information about the algorithm workflow. It is worth noting that in the correspondence of each time step (once the convergence is reached for both the iterative cycles) the evaluated total pressure field is used to calculate the linear wear rate field $\tau_w(\theta, \varphi, t)$ which updates the accumulation of wear depth $u_w(\theta, \varphi, t)$, modifying the gap geometry [6].

As stated in the introduction, the input for the developed lubrication model (i.e., the reference load $\boldsymbol{N}_{ref}(t)$ and the angular velocity $\boldsymbol{\omega}(t)$) is completely determined by

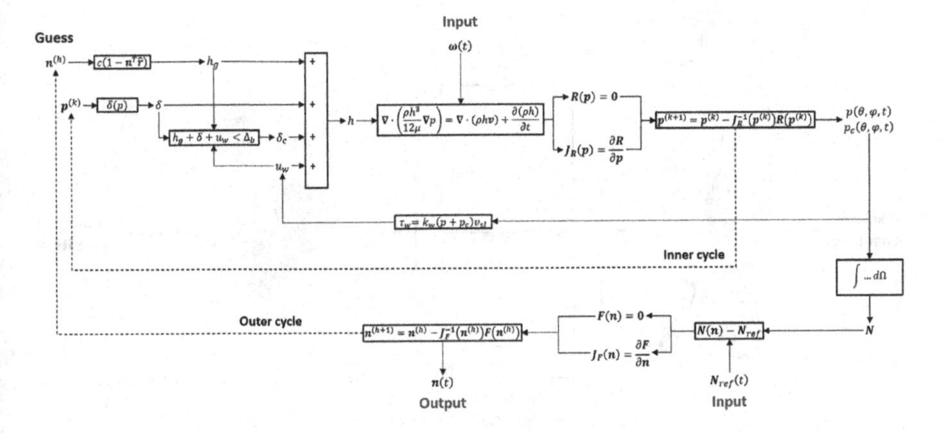

FIGURE 9.10 Numerical algorithm workflow.

the mechanics of the musculoskeletal system that includes the analysed hip joint. Therefore, the next section is dedicated to the description of a general multibody model, already developed by the authors, which elaborates the joint reactions and the relative motion needed to supply the described lubrication model and to evaluate the tribological state of the system's articulations.

9.3 MUSCULOSKELETAL MULTIBODY MODEL OF THE LOWER LIMB

The multibody approach constitutes a suitable technique to analyse the musculoskeletal systems, since they are composed by bodies (the bones) linked by joints (the articulations), moved by actuators (the muscles) and subjected to external environmental actions [29, 30]. In the framework of the wear prediction of an artificial joint belonging to the analysed system, the objective is to elaborate the loading and the relative velocity related to that joint due to a certain musculoskeletal kinematics. The specific aim of this chapter is to use the multibody output information to supply the mixed lubrication model described above, by logically connecting them as shown in the Figure 9.11.

9.3.1 Kinematical Analysis of Constrained Multibody Systems

Generally, the degrees of freedom of a musculoskeletal system are referred to the joint relative transformations; therefore, their time evolution represent the kinematical input needed to start the so-called *inverse dynamics* [31, 32]. The kinematics of each bone i belonging to the musculoskeletal system (which are assumed as rigid bodies) can be described along the time t by knowing its translation vector $\boldsymbol{t}_i$ (locating its mass centre with respect to the inertial reference frame, i.e., the ground) and its unit quaternion $\boldsymbol{\theta}_i$ (4-elements vector describing its orientation with respect to the ground reference frame): the described quantities are collected into the body's Lagrangian coordinates vector $\boldsymbol{q}_i$ [33].

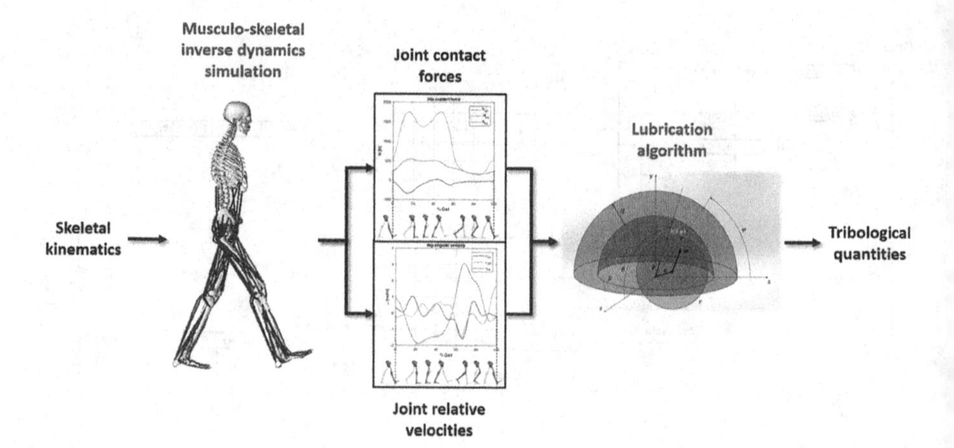

FIGURE 9.11 Multibody and lubrication models coupling.

$$q_i = \begin{bmatrix} t_i \\ \theta_i \end{bmatrix} \tag{9.21}$$

In order to perform a dynamical analysis which evaluates all the forces and torques acting on the bodies and on the joints, it is necessary to know the time evolution of all the Lagrangian coordinates $\boldsymbol{q}(t)$ (column ordering), as well as their time derivatives, i.e., the Lagrangian velocities $\dot{q}(t)$ and accelerations $\ddot{q}(t)$. Considering that each joint J linking the bodies i and j allows some relative movements and prevents others (out of total of six potential relative motion in the three-dimensional space), six constraint equations can be written in (9.22) for each joint:

- the *rheonomic* constraint equations $\boldsymbol{C}_{r_J}$, related to the allowed degrees of freedom $\boldsymbol{q}_{dof_J}(t)$, which force the linked bodies Lagrangian coordinates $\boldsymbol{q}_i$ and $\boldsymbol{q}_j$ to evolve along the time t in order to follow the known permitted movements;
- the *scleronomic* constraint equations $\boldsymbol{C}_{s_J}$, related to the locked degrees of freedom, which force the linked bodies Lagrangian coordinates $\boldsymbol{q}_i$ and $\boldsymbol{q}_j$ to avoid the prevented movements.

$$\begin{aligned} \boldsymbol{C}_{rJ}\left(\boldsymbol{q}_i, \boldsymbol{q}_j, t\right) = \boldsymbol{0} \\ \boldsymbol{C}_{sJ}\left(\boldsymbol{q}_i, \boldsymbol{q}_j\right) = \boldsymbol{0} \end{aligned} \tag{9.22}$$

Accounting for n_B bodies (including the ground) and considering the n_{dof} rheonomic equations and the n_c scleronomic ones, the number of constraint equations is $6(n_B - 1)$ while there are $7(n_B - 1)$ unknowns, since each body's Lagrangian coordinates $\boldsymbol{q}_i$ has 7 elements (because of the quaternion): the remaining $(n_B - 1)$ equations are related to the constraint equations C_{b_i} (9.23) which force each body i Langrangian coordinates $\boldsymbol{q}_i$ to evolve in order to keep unitary the norm of the unit quaternion $\boldsymbol{\theta}_i$.

$$C_{b_i}(q_i) = \theta_i^T \theta_i - 1 = 0 \tag{9.23}$$

Ordering all the constraints equations along the column direction, assembling all the rheonomic ones $\boldsymbol{C}_r$, the scleronomic ones $\boldsymbol{C}_s$ and the bodies' ones $\boldsymbol{C}_b$, the non-linear closed set of equations $\boldsymbol{C}$ needed to elaborate the position analysis is obtained in (9.24). It is solved through the Newton iterative method in the Lagrangian coordinates $\boldsymbol{q}$ unknowns, by evaluating the constraint equations $\boldsymbol{C}$ derivative with respect to the Lagrangian coordinates $\boldsymbol{q}$, i.e., the constraint Jacobian $\boldsymbol{C}_q$ [7].

$$\boldsymbol{C}(\boldsymbol{q},t) = \begin{bmatrix} \boldsymbol{C}_r(\boldsymbol{q},t) \\ \boldsymbol{C}_s(\boldsymbol{q}) \\ \boldsymbol{C}_b(\boldsymbol{q}) \end{bmatrix} = \boldsymbol{0} \rightarrow \boldsymbol{q}^{(k+1)} = \boldsymbol{q}^{(k)} - \boldsymbol{C}_q^{-1}\left(\boldsymbol{q}^{(k)},t\right)\boldsymbol{C}\left(\boldsymbol{q}^{(k)},t\right) \tag{9.24}$$

Once the position analysis problem has reached the convergence, the velocity analysis and the acceleration one are formulated by differentiating with respect to time the constraint equation, leading to two linear problems in (9.25) which are solved in the Lagrangian velocities $\dot{\boldsymbol{q}}$ and accelerations $\ddot{\boldsymbol{q}}$ unknowns by matrix inversion. The described time differentiation introduces other matrices and vectors (i.e., $\boldsymbol{C}_t$, $\dot{\boldsymbol{C}}_q$ and $\dot{\boldsymbol{C}}_t$).

$$\begin{aligned} \frac{d}{dt}\left(\boldsymbol{C}(\boldsymbol{q},t)\right) &= \dot{\boldsymbol{C}}(\boldsymbol{q},\dot{\boldsymbol{q}},t) = \boldsymbol{C}_q\dot{\boldsymbol{q}} + \boldsymbol{C}_t = \boldsymbol{0} \\ \frac{d}{dt}\left(\dot{\boldsymbol{C}}(\boldsymbol{q},\dot{\boldsymbol{q}},t)\right) &= \ddot{\boldsymbol{C}}(\boldsymbol{q},\dot{\boldsymbol{q}},\ddot{\boldsymbol{q}},t) = \boldsymbol{C}_q\ddot{\boldsymbol{q}} + \dot{\boldsymbol{C}}_q\dot{\boldsymbol{q}} + \dot{\boldsymbol{C}}_t = \boldsymbol{0} \end{aligned} \rightarrow \begin{aligned} \boldsymbol{C}_q\dot{\boldsymbol{q}} &= -\boldsymbol{C}_t \\ \boldsymbol{C}_q\ddot{\boldsymbol{q}} &= -\left(\dot{\boldsymbol{C}}_q\dot{\boldsymbol{q}} + \dot{\boldsymbol{C}}_t\right) \end{aligned} \tag{9.25}$$

The resolution of the position, the velocity and the acceleration analysis closes the whole kinematic problem in which, given the joint behavior and the time evolution of the musculoskeletal system degrees of freedom $\boldsymbol{q}_{dof}(t)$, the trajectories of the Lagrangian coordinates $\boldsymbol{q}(t)$ and their time derivatives are elaborated. In order to assemble the matrices and vectors needed (i.e., $\boldsymbol{C}$, $\boldsymbol{C}_q$, $\boldsymbol{C}_t$, $\dot{\boldsymbol{C}}_q$ and $\dot{\boldsymbol{C}}_t$), each joint J has to be described. Generally, each constraint equation $\boldsymbol{C}_J$ (and its time derivatives) is given in the form (9.26): therefore, the definition of the equation $\boldsymbol{C}_J(\boldsymbol{q}_i, \boldsymbol{q}_j, t)$ is all that is needed to build the Equations (9.24) and (9.25).

$$\begin{aligned} \boldsymbol{C}_J(\boldsymbol{q}_i,\boldsymbol{q}_j,t) &= \boldsymbol{0} \\ \boldsymbol{C}_{Jq_i}\dot{\boldsymbol{q}}_i + \boldsymbol{C}_{Jq_j}\dot{\boldsymbol{q}}_j + \boldsymbol{C}_{Jt} &= \boldsymbol{0} \\ \boldsymbol{C}_{Jq_i}\ddot{\boldsymbol{q}}_i + \boldsymbol{C}_{Jq_j}\ddot{\boldsymbol{q}}_j + \dot{\boldsymbol{C}}_{Jq_i}\dot{\boldsymbol{q}}_i + \dot{\boldsymbol{C}}_{Jq_j}\dot{\boldsymbol{q}}_j + \dot{\boldsymbol{C}}_{Jt} &= \boldsymbol{0} \end{aligned} \tag{9.26}$$

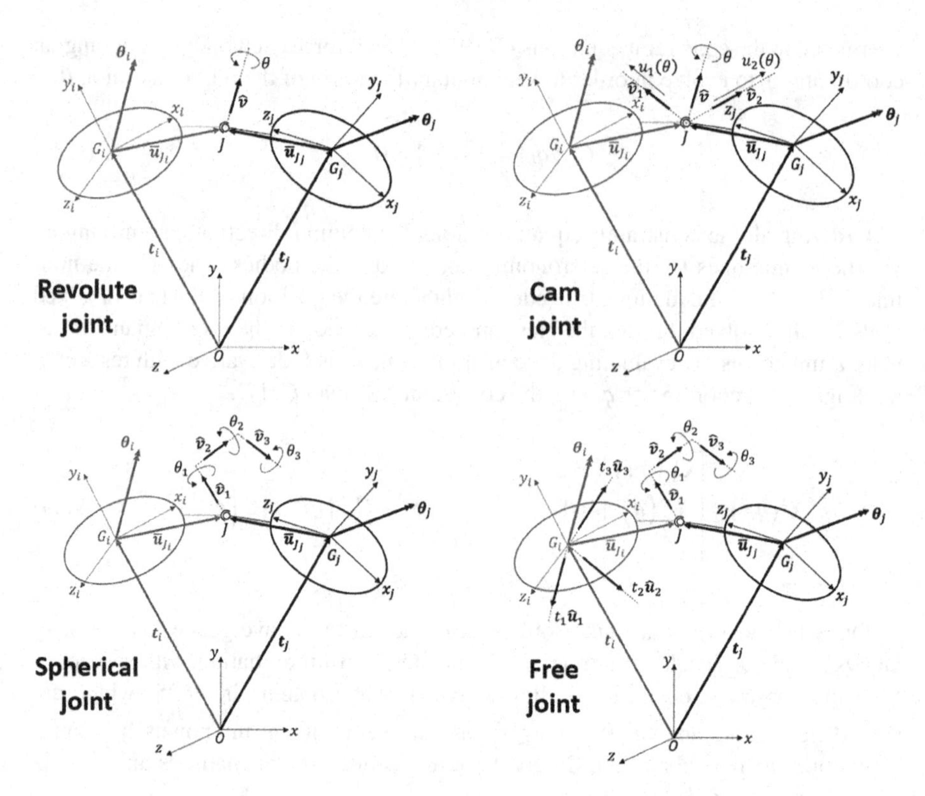

FIGURE 9.12 Lower limb musculoskeletal system considered joints.

The joints belonging to the lower-limb musculoskeletal system discussed in this chapter are listed in the following and shown in Figure 9.12 (for writing convenience, the constraint equation related to the joints are reported without their time derivatives analytical expressions):

- the *revolute joint*, which allows a single degree of freedom, i.e., the relative rotation $\theta(t)$ around the axis oriented by the unit vector $\hat{v}$ defined in the parent body i reference frame; then, it is characterized by one rheonomic constraint equation and by five scleronomic ones which lock the remaining two relative rotations and the three relative translations; the rheonomic part forces the linked bodies' quaternions $\boldsymbol{\theta}_i$ and $\boldsymbol{\theta}_j$ to combine in a relative quaternion with angle equal to $\theta(t)$; the scleronomic translational part constrains the linked bodies Lagrangian coordinates $\boldsymbol{q}_i$ and $\boldsymbol{q}_j$ to locate the position of the joint J through the given relative position defined in the body i ($\bar{\boldsymbol{u}}_{J_i}$) and in the body j ($\bar{\boldsymbol{u}}_{J_j}$), while the rotational one is defined in order to keep the mutual orthogonality between two vectors $\boldsymbol{s}_{i_1}$ and $\boldsymbol{s}_{i_2}$ perpendicular to the rotation axis belonging to the body i and a vector $\boldsymbol{s}_j$ parallel to the rotation axis belonging to the body j (Equation (9.27));

$$C_{r_J}(q_i,q_j,t) = \left\langle H(\theta_i)^T \theta_j \right\rangle - \theta(t) = 0$$

$$C_{s_J}(q_i,q_j) = \begin{bmatrix} t_i + R(\theta_i)\bar{u}_{J_i} - \left(t_j + R(\theta_j)\bar{u}_{J_j}\right) \\ s_{i_1}^T s_j \\ s_{i_2}^T s_j \end{bmatrix} = \mathbf{0} \tag{9.27}$$

- the *cam joint*, which is kinematically similar to the revolute one, is characterized by the same rotational degree of freedom, but it allows two additional relative translations between the linked bodies dependent on the angle $\theta(t)$ described by the functions $u_1(\theta)$ and $u_2(\theta)$ along the axes oriented by the unit vectors $\hat{v}_1$ and $\hat{v}_2$ defined in the parent body i reference frame; the constraint equations referred to this joint are reported in (9.28), by substituting the joint local position in the parent body reference frame $\bar{u}_{J_i}$ with the given cam displacement;

$$C_{r_J}(q_i,q_j,t) = \left\langle H(\theta_i)^T \theta_j \right\rangle - \theta(t) = 0$$

$$C_{s_J}(q_i,q_j) = \begin{bmatrix} t_i + R(\theta_i)\left[u_1(\theta)\hat{v}_1 + u_2(\theta)\hat{v}_2\right] - \left(t_j + R(\theta_j)\bar{u}_{J_j}\right) \\ s_{i_1}^T s_j \\ s_{i_2}^T s_j \end{bmatrix} = \mathbf{0} \tag{9.28}$$

- the *spherical joint*, which allows three relative rotations $\theta_1(t)$, $\theta_2(t)$ and $\theta_3(t)$ around three successive rotation axes oriented by the unit vectors $\hat{v}_1$, $\hat{v}_2$ and $\hat{v}_3$ defined in the parent body i reference frame, while locking the relative translation between the linked bodies i and j; while the scleronomic equations are equals to the ones referred to the translational term of the revolute joint, the rheonomic equations are defined in order to constrain the linked bodies' quaternions $\boldsymbol{\theta}_i$ and $\boldsymbol{\theta}_j$ to combine in three relative quaternions with angles respectively equals to $\theta_1(t)$, $\theta_2(t)$ and $\theta_3(t)$ (equation (29));

$$C_{r_J}(q_i,q_j,t) = \begin{bmatrix} \left\langle H_{\theta_1}^T H(\theta_i)^T \theta_j \right\rangle - \theta_1(t) \\ \left\langle H_{\theta_2}^T H(\theta_i)^T \theta_j \right\rangle - \theta_2(t) \\ \left\langle H_{\theta_3}^T H(\theta_i)^T \theta_j \right\rangle - \theta_3(t) \end{bmatrix} = \mathbf{0}$$

$$C_{s_J}(q_i,q_j) = t_i + R(\theta_i)\bar{u}_{J_i} - \left(t_j + R(\theta_j)\bar{u}_{J_j}\right) = \mathbf{0} \tag{9.29}$$

- the *free joint*, which is basically a spherical joint able to provide additionally the relative translation between the linked bodies; the translation motion is characterized by three relative translations $t_1(t)$, $t_2(t)$ and $t_3(t)$ along three unit vectors, respectively, $\hat{u}_1$, $\hat{u}_2$ and $\hat{u}_3$ defined in the parent body i reference frame; therefore, while the free joint does not impose scleronomic equations, it is characterized by six rheonomic equations in which the rotational term is equal to the one referred to the spherical joint and the translational one is given by enforcing the equality between the free joint local position in the parent body i reference frame $\bar{u}_{J_i}$ and the given displacement (Equation (9.30)).

$$\bar{u}_{J_i} = \begin{bmatrix} \hat{u}_1 & \hat{u}_2 & \hat{u}_3 \end{bmatrix} \begin{bmatrix} t_1(t) \\ t_2(t) \\ t_3(t) \end{bmatrix} = Ut(t)$$

$$C_{rJ}\left(q_i, q_j, t\right) = \begin{bmatrix} \left\langle H_{\theta_1}^T H(\theta_i)^T \theta_j \right\rangle - \theta_1(t) \\ \left\langle H_{\theta_2}^T H(\theta_i)^T \theta_j \right\rangle - \theta_2(t) \\ \left\langle H_{\theta_3}^T H(\theta_i)^T \theta_j \right\rangle - \theta_3(t) \\ U^{-1} R(\theta_i)^T \left(t_j + R(\theta_j)\bar{u}_{J_j} - t_i\right) - t(t) \end{bmatrix} = 0 \quad (9.30)$$

9.3.2 Muscular Action Modelling

Once the kinematical analysis has been performed, the elaborated time evolution of the Lagrangian coordinates $q(t)$, velocities $\dot{q}(t)$ and acceleration $\ddot{q}(t)$ are used to evaluate the forces and the torques acting on the bodies. The forces related to the case of a musculoskeletal system are generally referred to internal driving actions (i.e., the muscular ones), internal reactions, inertial and environmental (e.g., the ground reaction and the weight forces in the case related to this chapter).

The muscles play a key role in the framework of the musculoskeletal system mobility and the force exerted by them is generally unknown for a certain kinematics. The muscles' force is taken into account in the dynamical equilibrium of the multibody system through the virtual work principle. As shown in the Figure 9.13, a muscle can be viewed as a linear actuator linking two bodies i and j in correspondence of the two respective attachment points A and B: the muscle linear actuator is able to produce the muscular force F_m along the line joining the points A and B only in the contraction verse [7, 33]. Therefore, the virtual work produced by the muscle δW_m for the virtual displacement of the attachment points δr_A and δr_B is given in (31).

$$l = r_B - r_A \rightarrow \hat{l} = \frac{l}{\sqrt{l^T l}} \rightarrow \delta W_m = F_m \hat{l}^T \delta r_A - F_m \hat{l}^T \delta r_B \quad (9.31)$$

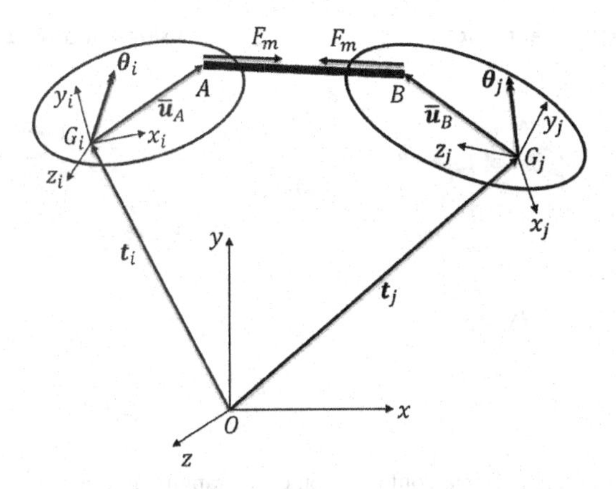

FIGURE 9.13 Muscle's action between two bodies.

Relating the virtual displacement of the attachment points A and B to the virtual variation of the respective bodies' Lagrangian coordinates $\delta \boldsymbol{q}_i$ and $\delta \boldsymbol{q}_j$, it follows that the vector of all the muscle forces $\boldsymbol{F}_m$ can be turned in the generalized force vector $\boldsymbol{Q}_m$ acting in the multibody space through the definition of a muscle Jacobian $\boldsymbol{\Phi}_q$, which can be assembled by summing the virtual work produced by all the muscles.

$$\begin{cases} \boldsymbol{r}_A = \boldsymbol{t}_i + \boldsymbol{R}(\theta_i)\bar{\boldsymbol{u}}_A \\ \boldsymbol{r}_B = \boldsymbol{t}_j + \boldsymbol{R}(\theta_j)\bar{\boldsymbol{u}}_B \end{cases}$$

$$\rightarrow \delta W_m = F_m \left(\hat{\imath}^T \begin{bmatrix} \boldsymbol{I} & -\boldsymbol{R}(\theta_i)\tilde{\bar{\boldsymbol{u}}}_A \bar{\boldsymbol{G}}(\theta_i) \end{bmatrix} \begin{bmatrix} \delta \boldsymbol{t}_i \\ \delta \boldsymbol{\theta}_i \end{bmatrix} - \hat{\imath}^T \begin{bmatrix} \boldsymbol{I} & -\boldsymbol{R}(\theta_j)\tilde{\bar{\boldsymbol{u}}}_B \bar{\boldsymbol{G}}(\theta_j) \end{bmatrix} \begin{bmatrix} \delta \boldsymbol{t}_j \\ \delta \boldsymbol{\theta}_j \end{bmatrix} \right) \tag{9.32}$$

$$\delta W_m = F_m \left(\Phi_{q_i} \delta \boldsymbol{q}_i + \Phi_{q_j} \delta \boldsymbol{q}_j \right) \rightarrow \boldsymbol{Q}_m^T \delta \boldsymbol{q} = \boldsymbol{F}_m^T \Phi_q \delta \boldsymbol{q}$$

The algorithm developed by the authors enables to consider also curved muscle actuators, in order to simulate the complex path of muscles that wrap around bony surfaces [7]. As shown in Figure 9.14, when the simulation detects the intersection between a straight line muscle AB and a wrapping object described by the surface explicit equation $\boldsymbol{x}(u, v)$, it varies the location of the intersection points P and Q on the surface until the split linear muscles AP and QB arrange tangentially with respect to the surface and the geodesic curves starting from that points become collinear in correspondence of the position where they are as close as possible: therefore, the definitive geodesics are joined in correspondence of the collinearity point, so that the curved muscle path $\boldsymbol{c}_{PQ}$ is obtained [34, 35, 36].

The wrapping model elaborates the geodesic curve $\boldsymbol{c}$ starting from a point on the surface located by the position vector $\boldsymbol{r}_0$ and with initial velocity $\hat{t}_0$ (i.e., the known tangential unit vector given by the split muscle's orientation), by solving the equivalent forward dynamics problem related to the free motion of a particle constrained to belong to the surface described by the implicit function $f(\boldsymbol{c})$ [7]. The forward dynamics problem provides the evolution along the curvilinear coordinate s of the geodesic

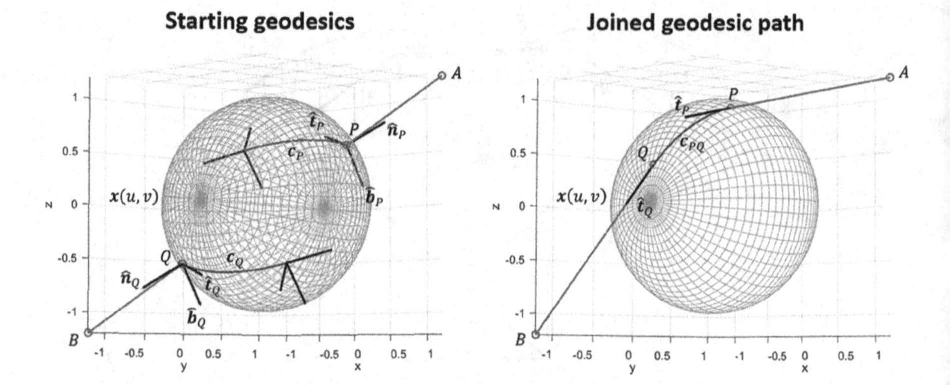

FIGURE 9.14 Starting and final configuration of a wrapping muscle.

curve $\boldsymbol{c}(s)$, introducing the particle's mass matrix $\boldsymbol{M}$, the constraint equation $\boldsymbol{C}(\boldsymbol{c})$ related to the surface constraint and the associated Lagrange multiplier $\boldsymbol{\lambda}$ which is the force orthogonal to the surface that keeps the particle on it. The described Equation (9.33) is solved numerically with marching finite differences.

$$\begin{cases} \boldsymbol{Mc}'' + \boldsymbol{C}_c^T \boldsymbol{\lambda} = \boldsymbol{0} \\ \boldsymbol{C}(\boldsymbol{c}) = f(\boldsymbol{c}) = \boldsymbol{0} \\ \boldsymbol{c}(0) = \boldsymbol{r}_0 \\ \boldsymbol{c}'(0) = \hat{\boldsymbol{t}}_0 \end{cases} \tag{9.33}$$

While elaborating, the wrapping algorithm evaluates the curve orthonormal basis along s (i.e., the tangential, normal and binormal unit vectors $\hat{\boldsymbol{t}}$, $\hat{\boldsymbol{n}}$ and $\hat{\boldsymbol{b}}$) in order to set to zero the function $\boldsymbol{f}(\boldsymbol{\xi})$, where the unknown $\boldsymbol{\xi}$ is composed by the surface coordinates u_0 and v_0 of the muscle's intersection points on the surface P and Q and the function $\boldsymbol{f}$ ensures the tangency of the split muscles on the surface and the collinearity in correspondence of the closest points P^* and Q^*. As written in Equation (9.34), the assembled non-linear problem is solved with the Newton iterative method by introducing the numerically obtained function Jacobian $\boldsymbol{J}_f$.

$$\begin{cases} \boldsymbol{r} = \boldsymbol{x}(u,v) \\ \hat{\boldsymbol{t}} = \begin{bmatrix} \boldsymbol{x}_u & \boldsymbol{x}_v \end{bmatrix} \begin{bmatrix} u' \\ v' \end{bmatrix} \\ \hat{\boldsymbol{n}} = \dfrac{\boldsymbol{x}_u \times \boldsymbol{x}_v}{\|\boldsymbol{x}_u \times \boldsymbol{x}_v\|} \\ \hat{\boldsymbol{b}} = \hat{\boldsymbol{t}} \times \hat{\boldsymbol{n}} \end{cases} \qquad \xi = \begin{bmatrix} u_{0P} \\ v_{0P} \\ u_{0Q} \\ v_{0Q} \end{bmatrix} \rightarrow \boldsymbol{f}(\xi) = \begin{bmatrix} \hat{\boldsymbol{t}}_P^T \hat{\boldsymbol{n}}_P \\ \hat{\boldsymbol{t}}_Q^T \hat{\boldsymbol{n}}_Q \\ \hat{\boldsymbol{b}}_{P^*}^T \left(\boldsymbol{r}_{P^*} - \boldsymbol{r}_{Q^*}\right) \\ \hat{\boldsymbol{b}}_{P^*}^T \hat{\boldsymbol{t}}_{Q^*} \end{bmatrix} \tag{9.34}$$

$$\xi^{(k+1)} = \xi^{(k)} - \boldsymbol{J}_f^{-1}\left(\xi^{(k)}\right) \boldsymbol{f}\left(\xi^{(k)}\right)$$

Also in the case of curved muscle actuator, the associated contribution to the muscle Jacobian can be found. In fact, with reference to Figure 9.14 showing a muscle linking the bodies i and j through the respective attachment points A and B while wrapping on a bony surfaces fixed to a third body k, once the muscle curved path $\boldsymbol{c}_{PQ}$ has been evaluated, the virtual work produced by the muscle δW_m, reported in (9.35), is related only to the split straight line segment AP and QB, since the geodesic dynamical problem in (9.33) assumes by definition that no frictional force is exchanged between the curve and the surface, so that the muscle force F_m keeps constant along the muscle from A to B.

$$\delta W_m = -F_m \hat{\boldsymbol{t}}_P^T\left(\delta \boldsymbol{r}_P - \delta \boldsymbol{r}_A\right) - F_m \hat{\boldsymbol{t}}_Q^T\left(\delta \boldsymbol{r}_B - \delta \boldsymbol{r}_Q\right) \tag{9.35}$$

As previously performed for the straight line muscle, relating the attachment and the splitting points virtual displacements to the virtual variation of the Lagrangian coordinates related to the involved bodies i, j and k the curved muscle's contribution to the muscle Jacobian $\boldsymbol{\Phi}_q$ is given in Equation (9.36).

$$\begin{cases} \boldsymbol{r}_A = \boldsymbol{t}_i + \boldsymbol{R}(\theta_i)\bar{\boldsymbol{u}}_A \\ \boldsymbol{r}_P = \boldsymbol{t}_k + \boldsymbol{R}(\theta_k)\bar{\boldsymbol{u}}_P \\ \boldsymbol{r}_Q = \boldsymbol{t}_k + \boldsymbol{R}(\theta_k)\bar{\boldsymbol{u}}_Q \\ \boldsymbol{r}_B = \boldsymbol{t}_j + \boldsymbol{R}(\theta_j)\bar{\boldsymbol{u}}_B \end{cases} \rightarrow \delta W_m = F_m\left(\Phi_{q_i}\delta \boldsymbol{q}_i + \Phi_{q_k}\delta \boldsymbol{q}_k + \Phi_{q_j}\delta \boldsymbol{q}_j\right) \tag{9.36}$$

In general, each muscle can be characterized by a heterogenous complex path connecting more than two bodies, which is modelled as a series of straight line and curved units along which the same muscle force is transmitted.

In order to evaluate the joint reaction in the successive paragraph, the muscle force F_m has to be divided in its active and passive parts, because, while the active term provides to the musculoskeletal motion, the passive one is going to load the joints. The approach used by the authors in the developed algorithm to separate the active contribution from the passive one, adopts the Hill muscle model [37, 38, 39, 40]. With reference to Figure 9.15, the Hill muscle-tendon model is composed by a series between:

- the tendon element *SE*, characterized by the given tendon slack length l_t considered as a constant in this work;
- the muscle unit, which is a parallel between the contractile element *CE*, i.e., the real active part of the muscle-tendon, and the passive one *PE*, representing the muscle-surrounding tissues stiffness.

As shown in Figure 9.15, the muscle unit of length l_m is inclined by the pennation angle α_p with respect to the muscle action line. Because of the series coupling, the force exerted by the muscle-tendon unit is constant along both the tendon and the muscle: therefore, the muscle force applied to the attachment points A and B is given

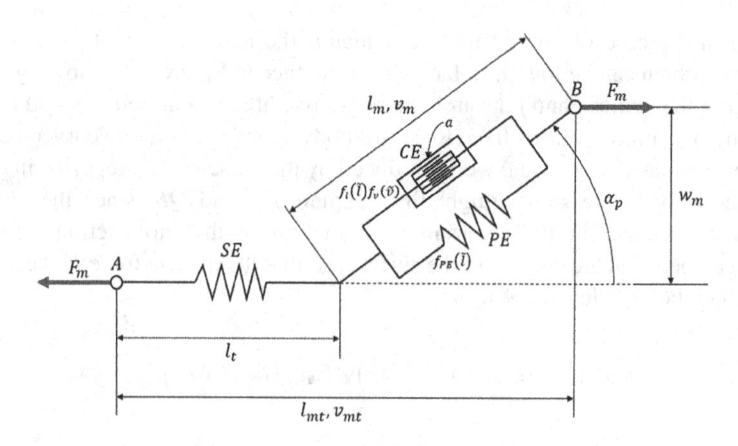

FIGURE 9.15 Hill muscle configuration.

in (9.37) by the sum between the contractive term F_{CE} and the passive one F_{PE} projected on the action line [7].

$$F_m = \left(F_{CE} + F_{PE}\right)\cos\alpha_p \tag{9.37}$$

The contractive force F_{CE} and the passive one F_{PE} are equal to a fraction of the known maximum isometrical force F_0 (Equation (9.38)) governed by the muscle activation level a and the force-length-velocity surface shown in Figure 9.16:

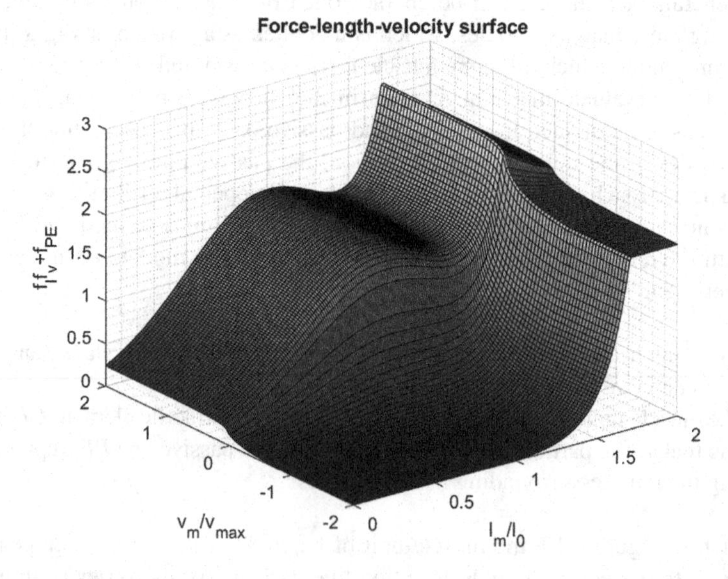

FIGURE 9.16 Hill muscle-tendon dependence on the muscle length and deformation velocity.

- for a given activation a, the generated force F_{CE} is maximum when the muscle length l_m is equal to the known optimal fiber length l_0 beyond which it extinguishes, according to the function $f_l(l_m/l_0)$; furthermore, it is greater than F_0 when the muscle deformation velocity v_m is positive (lengthening) and lower than F_0 when v_m is negative (shortening) until it reaches the maximum contraction velocity v_{max} beyond which the force cannot be generated, according to the function $f_v(v_m/v_{max})$;
- the passive force F_{PE} exists only when the muscle is lengthened beyond the optimal fiber length l_0, according to the function $f_{PE}(l_m/l_0)$.

$$\begin{cases} F_{CE} = af_l\left(\dfrac{l_m}{l_0}\right)f_v\left(\dfrac{v_m}{v_{\max}}\right)F_0 \\ F_{PE} = f_{PE}\left(\dfrac{l_m}{l_0}\right)F_0 \end{cases} \tag{9.38}$$

The muscle length l_m, deformation velocity v_m and pennation angle α_p can be evaluated starting from the muscle-tendon length l_{mt} and deformation velocity v_{mt} available from the kinematical analysis. In particular, for a complex muscle's series, the above quantities are composed by the sum of the single units contributions: considering a muscle-tendon consisting of n_s straight line units and n_c curved ones, its length l_{mt} and deformation velocity v_{mt} are given in (9.39).

$$\begin{cases} l_{mt} = \sum_{i=1}^{n_s} \|\boldsymbol{r}_{B_i} - \boldsymbol{r}_{A_i}\| + \sum_{j=1}^{n_c} \int \|\boldsymbol{c}_{P_jQ_j}{}'(s)\| ds \\ v_{mt} = \sum_{i=1}^{n_s} \hat{\boldsymbol{i}}_i^T\left(\boldsymbol{v}_{B_i} - \boldsymbol{v}_{A_i}\right) + \sum_{j=1}^{n_c}\left(\hat{\boldsymbol{t}}_{Q_j}^T \boldsymbol{v}_{Q_j} - \hat{\boldsymbol{t}}_{P_j}^T \boldsymbol{v}_{P_j}\right) \end{cases} \tag{9.39}$$

With reference to Figure 9.15, Equation (9.40) holds when the muscle width w_m is considered as a constant provided by the optimal configuration in terms of fiber length and pennation angle, i.e., the known parameters l_0 and α_0. Then, the Hill muscle states l_m, v_m and α_p are obtained.

$$\begin{cases} l_{mt} = l_t + l_m \cos\alpha_p \\ w_m = l_m \sin\alpha_p = l_0 \sin\alpha_0 \end{cases} \rightarrow \begin{cases} l_m = \sqrt{\left(l_0 \sin\alpha_0\right)^2 + \left(l_{mt} - l_t\right)^2} \\ \alpha_p = \arctan\left(\dfrac{l_0 \sin\alpha_0}{l_{mt} - l_t}\right) \\ v_m = v_{mt}\cos\alpha_p \end{cases} \tag{9.40}$$

Taking advantage of the Hill muscle description, each muscle k force F_{m_k} can be written in (9.41) as the sum of a contractive term $\bar{F}_{CE_k}$ scaled by the activation level a_k and a passive term $\bar{F}_{PE_k}$.

$$\begin{cases} \bar{F}_{CE_k} = f_l\left(\dfrac{l_{m_k}}{l_{0_k}}\right) f_v\left(\dfrac{v_{m_k}}{v_{\max k}}\right) F_{0_k} \cos\alpha_{p_k} \\ \bar{F}_{PE_k} = f_{PE}\left(\dfrac{l_{m_k}}{l_{0_k}}\right) F_{0_k} \cos\alpha_{p_k} \end{cases} \rightarrow F_{m_k} = a_k \bar{F}_{CE_k} + \bar{F}_{PE_k} \qquad (9.41)$$

Therefore, extending for the muscle force vector $\boldsymbol{F}_m$, the generalized muscle force vector $\boldsymbol{Q}_m$ can be defined through the definition of a contractile Jacobian $\boldsymbol{\Phi}_{q_{CE}}$, the muscle activation vector $\boldsymbol{a}$ and the generalized muscle passive force vector $\boldsymbol{Q}_{PE}$.

$$\begin{cases} \boldsymbol{F}_m = \operatorname{diag}\left(\bar{\boldsymbol{F}}_{CE}\right)\boldsymbol{a} + \bar{\boldsymbol{F}}_{PE} \\ \boldsymbol{Q}_m = \Phi_q^T \boldsymbol{F}_m \end{cases} \rightarrow \boldsymbol{Q}_m = \left(\Phi_q^T \operatorname{diag}\left(\bar{\boldsymbol{F}}_{CE}\right)\right)\boldsymbol{a} + \left(\Phi_q^T \bar{\boldsymbol{F}}_{PE}\right) = \Phi_{q_{CE}}^T \boldsymbol{a} + \boldsymbol{Q}_{PE} \qquad (9.42)$$

9.3.3 Static Optimization for the Inverse Dynamics

The kinematical analysis and the muscle modelling are needed to assemble the equation of motion that governs the musculoskeletal system dynamics [33], which is composed in (9.43) by:

- the inertial generalized term dependent on the mass matrix $\boldsymbol{M}$, the Lagrangian accelerations $\ddot{q}$ and the centrifugal and Coriolis force vector $\boldsymbol{Q}_v$ ($\boldsymbol{M}$ and $\boldsymbol{Q}_v$ are obtained by ordering the single bodies' contributions $\boldsymbol{M}_i$ and $\boldsymbol{Q}_{v_i}$ depending on the inertial properties of mass m_i and local inertia tensor $\bar{\boldsymbol{I}}_i$);
- the external generalized term $\boldsymbol{Q}_e$, where the particular force $\boldsymbol{F}_e$ acting on the body i with application point located by $\bar{\boldsymbol{u}}_e$ is turned in generalized force by the virtual work principle;
- the internal constraint term given by the product between the constraint Jacobian $\boldsymbol{C}_{\mathrm{q}}$ and the Lagrange multipliers $\boldsymbol{\lambda}$.

$$\begin{cases} \boldsymbol{M}_i = \begin{bmatrix} m_i \boldsymbol{I} & \boldsymbol{0} \\ \boldsymbol{0} & \bar{\boldsymbol{G}}^T(\theta_i)\bar{\boldsymbol{I}}_i\bar{\boldsymbol{G}}(\theta_i) \end{bmatrix} \\ \boldsymbol{Q}_{v_i} = \begin{bmatrix} \boldsymbol{0} \\ -2\bar{\boldsymbol{G}}^T(\dot{\theta}_i)\bar{\boldsymbol{I}}_i\bar{\boldsymbol{G}}(\theta_i)\dot{\theta}_i \end{bmatrix} \\ \boldsymbol{Q}_{ei} = \begin{bmatrix} \boldsymbol{I} & -\boldsymbol{R}(\theta_i)\tilde{\bar{\boldsymbol{u}}}_e\bar{\boldsymbol{G}}(\theta_i) \end{bmatrix}^T \boldsymbol{F}_e \end{cases} \rightarrow \boldsymbol{M}\ddot{\boldsymbol{q}} - \boldsymbol{Q}_v - \boldsymbol{Q}_e + \boldsymbol{C}_q^T\boldsymbol{\lambda} = \boldsymbol{0} \qquad (9.43)$$

After a rotational coordinate transformation is performed in order to replace the unit quaternion with the angular cartesian coordinates (so that the constraint Jacobian related to the quaternions' norm can be removed), the closed formulation in (9.44) allows to calculate the time evolution of the Lagrange multipliers $\boldsymbol{\lambda}$ by linear matrix inversion [7], where:

- the rheonomic Lagrange multipliers $\boldsymbol{\lambda}_r$ are the driving forces acting on the joint's degrees of freedom;
- the scleronomic Lagrange multipliers $\boldsymbol{\lambda}_s$ are referred to the imposition of the scleronomic constraint, but they are not fully related to the joint reactions, because the equation does not consider the muscles as the real actuators.

$$\begin{bmatrix} \boldsymbol{C}_{q_r}^T & \boldsymbol{C}_{q_s}^T \end{bmatrix} \begin{bmatrix} \boldsymbol{\lambda}_r \\ \boldsymbol{\lambda}_s \end{bmatrix} = -\left(\boldsymbol{M}\ddot{\boldsymbol{q}} - \boldsymbol{Q}_v - \boldsymbol{Q}_e \right) \tag{9.44}$$

Therefore, the introduction of the generalized muscle force vector $\boldsymbol{Q}_m$ in place of the rheonomic actions in (9.45) enables to refer the scleronomic multipliers $\boldsymbol{\lambda}_s$ to the joint reactions.

$$\boldsymbol{M}\ddot{\boldsymbol{q}} - \boldsymbol{Q}_v - \boldsymbol{Q}_e - \boldsymbol{Q}_m + \boldsymbol{C}_{q_s}^T \boldsymbol{\lambda}_s = \boldsymbol{0} \tag{9.45}$$

Relating $\boldsymbol{Q}_m$ to the muscle activations $\boldsymbol{a}$ in (9.46) using the Equation (9.42), the equation of motion is characterized by more unknowns (i.e., the muscle activations $\boldsymbol{a}$ and the joint reactions $\boldsymbol{\lambda}_s$) than equations since the complex human muscular actuation system is redundant: then, an optimization criterion has to be adopted in order to calculate a numerical solution.

$$\boldsymbol{M}\ddot{\boldsymbol{q}} - \boldsymbol{Q}_v - \boldsymbol{Q}_e - \left(\boldsymbol{\Phi}_{qCE}^T \boldsymbol{a} + \boldsymbol{Q}_{PE} \right) + \boldsymbol{C}_{q_s}^T \boldsymbol{\lambda}_s = \boldsymbol{0} \tag{9.46}$$

The so-called *static optimization* is actually addressed by several objective functions by the scientific researchers: in this work a simple cost function equal to the sum of the squared activations is minimized [7] (minimum muscle energy criterion) and coupled to the equality constraint represented by the equation of motion and to the boundaries on the muscles' activation being a scalar number between zero and one, as described in the Equation (9.47).

$$\begin{cases} \min\limits_{\boldsymbol{a},\boldsymbol{\lambda}_s} \boldsymbol{a}^T \boldsymbol{a} \\ \begin{bmatrix} \boldsymbol{\Phi}_{qCE}^T & -\boldsymbol{C}_{q_s}^T \end{bmatrix} \begin{bmatrix} \boldsymbol{a} \\ \boldsymbol{\lambda}_s \end{bmatrix} = \boldsymbol{M}\ddot{\boldsymbol{q}} - \boldsymbol{Q}_v - \boldsymbol{Q}_e - \boldsymbol{Q}_{PE} \\ \boldsymbol{0} \le \boldsymbol{a} \le \boldsymbol{1} \end{cases} \tag{9.47}$$

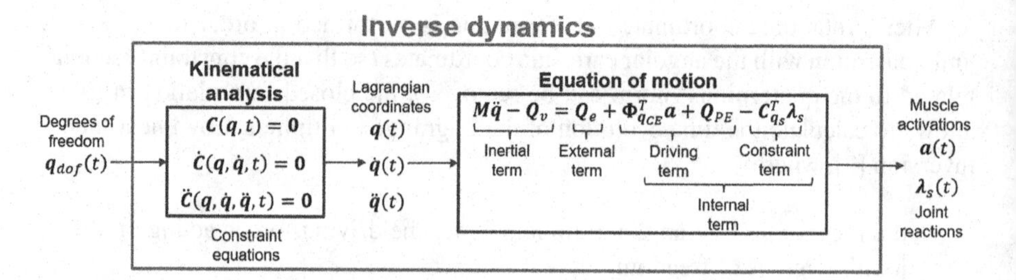

FIGURE 9.17 Multibody model workflow.

At this point, the developed multibody model is able to calculate the joint reactions, which are needed to supply the described above lubrication model as summed up in the scheme shown in the Figure 9.17. The merging between the multibody and the lubrication models gives the possibility to evaluate the tribological performance of a joint by knowing the kinematics of the whole analysed musculoskeletal system, therefore the tribological characterization of a prosthesis can be conducted for several physical activities and for different subjects.

9.4 APPLICATION TO THE THR WEAR ASSESSMENT DURING THE GAIT

The authors adopted the developed models, written totally in MatLab computational environment, to evaluate the wear evolution of an artificial hip joint Ceramic-on-UHMWPE belonging to the lower limb musculoskeletal system subjected to the gait kinematics. The analysed lower limb is constituted by input data (bodies' and joints' properties, muscles' characteristics, degrees of freedom trajectories and applied ground reaction forces) available on the "3DGaitModel2392" OpenSim model [41, 42, 43, 44]: the system is composed by 23 bodies (three kinematical chains being the back joined with the two legs by the pelvis) and 92 muscles. While the OpenSim model elaborates the muscle wrapping through moving muscles' attachment points, in this chapter some wrapping objects taken from a second OpenSim model ("Arnold Hamner Running Hybrid v2.1" [45]) are introduced. All the input data needed to perform the whole inverse dynamics of the analysed case, including the topology of all the involved objects, are collected and listed in the Excel file "gait-2392.xlsx" [7], which is called by the main MatLab script and elaborated. In Figure 9.18, the lower-limb musculoskeletal system is shown with the black lines representing the bodies, the joints black circles, the bodies mass centres green circles together with the local reference frames, the red muscles and the cyan wrapping objects (cylinders or ellipsoids). The four types of joint described are used:

- the revolute joint is adopted for the ankle, the subtalar and the metatarso-phalangeal joints;
- the cam joint is adopted for the knee joints, which allows a relative translation between the femur and the tibia bones in the sagittal plane dependent on the knee flexion angle;

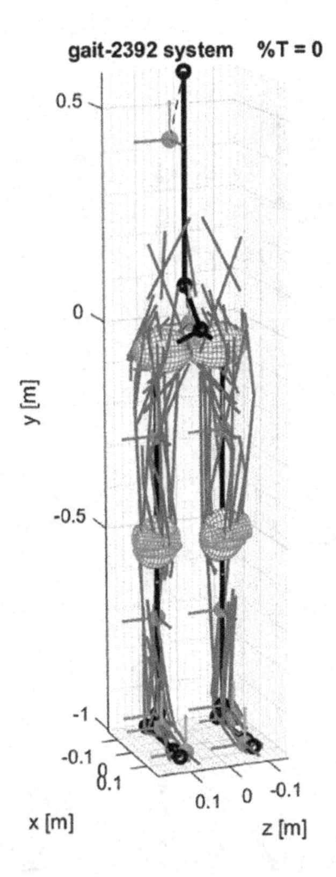

FIGURE 9.18 MatLab lower limb musculoskeletal system.

- the spherical joint is adopted for the hip and the back joints;
- the free joint is adopted for the ground-pelvis connection, in order to give the mobility to the whole system in terms of translation and orientation.

As usual, the gait cycle starts with the *stance* phase, when the right heel strikes the ground; then, when the right foot toes leave the ground, in correspondence of about 60% of the gait cycle, the *swing* phase takes place, until the end of the cycle identified by the successive right heel strike [46]. The gait cycle kinematics evaluated through the multibody model is shown in Figure 9.19, together with the ground inertial reference frame and the ground reaction force vector depicted with a green arrow moving over the ground, while the muscles take on a colour shade between blue (zero activation) and red (total activation) as a result of the static optimization.

In order to show the reliability and the capability of the developed multibody algorithm, in Figure 9.20 the driving torques referred to the right hip, knee and ankle in the sagittal plane during the gait are shown, while Figure 9.21 reports the muscle states referred to the right semitendinosus muscle wrapping around an ellipsoid bony surface fixed to the tibia bone.

The most interesting output, from a tribological point of view, elaborated by the multibody algorithm are surely the joint reactions. Since the objective is to evaluate

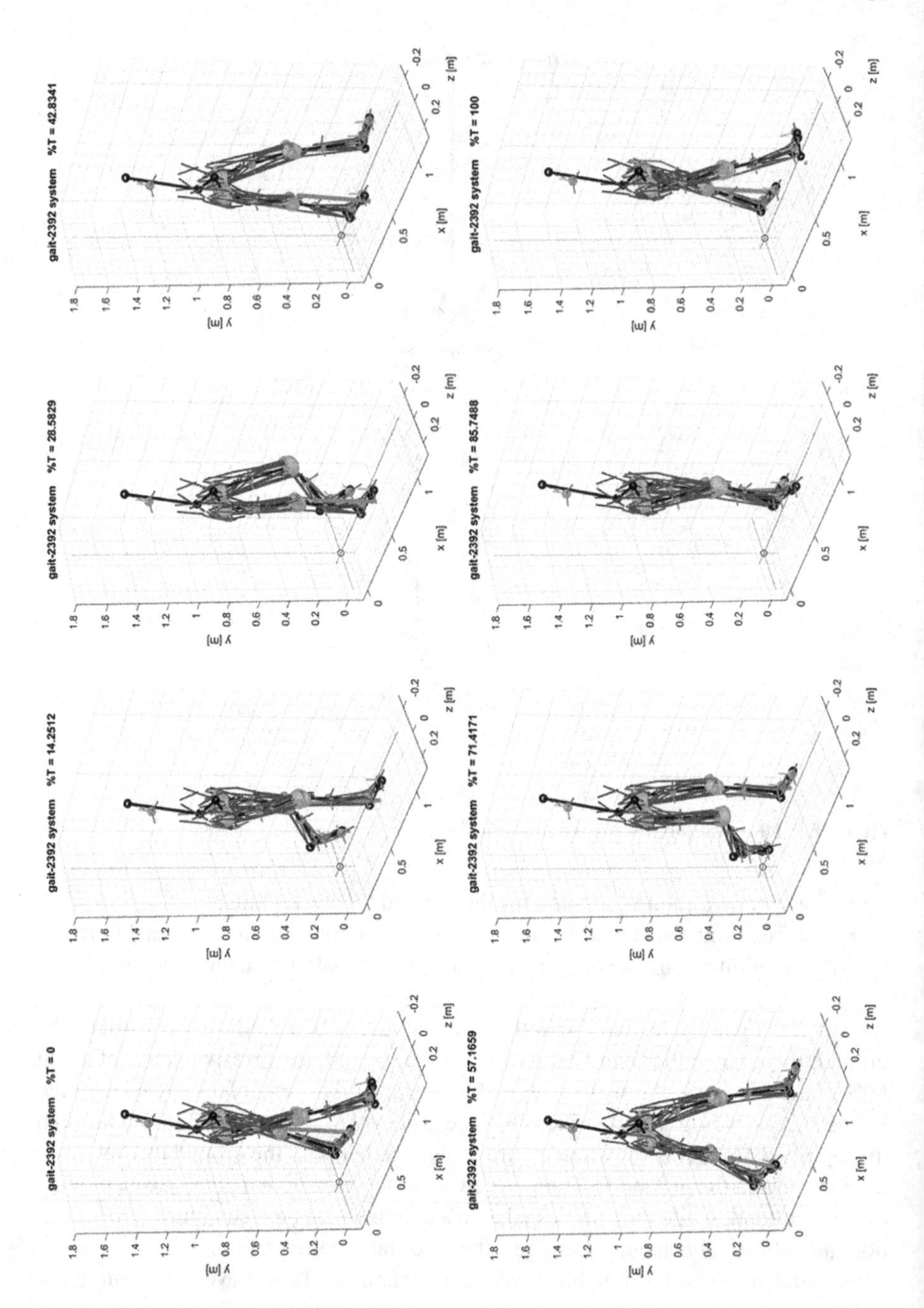

FIGURE 9.19 Multibody visual output.

the tribological state of the right hip joint, the associated joint reactions λ_s are rotated in the femur reference frame in order to compare them with the established *in vivo* measurements performed by Bergmann [47] in Figure 9.22. Considering that the computed *in silico* results are referred to a particular subject, while the *in vivo*

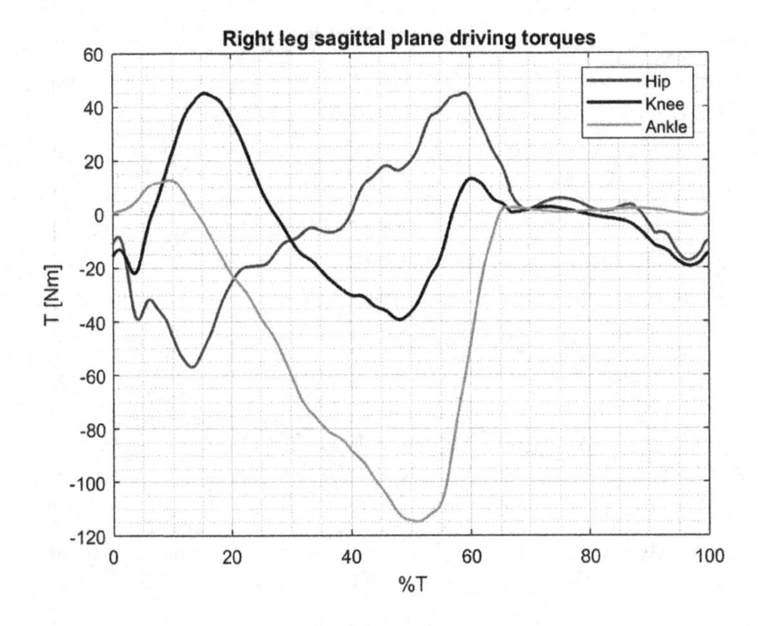

FIGURE 9.20 Sagittal driving torques referred to the right leg joints.

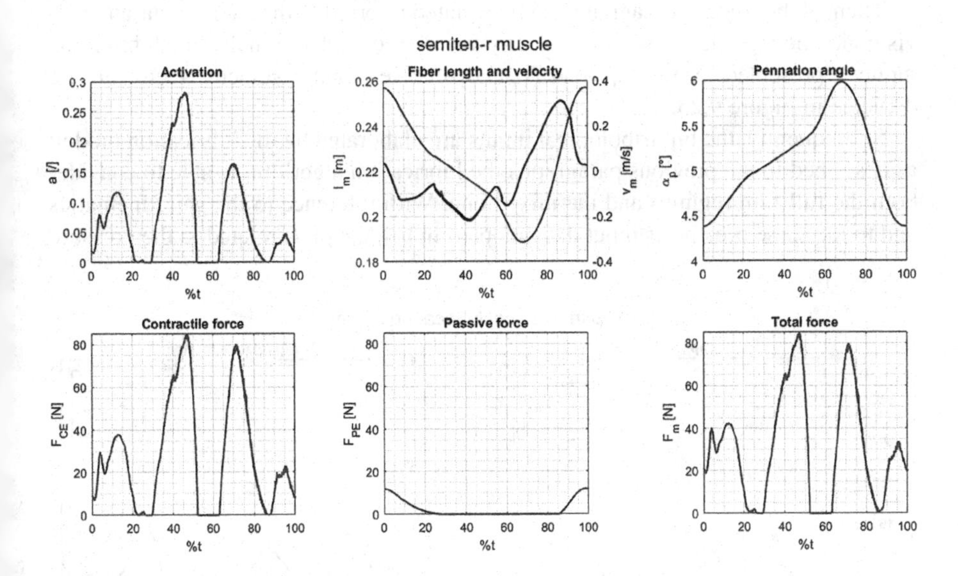

FIGURE 9.21 Right semitendinosus muscle states.

Bergmann loads are referred to the average applied to several people, the calculated hip joint reactions constitute a very satisfactory outcome, which encourages the utilization of the multibody algorithm as a synovial joint tribological configurations generator.

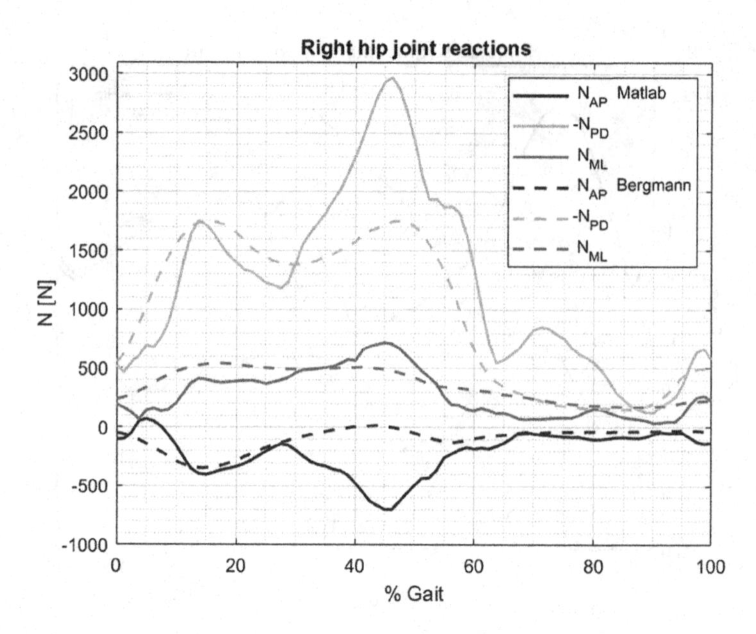

FIGURE 9.22 Comparison between the simulated and the Bergmann hip loads.

Then, elaborating the Lagrangian coordinated referred to the right femur and pelvis bone, the hip relative angular velocity vector is obtained. Finally, the lubrication inputs represented by the hip load $\boldsymbol{N}_h$ and relative angular velocity vector $\boldsymbol{\omega}_h$ are depicted in Figure 9.23.

Subsequently, the hip tribological inputs are elaborated by the lubrication model: as described in the previous paragraphs the lubrication algorithm is able to consider both the full film regimes and the mixed one. With reference to the gait kinematics and to the same prosthesis input data adopted in [6], the plots related to the pressure

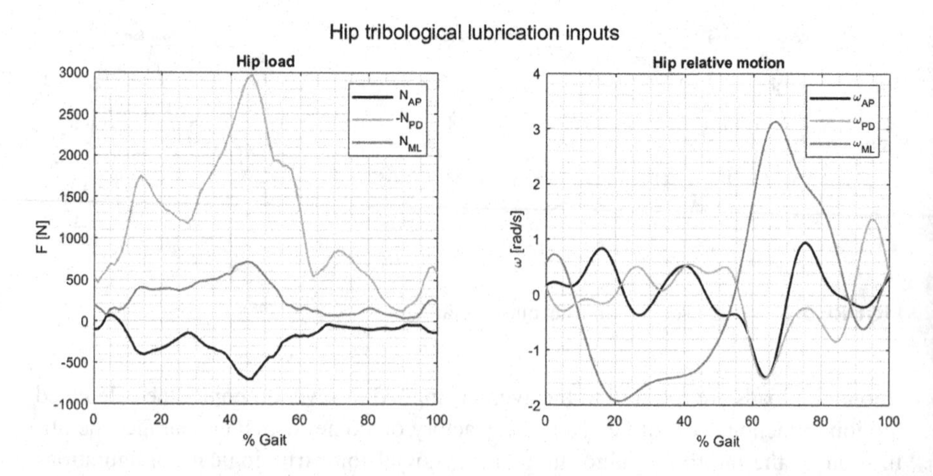

FIGURE 9.23 Right hip tribological state during the gait.

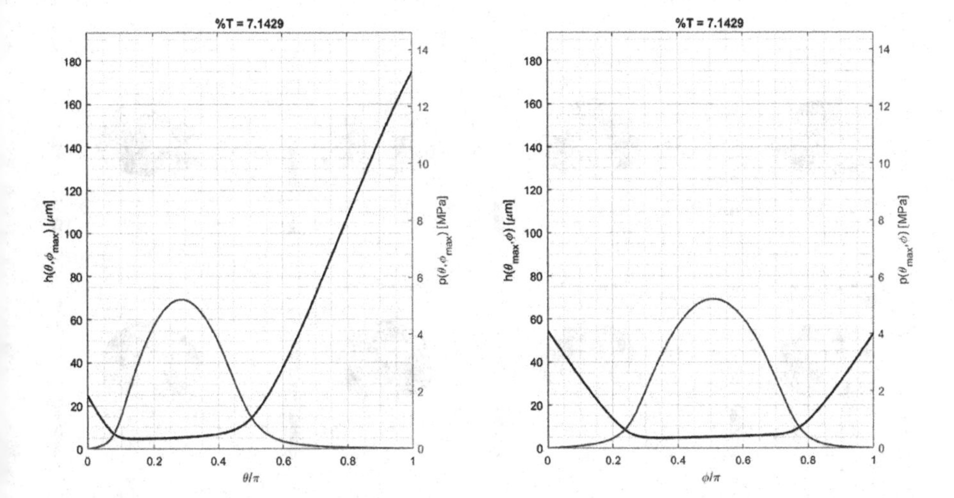

FIGURE 9.24 Full lubrication condition.

and the separation shapes within the synovial gap in the pressure peak position are shown: Figure 9.24 is related to an instant in which the surfaces are completely separated by the synovial fluid meatus, while Figure 9.25 is depicted in correspondence of a time instant in which the load is supported by both lubricated areas and contact ones, where the latter are characterized by zero fluid film thickness.

The transition from the lubricated areas to the contact ones over the domain is a challenging numerical issue in the framework of the mixed lubrication algorithms. Regarding the artificial hip joint during the gait, the outcome referred to the coexistence of the synovial fluid pressure with the contact pressure is shown in Figure 9.26: it shows that the full film mode is established only in the early steps of the cycle, after which the contact pressure gradually replaces the fluid one due to the high load

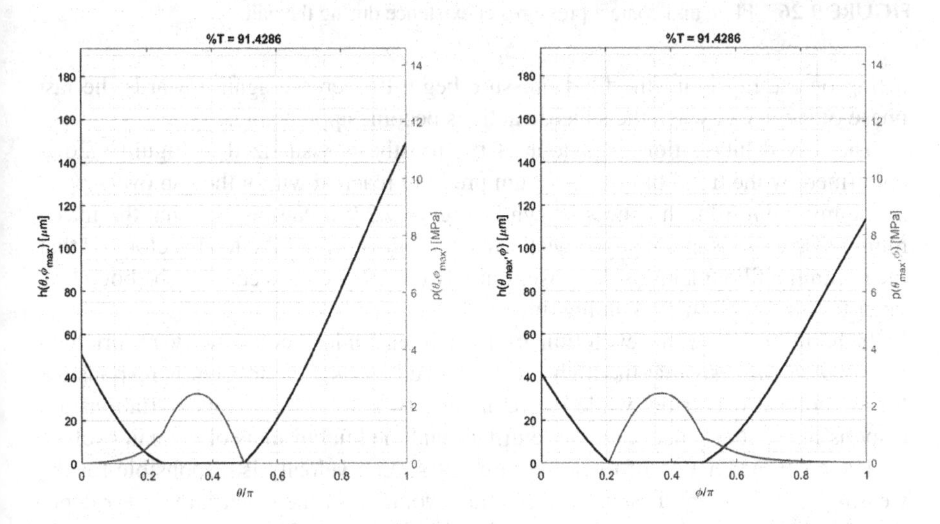

FIGURE 9.25 Mixed lubrication condition.

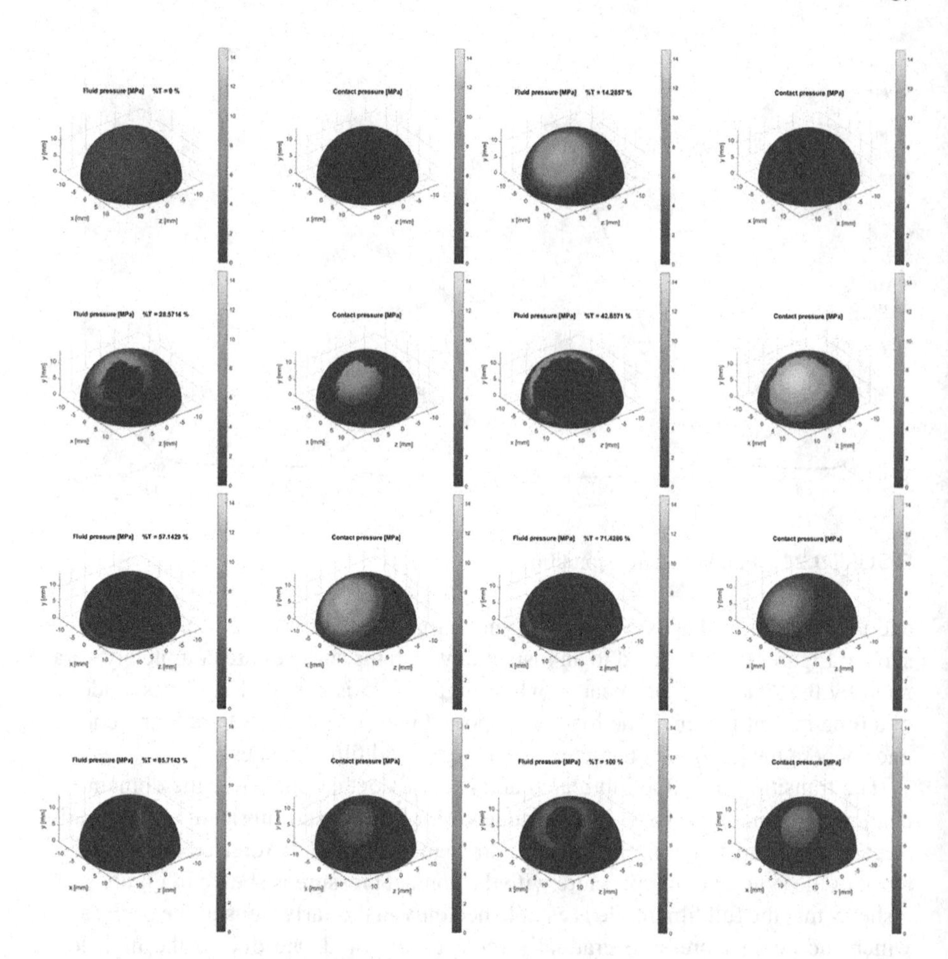

FIGURE 9.26 Fluid and contact pressures coexistence during the gait.

acting on the hip joint; the fluid pressure begin to increase again towards the last phase of the cycle, when the squeeze actions become appreciable.

The mixed lubrication behaviour of the prosthesis visualized in Figure 9.26 is confirmed by the trend of the maximum pressure reached within the gap overlapped to the minimum film thickness shown in Figure 9.27: it can be seen that the maximum pressure substantially follows the shape of the norm of the load vector and that the minimum film thickness decreases quickly until it becomes equal to the boundary layer for almost all the remaining time.

In addition, the time evolution of the dimensionless eccentricity reported in Figure 9.28 suggests that the high level achieved by its y-component is responsible for the contact occurring, while the remaining two components time variations are responsible for the squeeze action return around the final instants of the gait cycle.

The combined action of total pressure and sliding velocity is responsible for the wearing of the prosthesis surfaces: the time evolution of the penetration wear depth field for this case is shown in Figure 9.29: this type of analysis can be useful while

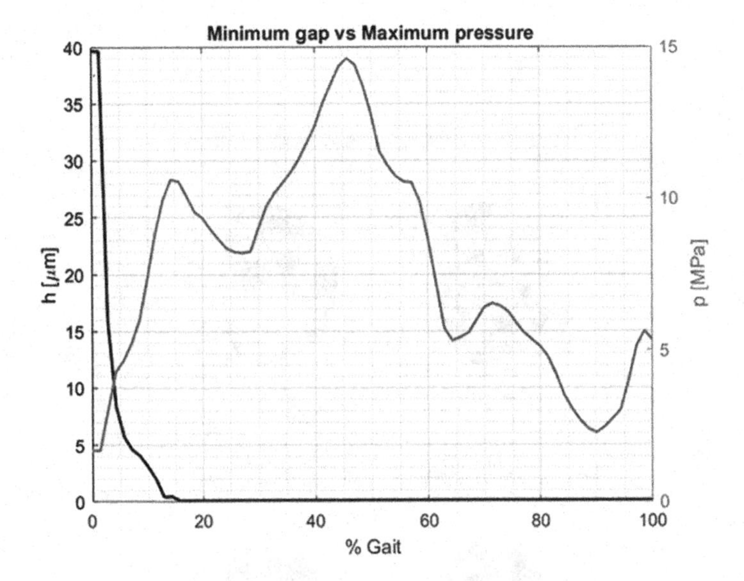

FIGURE 9.27 Maximum pressure and minimum surfaces' separation within the synovial gap.

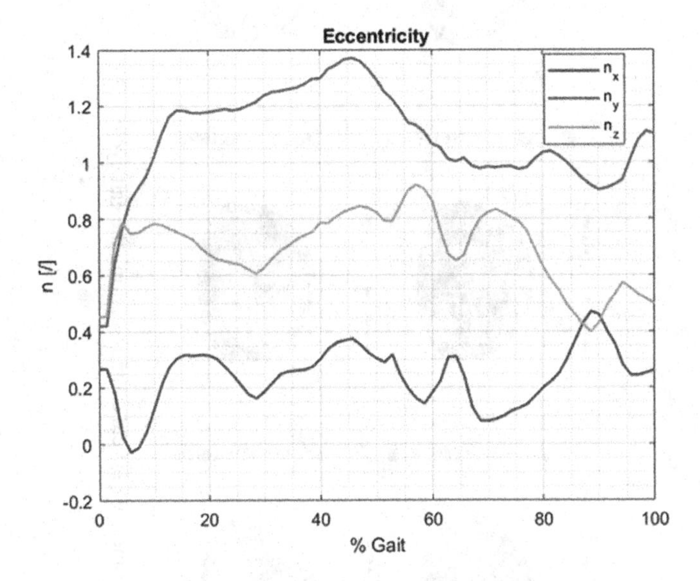

FIGURE 9.28 Dimensionless eccentricity vector evolution over time.

evaluating which area of the implant is the most critical in terms of material loss for a certain kinematics.

The final output of the wear volume is obtained by integrating over the surface's domain the penetration wear depth and shown in Figure 9.30: its trend can be relevant in the framework of the implant duration estimation, in fact, for this case, it gives information about the quantity of volume removed for a single gait cycle.

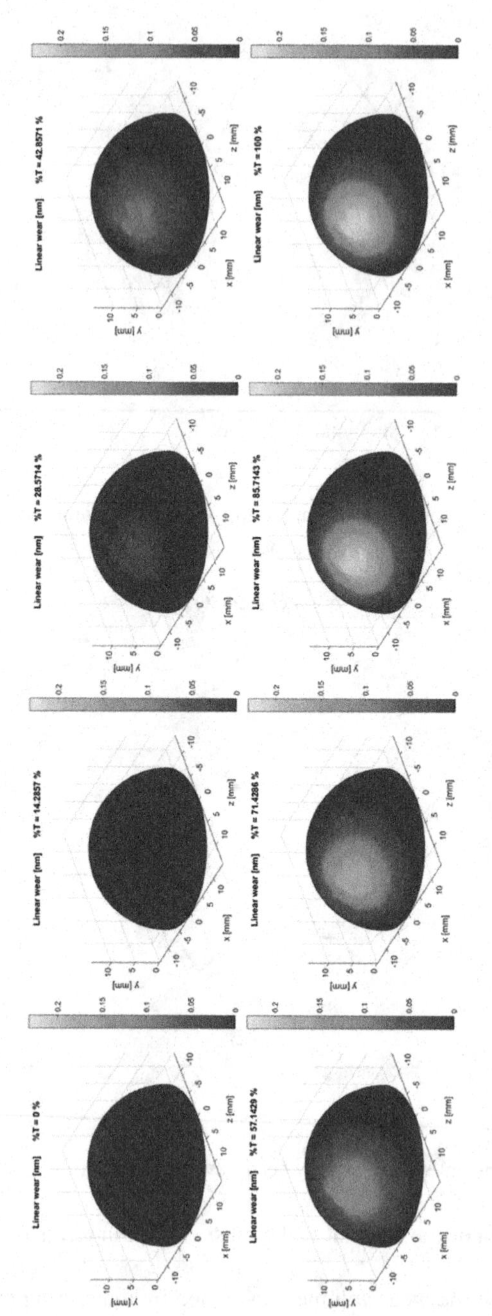

FIGURE 9.29 Linear wear field evolution during the gait cycle.

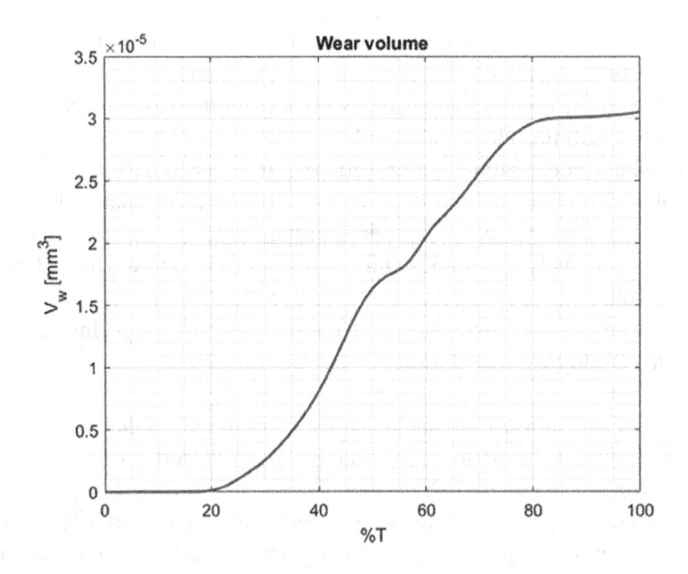

FIGURE 9.30 Wear volume time evolution over the gait cycle.

9.5 CONCLUSIONS

The aim of this chapter was to present works already developed by the authors in the framework of the biotribology of the hip implants. In particular, the main objective was to estimate the artificial hip joint duration through numerical models developed in MatLab computational environment able to simulate the complex phenomena evolving within the synovial cavity of the prosthesis.

A numerical model simulating the mixed elasto-hydrodynamic lubrication of a general spherical joint was developed and applied to the Ceramic-on-UHMWPE hip prosthesis case. The model takes into account the deformability of the softer acetabular cup through the column model and the variability of the rheological properties of the synovial fluid with the fluid pressure. Furthermore, it consider the non-Newtonian behaviour of the synovial fluid due to the hyaluronic acid macromolecules within it and the modification over time of the synovial gap geometry because of the wear cumulation. Above all, the developed numerical model allows to distinguish the lubricated areas by the contact one by dividing the domain's area in those zones in which the surfaces' interpenetration occurs. The workflow of the lubrication model starts from the knowledge of the hip load and relative motion conditions during a certain kinematics and it ends with the calculation of the time evolution of the dimensionless eccentricity vector of the femoral head with respect to the acetabular cup.

The tribological inputs for the lubrication algorithm are evaluated by a second developed model, which studies the multibody mechanics of general musculoskeletal systems. In particular, the developed procedure is able to perform the inverse dynamics of musculoskeletal systems, by assembling and solving the constraint equations (kinematical analysis) and by characterizing an optimization problem able to solve the equation of motion in terms of the unknown muscle activations and joint reactions. The multibody approach considers the complex muscle paths through a wrapping algorithm based on the geodesic curves generated around bony surfaces and distinguishes the contractile muscle actions from the passive ones by using the well-established Hill muscle-tendon model.

The two macro-codes are joined and coupled in order to have a complex computational tool which estimates the *in silico* wear of the hip implants belonging to a particular musculoskeletal system moving with a certain kinematics. The algorithm was applied to gait cycle kinematics and it turns out to be a powerful indicator of the implant duration.

In order to have an increasingly accurate computational instrument, some improvements could be devoted to:

- the boundary lubrication mechanisms modelling, so that more detailed information can be obtained in that zones characterized by no surfaces' separation;
- the introduction of the surface roughness fields, which can affect the pressure distribution significantly by preserving some lubricated areas in the valleys between the asperities;
- the implementation of a finite element model able to compute the deformation of the whole body of the prosthesis, in order to analyse hard-on-hard implants or to evaluate the stress state of the joint parts;
- the modelling of different wear mechanisms, such as the delamination or the tribocorrosion due to the active behaviour of biological fluids;
- the introduction of kinematically more articulated joints;
- the modelling of the tendon dynamics, which can be very relevant for kinematics faster and/or heavier than the gait (e.g., running, swimming, climbing, cycling, etc.).

FUNDING

This research was funded by MIUR, PRIN 2017 BIONIC.

BIBLIOGRAPHY

1. M. Nordin and V. H. Frankel, *Basic biomechanics of the musculoskeletal system*, Lippincott Williams & Wilkins, 2001.
2. J. P. Gleghorn and L. J. Bonassar, "Lubrication mode analysis of articular cartilage using Stribeck surfaces," *Journal of Biomechanics*, vol. 41, no. 9, pp. 1910–1918, 2008.
3. L. Mattei, F. Di Puccio, B. Piccigallo and E. Ciulli, "Lubrication and wear modelling of artificial hip joints: A review," *Tribology International*, vol. 44, no. 5, pp. 532–549, 2011.
4. J. Fisher and D. Dowson, "Tribology of total artificial joints," *Proceedings of the Institution of Mechanical Engineers, Part H: Journal of Engineering in Medicine*, vol. 205, no. 2, pp. 73–79, 1991.
5. S. Scholes and A. Unsworth, "Comparison of friction and lubrication of different hip prostheses," *Proceedings of the Institution of Mechanical Engineers, Part H: Journal of Engineering in Medicine*, vol. 214, no. 1, pp. 49–57, 2000.

6. A. Ruggiero and A. Sicilia, "A mixed elasto-hydrodynamic lubrication model for wear calculation in artificial hip joints," *Lubricants*, vol. 8, no. 7, p. 72, 2020.
7. A. Ruggiero and A. Sicilia, "A Novel Explicit Analytical Multibody Approach for the Analysis of Upper Limb Dynamics and Joint Reactions Calculation Considering Muscle Wrapping," *Applied Sciences*, vol. 10, no. 21, p. 7760, 2020.
8. A. Ruggiero, E. Gòmez and D. Roberto, "Approximate closed-form solution of the synovial fluid film force in the human ankle joint with non-Newtonian lubricant," *Tribology International*, vol. 57, pp. 156–161, 2013.
9. A. Ruggiero and A. Sicilia, "Lubrication modeling and wear calculation in artificial hip joint during the gait," *Tribology International*, vol. 142, p. 105993, 2020.
10. A. Ruggiero, A. Sicilia and S. Affatato, "In silico total hip replacement wear testing in the framework of ISO 14242-3 accounting for mixed elasto-hydrodynamic lubrication effects," *Wear*, vol. 460, p. 203420, 2020.
11. A. Ruggiero, R. D'Amato and S. Affatato, "Comparison of meshing strategies in THR finite element modelling," *Materials*, vol. 12, no. 14, p. 2332, 2019.
12. A. Ruggiero, M. Merola and S. Affatato, "Finite element simulations of hard-on-soft hip joint prosthesis accounting for dynamic loads calculated from a musculoskeletal model during walking," *Materials*, vol. 11, no. 4, p. 574, 2018.
13. D. Jalali-Vahid, M. Jagatia and Z. A. D. D. Jin, "Prediction of lubricating film thickness in a ball-in-socket model with a soft lining representing human natural and artificial hip joints," *Proceedings of the Institution of Mechanical Engineers, Part J: Journal of Engineering Tribology*, vol. 215, no. 4, pp. 363–372, 2001.
14. A. Ruggiero and A. Sicilia, "Mathematical development of a novel discrete hip deformation algorithm for the in silico elasto-hydrodynamic lubrication modelling of total hip replacements," *Lubricants*, vol. 9, no. 4, p. 41, 2021.
15. F. Wang and Z. Jin, "Lubrication modelling of artificial hip joints: From fluid film to boundary lubrication regimes," *Engineering Systems Design and Analysis*, vol. 41731, pp. 605–611, 2004.
16. F. Wang, C. Brockett, S. Williams, I. Udofia, J. Fisher and Z. Jin, "Lubrication and friction prediction in metal-on-metal hip implants," *Physics in Medicine & Biology*, vol. 53, no. 5, p. 1277, 2008.
17. D. Zhu and Y.-Z. Hu, "The study of transition from elastohydrodynamic to mixed and boundary lubrication," *The Advancing Frontier of Engineering Tribology, Proceedings of the 1999 STLE/ASME HS Cheng Tribology Surveillance*, pp. 150–156, 1999.
18. D. Zhu, "On some aspects of numerical solutions of thin-film and mixed elastohydrodynamic lubrication," *Proceedings of the Institution of Mechanical Engineers, Part J: Journal of Engineering Tribology*, vol. 221, no. 5, pp. 561–579, 2007.
19. L. Gao, D. Dowson and R. W. Hewson, "A numerical study of non-Newtonian transient elastohydrodynamic lubrication of metal-on-metal hip prostheses," *Tribology International*, vol. 93, pp. 486–494, 2016.
20. H. van Leeuwen, "The determination of the pressure—viscosity coefficient of a lubricant through an accurate film thickness formula and accurate film thickness measurements," *Proceedings of the Institution of Mechanical Engineers, Part J: Journal of Engineering Tribology*, vol. 223, no. 8, pp. 1143–1163, 2009.
21. D. Jalali-Vahid, Z. Jin and D. Dowson, "Elastohydrodynamic lubrication analysis of hip implants with ultra high molecular weight polyethylene cups under transient conditions," *Proceedings of the Institution of Mechanical Engineers, Part C: Journal of Mechanical Engineering Science*, vol. 217, no. 7, pp. 767–777, 2003.
22. J. F. Archard and W. Hirst, "The wear of metals under unlubricated conditions," *Proceedings of the Royal Society of London. Series A. Mathematical and Physical Sciences*, vol. 236, no. 1206, pp. 397–410, 1956.

23. L. Gao, D. Dowson and R. W. Hewson, "Predictive wear modeling of the articulating metal-on-metal hip replacements," *Journal of Biomedical Materials Research Part B: Applied Biomaterials*, vol. 105, no. 3, pp. 497–506, 2017.
24. S. Affatato, A. Ruggiero, M. Merola and S. Logozzo, "Does metal transfer differ on retrieved Biolox® Delta composites femoral heads? Surface investigation on three Biolox® generations from a biotribological point of view," *Composites Part B: Engineering*, vol. 113, pp. 164–173, 2017.
25. S. Affatato, A. Ruggiero and M. Merola, "Advanced biomaterials in hip joint arthroplasty. A review on polymer and ceramics composites as alternative bearings," *Composites Part B: Engineering*, vol. 83, pp. 276–283, 2015.
26. A. Aherwar, A. K. Singh and A. Patnaik, "Current and future biocompatibility aspects of biomaterials for hip prosthesis," *AIMS Bioeng*, vol. 3, no. 1, pp. 23–43, 2016.
27. L. Mattei, F. Di Puccio and E. Ciulli, "A comparative study of wear laws for soft-on-hard hip implants using a mathematical wear model," *Tribology International*, vol. 63, pp. 66–77, 2013.
28. M. Jagatia and Z. Jin, "Elastohydrodynamic lubrication analysis of metal-on-metal hip prostheses under steady state entraining motion," *Proceedings of the Institution of Mechanical Engineers, Part H: Journal of Engineering in Medicine*, vol. 215, no. 6, pp. 531–541, 2001.
29. P. A. Varady, U. Glitsch and P. Augat, "Loads in the hip joint during physically demanding occupational tasks: A motion analysis study," *Journal of biomechanics*, vol. 48, no. 12, pp. 3227–3233, 2015.
30. J. Hu, Z. Chen, H. Xin, Q. Zhang and Z. Jin, "Musculoskeletal multibody dynamics simulation of the contact mechanics and kinematics of a natural knee joint during a walking cycle," *Proceedings of the Institution of Mechanical Engineers, Part H: Journal of Engineering in Medicine*, vol. 232, no. 5, pp. 508–519, 2018.
31. M. T. Silva, A. F. Pereira and J. M. Martins, "An efficient muscle fatigue model for forward and inverse dynamic analysis of human movements," *Procedia IUTAM*, vol. 2, pp. 262–274, 2011.
32. M. P. Silva and J. A. Ambrosio, "Solution of redundant muscle forces in human locomotion with multibody dynamics and optimization tools," *Mechanics Based Design of Structures and Machines*, vol. 31, no. 3, pp. 381–411, 2003.
33. A. A. Shabana, *Dynamics of multibody systems*, Cambridge: Cambridge University Press, 2005.
34. O. Zarifi and I. Stavness, "Muscle wrapping on arbitrary meshes with the heat method," *Computer Methods in Biomechanics and Biomedical Engineering*, vol. 20, no. 2, pp. 119–129, 2017.
35. I. Stavness, M. Sherman and S. Delp, *A general approach to muscle wrapping over multiple surfaces*, 2012.
36. A. Scholz, I. Stavness, M. Sherman, S. Delp and A. Kecskeméthy, "Improved muscle wrapping algorithms using explicit path-error Jacobians," *Computational Kinematics*, pp. 395–403, 2014.
37. K. Jovanovic, J. Vranic and N. Miljkovic, "Hill's and Huxley's muscle models: Tools for simulations in biomechanics," *Serbian Journal of Electrical Engineering*, vol. 12, no. 1, pp. 53–67, 2015.
38. M. Millard, T. Uchida, A. Seth and S. L. Delp, "Flexing computational muscle: Modeling and simulation of musculotendon dynamics," *Journal of Biomechanical Engineering*, vol. 135, no. 2, 2013.
39. F. Romero and F. Alonso, "A comparison among different hill-type contraction dynamics formulations for muscle force estimation," *Mechanical Sciences*, vol. 7, no. 1, pp. 19–29, 2016.

40. D. G. Thelen, "Adjustment of muscle mechanics model parameters to simulate dynamic contractions in older adults," *Journal of Biomechanical Engineering*, vol. 125, no. 1, pp. 70–77, 2003.
41. S. L. Delp, J. P. Loan, M. G. Hoy, F. E. Zajac, E. L. Topp and J. M. Rosen, "An interactive graphics-based model of the lower extremity to study orthopaedic surgical procedures," *IEEE Transactions on Biomedical Engineering*, vol. 37, no. 8, pp. 757–767, 1990.
42. G. T. Yamaguchi and F. E. Zajac, "A planar model of the knee joint to characterize the knee extensor mechanism," *Journal of Biomechanics*, vol. 22, no. 1, pp. 1–10, 1989.
43. F. C. Anderson and M. G. Pandy, "A dynamic optimization solution for vertical jumping in three dimensions," *Computer Methods in Biomechanics and Biomedical Engineering*, vol. 2, no. 3, pp. 201–231, 1999.
44. F. C. Anderson and M. G. Pandy, "Dynamic optimization of human walking," *Journal of Biomechanical Engineering*, vol. 123, no. 5, pp. 381–390, 2001.
45. E. M. Arnold, S. R. Ward, R. L. Lieber and S. L. Delp, "A model of the lower limb for analysis of human movement," *Annals of Biomedical Engineering*, vol. 38, no. 2, pp. 269–279, 2010.
46. A. Kharb, V. Saini, Y. Jain and S. Dhiman, "A review of gait cycle and its parameters," *IJCEM International Journal of Computational Engineering & Management*, vol. 13, pp. 78–83, 2011.
47. G. Bergmann, A. Bender, J. Dymke, G. Duda and P. Damm, "Standardized loads acting in hip implants," *PLoS One*, vol. 11, no. 5, p. e0155612, 2016.

10 The Role of Surface Engineering in Tribology

P. Kumaravelu, Sudheer Reddy Beyanagari, S. Arulvel and Jayakrishna Kandasamy
Vellore Institute of Technology, Vellore, India

CONTENTS

10.1 INTRODUCTION

Tribology is the study of the science and technology of interacting surfaces in relative motion, including friction, wear, lubrication, and related design factors [1]. Because tribology is so closely linked to practical applications, elaborative research and empirical experience are extremely important in today's circumstances. The operating environment and contact mechanism are two of the most critical factors that affect the investigation of tribology. To understand the tribological behaviour, it is necessary to have a knowledge of physics, chemistry, metallurgy, and mechanics, which makes tribology an interdisciplinary science. Tribology deals, in particular, with friction, wear, and lubrication. Friction is defined as a body's resistance to

DOI: 10.1201/9781003243205-10

a movement against another body. Friction is a system response in the form of a reaction force, not a material property which is mainly influenced by the parameters such as temperature, moisture, load, mechanical characteristics, and surface topography [2].

The interaction between the two contacting surfaces causes tribological wear, which results in the gradual removal of the surface materials, i.e. material loss. Wear of the materials in contact is a system parameter, just as friction. The shearing contact between the asperities of two solids in relative motion causes adhesive wear [3]. The asperities deform elastically and plastically during sliding, resulting in a contact area where the bonding forces provide high adhesion and the surfaces are welded together. When the tangential relative motion causes a separation in the bulk of the asperities in the softer material rather than the interface, adhesive wear occurs. When one of the surfaces in contact is significantly harder than the other, abrasive wear develops, causing severe plastic deformation of the surface material [4]. Fatigue wear is vital in dies and instruments that are loaded regularly, such as rolls. The surface of loaded tools is compressed, and shear stresses are generated underneath the surface. Microcracks form at a site of weakness, such as an inclusion or second-phase particle because of repeated loading and unloading. Which the crack reaches the critical size, a flat sheet-like particle is detached from the surface, which is also known as delamination wear.[5]

10.2 ROLE OF SURFACE ENGINEERING

Friction-related energy losses cost the industrialized world between 5% and 7% of their GDP each year, and friction response for around one-third of the world's energy resources are being used. Every one of hundreds of components manufactured fails due to excessive wear, and it is also estimated that 10% of oil consumption wear is simply used to overcome the friction caused between the components.[6] Tribological contacts account for 23% of the world's total energy consumption; 20% is used to counteract friction, and 3% is used to remanufacture worn components and spare equipment owing to wear. The goal of maintaining greenhouse gas emissions in 2004 is to meet the climate targets of 500 ppm CO_2 in the atmosphere and a 2°C increase in mean surface temperature by 2054 [7]. In the automobile industry, the reduction of friction and wear among components such as the engine and gearbox can lower total fuel consumption by around 6%. This could save around 3000 crores liters per year. On the other hand, the 25 crores tons of CO_2 emission can be decreased when there is a reduction in the friction and wear reduction of automobile components [8].

The surfaces could wear in different ways given the variations in operating conditions. The surface finish and dimensions of the materials may alter and, hence, the material will ultimately fail to meet the customer's needs [9]. For example, abrasive and adhesive wear are the two major wear mechanisms of steel wire drawing dies. If adhesive wear occurs when the wire material attaches to the die surface and minute pieces are detached from the die surface by adhesive shear and fracture, it is important to understand the causes of adhesive wear [10]. When the engine parts are in contact with each other without a proper lubricant, the asperities could cause 'rough patches' on the material surface. Even the low-height asperities can cause the material to become extremely hot when rubbing against the other surface. The piston rings tightly pressed on the cylinder bore could be subject to friction and wear when the piston moves up and down in the cylinder. This is because of high elasticity, high combustion pressure, and side force. In such cases, the wear mechanisms include adhesion, scuffing, and abrasion [11] are dominant in piston rings and cylinder liners. Lubricating conditions, material nature, and surface roughness are the crucial factors that impact adhesive wear. Scuffing occurs at high temperatures when two relatively sliding metallic surfaces come into direct contact with insufficient lubrication.

Tribology could have an impact on factors such as the selection of materials, special structural design, lubrication, metal matrix composite, surface engineering, reducing the weight of the component. To improve tribology properties, surface engineering is one effective and flexible way handled by many industries [12]. Surface engineering has gained a lot of interest in recent years because of the production of low wear and friction phenomena. Modern surface engineering is seen in various industrial areas such as automotive, aviation, power engineering, medical engineering, micro-systems technology, microelectronics, etc. [13]. Surface engineering approaches are mainly used to optimize surface attributes and bulk materials. Surface coatings and surface treatments are two main components of surface engineering. Surface engineering can modify the surface properties like residual stresses, friction coefficients, surface hardness, wear, surface roughness, wettability, and corrosion rate (Figure 10.1).

Over recent decades, several surface engineering technologies have been successfully developed and deployed to minimize friction and protect surfaces from damage [14]. For example, the grinding tools required a high wear performance, which cannot be achieved by conventional materials. In such conditions, high hardness ceramic/diamond-coated materials can be used to improve the tool performance. In many engineering components, wear failures are related to surface properties. Surface coatings could introduce the needed phase composition on the surface of the

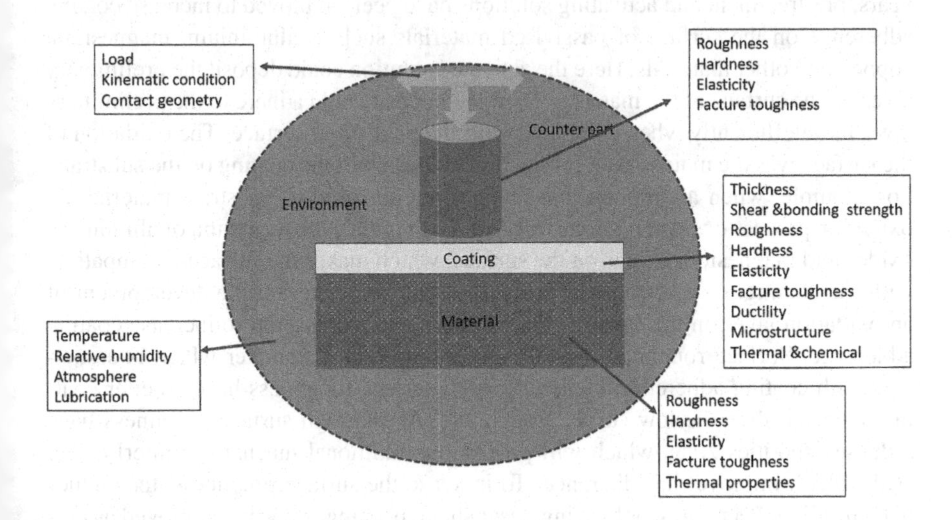

FIGURE 10.1 Influencing parameter of a coating system tribological performance.

materials, which eventually increases the friction resistance, wear resistance, corrosion resistance, and other functional features. The various coating processes include Physical Vapour Deposition (PVD), Chemical Vapour Deposition (CVD), Laser, plasma and thermal spraying, high-velocity oxyfuel, sol-gel, cladding, Chemical solution deposition, Electrochemical deposition, and electroplating were employed in recent years [15].

10.3 SURFACE MECHANISM

Surface atoms can never be confused with their bulk counterparts. The fundamental explanation for this is that atoms in bulk materials are surrounded by other atoms on all sides, while surface atoms are linked to other atoms on just one side. As a consequence, the surface atom transfers the attractive force of other atoms in one direction, resulting in surface energy. Furthermore, surface atoms may attract to any form, such as liquid-phase or gas-phase atoms. This is why the materials oxidized in the environment. When the surface functional properties are insufficient, the ability to connect foreign atoms and molecules is helpful [16].

10.4 IMPORTANCE OF PRE-TREATMENTS BEFORE THE COATING PROCESS

Coating adherence is one of the most serious issues in surface coating, which is mainly influenced by factors such as oxidation, poor surface finish, foreign elements, etc. To increase coating adhesion, the surface of the material must be mechanically cleaned (through polishing, sandblasting, etc.), degreased, and then activated by a physical or chemical technique. In certain circumstances, an undercoat that is intended to accomplish at least one of the tasks (thermal expansion coefficients) in addition to increasing coating adherence to the substrate is necessary [17]. In recent years, pre-treatment and activating solutions have been employed to increase coating adherence on the surface of passivised materials such as aluminium, magnesium, copper, and other materials. Here the activated solution could deposit the preliminary layer on the surface of the material. Then the deposit could adhere to the preliminary layer more efficiently when compared with the passivized surface. The oxidation of the surface was the main reason for the poor adhesion of the coating on the substrate. For example, when aluminium and magnesium are used as substrate material, an oxidation process occurs on the surface. This will induce the formation of aluminium oxides and magnesium oxides on the surface, which makes the surface incompatible with the coating. As a result, it is critical to remove or prevent the development of an oxidation layer on the surface of aluminium and magnesium to increase coating adherence. Surface roughness was the other important parameter which is associated with coating adherence. Generally, high surface roughness have a better coating adhesion then the low surface roughness. As the high surface roughness have a dense asperities zone, which will provide an additional mechanical interlocking aside which improve the adherence. To increase the surface roughness, techniques like Grinding, Peening, sandblasting, laser shock peening, etc were employed as pre-treatment techniques in recent years.

10.5 COATING TECHNIQUES

Surface coating methods are used in various industries due to the enormous diversity of applications. Plasma sprayed coating, cold spray coating, thermal spray coating, plasma electrolytic oxidation, electroless Ni-P coating, and chemical conversion coating are among the coating technologies that are available in recent years (Figure 10.2). These approaches include a variety of parameters that help in the improvement of materials' microstructure, biocompatibility, friction resistance, wear resistance, corrosion resistance, suitability, and durability [18], which is briefly discussed below. Each technique has its uniqueness, advantages, and disadvantages. Among the disadvantages of the processes are: negative heat effects (e.g., distortion, crack, delamination, etc.), atmospheric condition, and poor coating adhesion [19]. This limits the selection of coating material in various technologies. For example, pure ceramic coating can't be done using the Electroless process, which restricts its use in high-temperature applications. In recent years, the protective layer has been made up of a variety of materials, such as metals, ceramics, and polymers. Materials such as Al, Ti, Hf, Zr, Ni, Co, Pt, MgO, ZrO_2, Al_2O_3, Y_2O_3, BeO, PEEK, and PTFE were often utilized in coating technology [18], to improve the resistance towards the friction, wear, and corrosion.

10.5.1 Cold Spray Coating

In cold spray coating, the powder is passed through a de Level nozzle, which accelerates it to reach supersonic velocities. Because there is no heat source (plasma or electrical discharge), the powder remains solid throughout the process. The mechanical

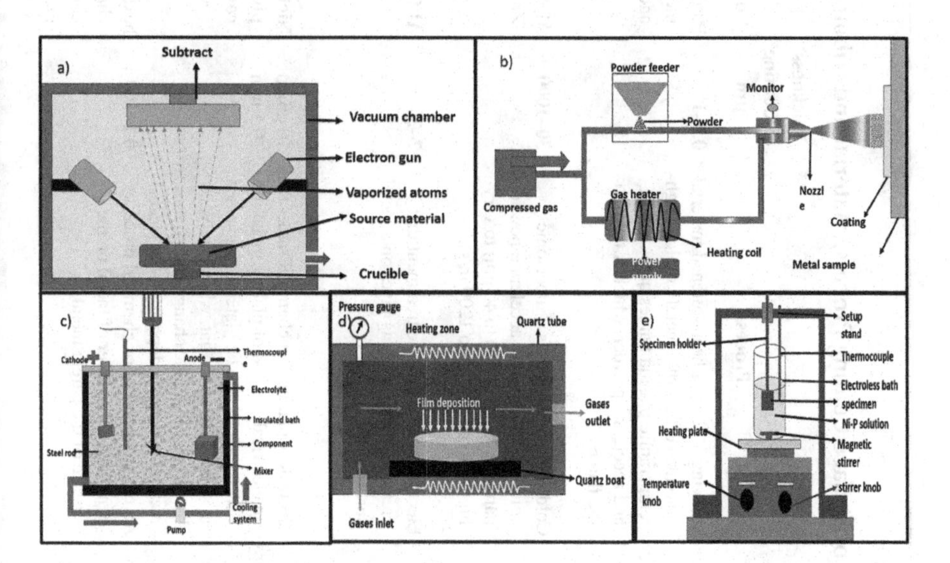

FIGURE 10.2 Schematic of different surface coating processes a) Physical vapor deposition, b) Cold spray coating, c) Plasma electrolytic oxidation, d) chemical vapor deposition, e) Electroless Ni-P coating (Table 10.1).

TABLE 10.1
Summary of Surface Coating Process, Material and Tribological Properties

Surface Coating Process	Process	Thickness of Coating µm	Coated Material	Parent Material	Tribological Properties	Reference
Plasma sprayed coating	Spraying molten or heat-softened material onto the surface of a metal or alloy High-temperature plasma flames quickly heat and accelerate powder, which is fed into the flames	**0.5–1**	Pure metals, alloy metals, ceramics, and composites	All metal, ceramics, some plastics, and carbon fiber composite	Easily worn off when in contact with the counter face, eliminating structural damage while providing adequate erosion and permeability resistance to maintain tight clearance over the engine service life.	[26]
Cold spray coating	Cold spray coating uses a supersonic compressed gas jet to accelerate powder particles (typically 10 to 40 µm) to very high velocities (200 to 1200 m/s).	**30–1000**	Al, Cu, Ni, Ti, Ag, Zn, Ta, Nb	Ti-6Al-4V, Metal alloy	The low-temperature method that yields cold-worked microstructure metallic coatings with minimal porosity, little thermal residual stress, and a high thickness for use in repair applications.	[27]
Plasma electrolytic oxidation	Electrochemical surface treatment that creates an oxide coating layer on a metal surface	**2–20**	**Al_2O_3,**	Aluminium, Titanium, Magnesium, Zirconium	The ability to resist corrosion, as well as its photocatalytic activity and aesthetic appeal when exposed to a variety of environmental stressors, are all taken into account when designing a PEO coating	[28]
Electroless Ni-P coating	Electroless plating is a chemical rather than an electrical method of plating metal in which the piece to be plated is immersed in a reducing agent that, when catalyzed by certain materials, changes metal ions to metal that forms a deposit on the piece.	10 µm to 0.5 mm	**Nickel, phosphorous, copper, ceramics**	Most ferrous and non-ferrous alloy	The corrosion resistance of the highest order, Co-deposits with high hardness and low friction are offered. Good coating durability and low wear are some of the benefits of a uniform deposit. A coating that is self-lubricating. The post-plate finish is eliminated.	[29]
Chemical conversion coating	An electrochemical or chemical process generates a thin layer of metal on the surface of the parent metal.	**0.1–100**	hexavalent chromium compounds, graphite, MoS_2, polymer, Al_2O_3	**aluminium, magnesium, titanium, and zinc alloys**	To protect enclosures and metal parts from corrosion, and to ensure that reliable electrical connections can be made to aluminum chassis and components. It is also an effective pre-treatment for powder coats.	[30]

PVD	At high vacuum conditions solid or liquid materials transfer into the vapor phase than by metal vapor condensation which creates the dense and thin film on the material	0.05 to 100	aluminum, copper, nickel, and zirconium.	most metals, plastics, ceramics, and glass	Components in an IC engine benefit greatly from PVD coatings because it improves wear resistance, scratch resistance, surface roughness, and friction	[31]
CVD	A chemical reaction would generate a deposition layer on the substrate's surface when it was exposed to a mixture of volatile material precursors.	1 μm to 2 mm	aluminium, cobalt, copper, iridium, titanium, tungsten; carbides,	ceramics, glass, metals, and metal alloy	tribological characteristics are improved through abrasion resistance, crack resistance/ coating cohesiveness, and adhesion.	[32, 33, 34]
Electroplating	One metal is negatively charged, while the other is positively charged, Over time, the positively charged metal molecules slowly migrate to the surface of the negatively charged metal, forming a very thin layer.	**0.1 to 85**	Nickel, Chromium, Brass, Cadmium, Copper, Gold, Palladium, Platinum, Ruthenium, Silver, Tin Zinc	**All metal**	Forms a Protective Barrier, Enhances Appearance, Reduces Friction Conducts Electricity, Prevents Formation of Whiskers, Plated surfaces are less susceptible to damage when struck or dropped, which can increase their lifespan. Nickel plating improves performance and reduces premature wear and tear.	
HVOF	Coating materials in powder form are injected into this hot jet stream and partially melted as they exit the nozzle tip.	**70–100**	Tungsten Carbide-Nickel Superalloy, Chromium Carbide, Aluminum Bronze, molybdenum		Wear protection, Low friction surfaces, Corrosion protection, High-temperature oxidation resistance, Improved cohesive strength inside the covering, and a stronger connection to the underlying substrate because of increased impact velocities and smaller powder particles, the as-sprayed surface is smoother. Wear resistance is improved as a result of stronger, tougher coatings.	[35]

deformation of the particles is the mechanism for coating deposition. As a consequence, cold spray is especially suited to metals, which bend and experience adiabatic shear at the particle extremities as a result of the high-speed impact, leading to rapid local temperature spikes. Cold spray coatings provide desired properties such as virtually full density and minimum chemical oxidation. The operation is rather rapid and may be cost-effective. It has been utilized in the past to deposit protective coatings for common metals used in corrosive conditions (e.g. Al, Ta, and Ti). Metallic coatings, on the other hand, are not the ideal answer for wear resistance, as previously stated. This naturally inspired cold spray researchers to look into the possibility of developing composite coatings with a metal matrix, ceramic reinforcements, solid lubricant inclusions. These coatings may be created using the cold spray process, but there are substantial hurdles in tailoring the process such that the coatings have a strong structure and qualities that give high wear resistance. For example, even at 800°C, the wear resistance of cold-sprayed (Fe) Al alloy coating was greater than that of heat-resistant 2520 stainless steel [20]. Hence, the cold spray technique has been used to improving the wear resistance.

10.5.2 Plasma Electrolytic Oxidation (PEO)

Plasma electrolytic oxidation (PEO) is the technique generally used for coating metal alloys. Here, the plasma discharge is created in the metal electrolyte interface, allowing the surface to turn into a dense hard ceramic oxide layer without causing damage to the substrate surface. The Keronite PEO method uses a complex combination of oxide growth, fusing, re-crystallization of the oxide film, and partial metal dissolution at tiny levels to create an oxide layer. Extensive plasma discharges on the surface of Keronite PEO might make it an aggressive process, causing the oxide layer to explode at tiny levels at very high local pressures. The operating temperature for the PEO method is between 12 and 30 degrees Celsius. Additionally, the electrolyte's elemental co-deposition may result in the creation of a ceramic layer on the surface, resulting in crystalline and amorphous phases in addition to oxidation. Compared to hard-anodized coating, the PEO coating was two or three times greater in elastic modulus and hardness. This eventually increased the wear resistance of PEO coating over the hard anodized coating [21].

10.5.3 Electroless Coating

Electroless coating is a technique that uses redox reactions for the deposition of metals, non-metals, and composites on various conducting and non-conducting substrates. The main use of coatings is to reduce friction, wear, and corrosion. In addition, the recent applications of the electroless coating were seen in Microelectromechanical systems, electromagnetic interference, powder metallurgy, membrane reactors, heat exchanger reduction, and bacterial adhesion reduction. Because of their distinctive nodular microstructure (cauliflower), the electroless nickel coatings are smooth and lubricious [22]. However, in recent years, various surface treatments and reinforcement were used to change the microstructure, which eventually increases the friction and wear resistance. Heat treatments at 400°C generally reduce the friction coefficient and wear rate as compared to the as-deposited Nickel coatings [23]. Electroless composite coatings can be distinguished into two groups of tribology: solid lubricant coatings, hard-wear resistance coatings induce corrosion-resistance coatings. Solid lubricants, such as WS_2, MoS_2, PTFE (polytetrafluoroethylene) and graphite [24], are commonly used in electroless composite coatings, which typically have a lower friction coefficient than electroless NiP coatings. The hard wear-resistant composite coatings include the co-deposition of WC, SiC, Al_2O_3, B_4C, and diamond, which has greater hardness and wear resistance compared to the electroless NiP coatings [24]. Similar to other coating techniques, electroless coating has an adhesion issue over the passivized surface. Thus, chemical treatments were devised to increase the adherence of the coating to the substrate. In addition, the post-treatment process after the electroless process has improved the coating adhesion; however, researchers are still finding a way to improve the coating adhesion to make it better.

10.5.4 HOVF

In this process, the fuel (propylene, propane, or hydrogen) is combined with oxygen and burnt in a chamber, after which the products of combustion are allowed to expand with the use of a nozzle, where the gas velocity accelerates to supersonic levels. The powder is delivered into the nozzle along the axis, which is heated and accelerated through the nozzle. During the firing process, the powder is exposed to the state of combustion, where it becomes melted or oxidized.

At The oxidation of metallic and carbide is also conceivable in this process and found to be advantageous to the creation of any metallic or cermet material [25].

10.6 SURFACE MODIFICATION

Surface modification is the act of modifying the surface of a material by introducing physical, chemical, or biological qualities that vary from those present on the material's surface. Surface modification methods are employed in a variety of industries, including automotive, aerospace, power, electronic, biomedical, textile, petroleum, petrochemical, chemical, steel, power, cement, machine tools, and so on [36]. Surface modifications are required for additively made metals to increase mechanical, chemical, and physical qualities such as wear resistance, corrosion resistance, biocompatibility, and surface wettability. The post-treatments cause surface plastic deformation without removing any material, improving the surface and mechanical characteristics of metallic objects. Deformation is induced by the application of loads or shocks, kinetic energy, and thermal energy, which may alter the whole body's surface layer.[37]

10.7 IMPORTANCE OF POST-TREATMENT PROCESS IN COATING

Post-treatments are an important process that improves the efficiency of coatings include: adhesion, hardness, wettability, friction, wear, and corrosion. Generally, peening, heat treatment, cryogenic treatment, and laser treatment process are some

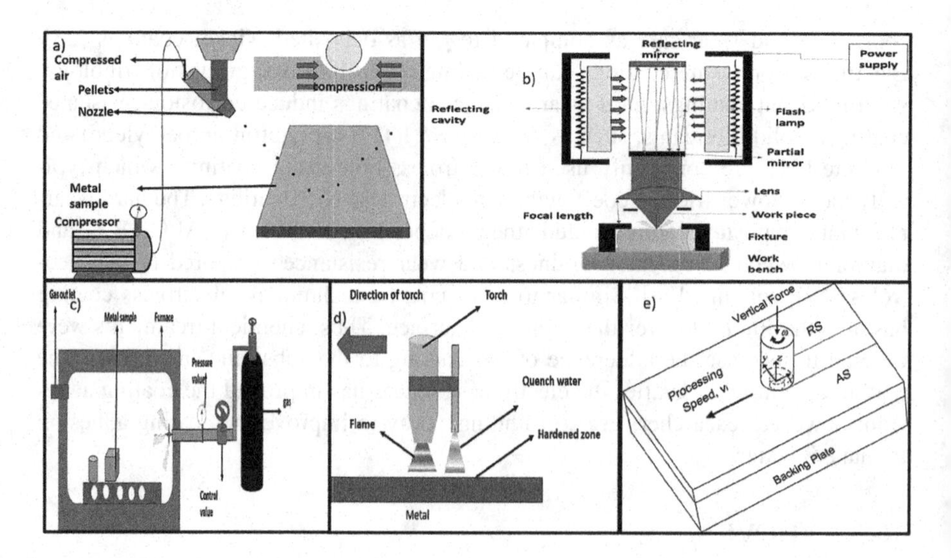

FIGURE 10.3 Schematic diagram of different surface modification processes (a) Shot peening, (b) Laser surface treatment, (c) Nitriding Treatment, (d) flame hardening, and (e) Friction stir processing.

of the Post-treatments process (Figure 9.3). Peening is the process, where the cold work process is employed to induce the compressive residual stress on the surface. The compressive residual stress may cause mechanical deformation, affecting the surface characteristics. Here, the deformation is caused only by the application of stresses [38] and, hence, there is no phase transformation in the surface of materials. In addition, it is observed that, based on the thermochemical diffusion treatments, the coatings can have either compressive or tensile residual stresses [39].

Laser treatments and heat treatments are the processes, which are the surface properties employing changing the microstructure as well as the phase composition. For example, in electroless Ni-P deposit, the phase structure is amorphous and has the NiP phase composition. After the heat treatment process, the formation of metastable phases like Ni_3P and $Ni_{12}P_5$. The formation of these phases was found advantageous in improving the hardness as well as the friction and wear resistance. Similarly, in the laser treatment, the phase composition can be altered effectively compared to the heat treatments.

10.7.1 Shot Peening

Shot penning is a cold working method that modifies the surface and mechanical characteristics of materials by using compressive residual stress. Following the shot-peening procedure, dome-shaped structures emerge on the surface. Shot peening, in general, produces a high level of surface roughness on the surface of the materials. As the diameter or projection speed of the shots increases, so does the degree of roughness, and vice versa. Executing a finishing shot peening operation and optimizing the settings is a straightforward technique to eliminate shot peening-induced roughness. For example, the surface and the tribological reaction of Ti6Al4V alloy shot-peened under varied parameters (shot size and peening duration) is investigated, and concluded the parameter impact on the wear resistance. It helps in processing gears, bars landing gears, and turbine blades in the aerospace and automotive industries. In addition, it is used to process hip implants in the biomaterials industry [40].

10.7.2 Flame Hardening

A gas flame glides over the component's surface, raising the temperature to 850°C, followed by a water quenching nozzle that sprays water over the previously heated surface. There are several fuels available, including acetylene, propane, and natural gas. Another method version involves the flame head being stationary while the component moves at a predetermined rate, or the part spinning in front of the flame head before being quenched, as in spin gardening. The quenching mechanism might be incorporated directly inside the torch. This is the best approach for larger pieces and smaller amounts. The mechanical assessment of treated samples demonstrated a significant increase in microhardness compared to the untreated condition, reaching 76% for the treated sample after two hours at 850°C. Under a load of 10N, the geometry of the wear tracks shows that friction produces main adhesive wear.[41]

10.7.3 Induction Hardening

An alternating current electric current is provided near the component surface using water-cooled, prepared copper coils, generating an electromagnetic field that heats the part surface through eddy currents. Temperatures as high as 850°C have been recorded. At 500 kHz for 0.5 mm depth, low frequencies create a larger depth of heat, while high frequencies provide a smaller depth of heat at 500 kHz for 0.5 mm depth or 1 kHz for 5 mm depth. It generates a shallower hardness layer than, for example, flame hardening and has a shorter cycle time (seconds to minutes), making it suitable for small objects. Fast quench with water, oil, or air using a separate quench ring integrated with the coil assembly. Part complexity to fit a variety of inductor shapes, including pancake and coil (both internal and exterior). The impact of cryogenic treatments on the tribological properties of high-carbon chromium alloy steel after induction hardening The friction coefficient of general power induction with cryogenic treatment is much lower than that of high power induction. Wear resistance and reduced friction may both be provided through cryogenic treatment. A cryogenic treatment boosts abrasion resistance by 34 times. Another significant effect of cryogenic treatment is that the experimental variation in wear depth is reduced, meaning that material organization is more uniform.[42]

10.7.4 Carburizing & Carbonitriding

Carburizing is a thermochemical process in which carbon is diffused into the surface of low-carbon steels to raise the carbon content to an adequate level. Vacuum carburizing is described as a brief immersion in carbon particles from the steel surface layer. The particles arrive due to the reactive influence of the steel's exterior on the carburizing environment. Carbon iotas are digested on the steel surface and subsequently diffuse toward the core as a consequence [43]. The carbon potential, as well as the capacity of the carburizing environment to provide carbon particles as a consequence of heated disintegration of air segments, is substantial. To boost the mechanical conductivity of designed segments, vaporous carbonitriding has long been utilized in combination with the carburizing cycle and other case-solidifying techniques. The vaporous carbonitriding process is often performed in carburizing heaters that include an endothermically determined carrier gas with alkali increases. The cycle temperature varies from 800 to 900 degrees Fahrenheit, which is 50 to 100 degrees Fahrenheit lower than the temperature for carburizing. The inter-diffusion of iron and carbon at the interface affects the carburized layer–substrate adhesion behavior. Tribological and adhesion qualities increased after applying UNSM technology to a carburized heat-resistant KHR 45A steel cracking tube, which may be attributable to the decrease in surface roughness and the increase in surface hardness of both the carburized layer and the KHR 45A steel cracking tube. Because of the hardened surface layer and variance in surface roughness of the surface of the KHR 45A steel cracking tube, both pre-and post-carburizing UNSM treatments increased wear enhancement and scratch resistance.[44]

10.7.5 Nitriding

Nitriding is a heat-treating procedure that introduces nitrogen into a metal's surface to produce a case-hardened surface. The nitriding process starts at the surface of the component with the creation of a compound layer, the composition of which is highly dependent on the steel composition (in particular, the carbon content) [45]. The epsilon and gamma prime phases that develop inside this surface layer are the most common. Each possesses characteristics that improve the mechanical qualities of the casing, and the proportion of each may be adjusted through temperature and gas chemical composition. The thickness of the compound layer is a function of the material (plain-carbon steel forms a thicker compound layer than alloy steels), the temperature, the duration, and the composition of the process-gas mixture used. Nitrocarburizing operations on medium-carbon steels such as 1045 in the pre-hardened stage (quench + temper). In particular, the epsilon (e) phase generated on the surface has a considerable impact on the reduction of the sliding friction coefficient and improvement of the wear resistance of a surface. It was also shown that nitrocarburizing is followed by oxidation, which has minimal influence on the friction coefficient but has a considerable impact on wear resistance. The wear resistance of nitrocarburized specimens is increased even more when they are subjected to oxidation, polishing, and further oxidation treatments. According to studies, ferritic nitrocarburizing of high-performance tool steels has been demonstrated to increase rolling-contact fatigue life, especially in high-stress, minimally lubricated circumstances. The iron-carbonitride compound layer is believed to be deleterious, but the diffusion zone under the surface is thought to be advantageous. M50 NiL is superior to ferritic nitrocarburized M50 NiL and untreated M50 NiL in terms of sliding wear performance. FNC does not influence this performance. The hardness of the diffusion zone and the formation of substantial compressive residual stresses are the primary factors that contribute to the improvement in mechanical characteristics of the diffusion zone [46] (Table 10.2).

10.7.6 Ion Implantation

Ion implantation is a surface modification process that involves injecting ions into a substrate's surface area. These techniques can increase the hardness, wear resistance, corrosion resistance, and fatigue resistance of substrate materials, extending the product's service life. In an accelerator, high-energy ions with energies ranging from 10 to 200 keV are created and focused as a beam onto the substrate's surface [53]. The kinetic energies created by the ions impingement on the substrate are more than four times more than the binding energy of formation of the solid substrate with the surface upon impact. In the near-surface area of any solid substrate, virtually any element may be injected. Plasma source ion implantation is the most current form of implantation. Plasma is used as a source in this type of ion implantation, which is normally stimulated from gas using an RF antenna. Products such as nitrides, borides, and carbides are often used in ion implantation operations [54].

10.7.7 FSP

Friction stir processing (FSP) is an evolution of friction stir welding, which is a method that uses extreme plastic deformation and mechanical mixing to combine metallic materials in a solid state. FSP intends to change the microstructure of the surface layer of a metallic material workpiece, not to combine two pieces of material [55]. FSP has been used to increase the ductility, fatigue resistance, and remove microstructural flaws in a variety of metallic materials and alloys [56]. In the FSP process, a non-consumable spinning tool is plunged to the surface being treated until the tool shoulder touches the workpiece. The lateral translation is then utilized to cover the area of interest. The rotation of tool against the workpiece material delivers a substantial heating and severe plastic deformation in the processed zone. This allows to process the low-melting-point and soft alloys such as aluminium, magnesium, and bronze in recent years [57].

10.8 SUMMARY

The present chapter explore the various surface engineering technologies used to improve tribological behaviour. Physical Vapour Deposition (PVD), Chemical Vapour Deposition (CVD), plasma and thermal spraying, sol-gel, cladding, and electroplating, and electroless are among the commercially employed coating techniques. Each technique has its own unique properties, advantages and disadvantages. Different types of surface modification techniques include flame hardening, induction hardening, laser hardening, ion implantation, cladding, carburizing and nitriding, all of which were used to improve surface properties such as friction, wear and corrosion resistance. Of these techniques, laser cladding and texturing have been widely used to create a complex surface on the substrate. Some of the processes use

TABLE 10.2
Summary of Surface Modification Process, Material and Tribological Behaviour

Surface Treatment Process	Process	Treated Thickness (mm)	Surface Hardness Maximum (μm)	Parent Material	Tribological Behaviour	Reference
Shot-peening	SP is a cold working procedure that uses high pressure to blast the component surface with spherical beads.		481	Al alloy, steels, irons, Ni and Ti alloy	shot peening exerts a beneficial effect on tribological behaviour reducing wear and friction coefficient	[47]
Flame hardening	Heat treatment procedure in which oxyfuel gas flames have immediately impinged on the hardened surface area, which is subsequently quenched.	**1–6**	600	Harden able steels and cast irons	Improved wear resistance	[48]
Induction hardening	Induction hardening is the process of heating a material to a certain temperature using induction heating. A quenching media rapidly cools the temperature after it has been attained. This quick cooling creates a rigid and robust material microstructure.	**0.2–2**	**700**	Hardenable steels and cast irons	Correlation between hardness and wear resistance. The wear resistance of a part increases significantly with induction hardening,	[49]
Laser hardening	The energy from the laser beam is applied directly to the component surface during surface layer hardening. In a small area, the surface layer is heated to the hardening temperature (>1000°C) in a short amount of time.	**0.1–0.6**	**700**	**Carbon and alloy steels, mg, al, Ti**	After laser surface treatment, the specimens' microhardness rises to a higher degree than enhanced wear resistance of laser-treated specimens over untreated specimens	[50]
Carburizing and carbonitriding	Carburizing is a method of hardening metal surfaces while leaving the metal beneath soft. The surface case, a thinner and harder layer of metal, is formed as a result of this.	**0.05–1.5**	**900**	Low carbon steels	Improved tribological performance, scratch resistance, load-bearing capacity, and metal adhesion	[44]

Nitriding	A heat treatment procedure introduces nitrogen into the surface of a metal to produce a case-hardened surface.	**0.025–0.5**	**1200**	Steels Tool steels (hot working and HSS)	Nitriding can increase wear resistance and improve bending and contact-fatigue properties. The ability of the nitrided layer to withstand thermal stresses improves part stability.	[45]
Ion implantation	A surface treatment method in which nitrogen or carbon ions are accelerated and forced to enter a component's surface to impart wear resistance. Electron collisions in a plasma turn nitrogen or carbon atoms into ions, which are then focussed into a stream using magnets and propelled towards the substrate by a voltage gradient.	**0.001**	**1100**	Al alloys, stainless steel, tool steel and die steel, ceramics, and polymers	Reduces sample wear and, to a lesser degree, decreases the friction coefficient.	[51]
Friction stir processing	Pressing a non-consumable tool into the workpiece and spinning it as it is driven laterally through it.	**1–2**		**All lightweight metal**	Enhancement of tribological qualities as a consequence of refining of the hard phases existing in the alloy after FSP modification.	[52]

heat sources to transform the feedstock into liquids and semisolids in the form of particles, droplets, and clusters, while the other techniques rely on the chemical sources. The thickness, microstructure, and functionality of the plasma/cold spray coating depend upon various parameters such as substrate materials, feedstock, and method of deposition. The conductivity was also the important factor to restrict the coating techniques to deposit polymers and ceramics. Multiple-layered coating has been recognised as being the most effective technique for improving the surface properties at extreme environment.

When it comes to engineering materials in terms of quality, performance, and lifecycle cost (among other factors), surface engineering is one of the most essential methodologies to be applied. The surface characteristics of materials have a substantial impact on the serviceability and durability of a component, and, as a result, they should not be overlooked throughout the engineering design process. The surface engineering technique delivers practical answers for challenging circumstances and, hence, surface engineering can be applied in various challenging areas such marine, automobile as aerospace and defence applications.

REFERENCES

1. Dohda, K.; Boher, C.; Rezai-Aria, F.; Mahayotsanun, N. Tribology in metal forming at elevated temperatures. *Friction* 2015, *3*, 1–27, doi:10.1007/s40544-015-0077-3.
2. Nilsson, M.; Olsson, M. An investigation of worn work roll materials used in the finishing stands of the hot strip mill for steel rolling. *Proc. Inst. Mech. Eng. Part J J. Eng. Tribol.* 2013, *227*, 837–844, doi:10.1177/1350650113478333.
3. Surfaces, I.; Relative, I.N. Tribology of coatings. *Tribol. Ser.* 1994, *28*, 33–124, doi:10.1016/S0167-8922(08)70753-3.
4. Dimaki, A.V.; Shilko, E.V.; Dudkin, I.V.; Psakhie, S.G.; Popov, V.L. Role of adhesion stress in controlling transition between plastic, grinding and breakaway regimes of adhesive wear. *Sci. Rep.* 2020, *10*, 1–14, doi:10.1038/s41598-020-57429-5.
5. Swain, B.; Bhuyan, S.; Behera, R.; Sanjeeb Mohapatra, S.; Behera, A. Wear: A serious problem in industry. *Tribol. Mater. Manuf. - Wear, Frict. Lubr.* 2021, doi:10.5772/intechopen.94211.
6. IEA World energy outlook 2021 – Revised version October 2021. 2021.
7. Woydt, M. The importance of tribology for reducing CO_2 emissions and for sustainability. *Wear* 2021, *474–475*, 203768, doi:10.1016/j.wear.2021.203768.
8. Holmberg, K.; Erdemir, A. The impact of tribology on energy use and CO_2 emission globally and in combustion engine and electric cars. *Tribol. Int.* 2019, *135*, 389–396, doi:10.1016/j.triboint.2019.03.024.
9. Nilsson, M.; Olsson, M. Tribological testing of some potential PVD and CVD coatings for steel wire drawing dies. *Wear* 2011, *273*, 55–59, doi:10.1016/j.wear.2011.06.020.
10. Podgornik, B.; Hogmark, S. Surface modification to improve friction and galling properties of forming tools. *J. Mater. Process. Technol.* 2006, *174*, 334–341, doi:10.1016/j.jmatprotec.2006.01.016.
11. Rahmani, R.; Rahnejat, H.; Fitzsimons, B.; Dowson, D. The effect of cylinder liner operating temperature on frictional loss and engine emissions in piston ring conjunction. *Appl. Energy* 2017, *191*, 568–581, doi:10.1016/j.apenergy.2017.01.098.
12. Neville, A.; Morina, A.; Haque, T.; Voong, M. Compatibility between tribological surfaces and lubricant additives-How friction and wear reduction can be controlled by surface/lube synergies. *Tribol. Int.* 2007, *40*, 1680–1695, doi:10.1016/j.triboint.2007.01.019.

13. Meng, Y.; Xu, J.; Jin, Z.; Prakash, B.; Hu, Y. *A Review of Recent Advances in Tribology*; 2020; vol. 8; ISBN 4054402003.
14. Sultana, A.; Zare, M.; Luo, H.; Ramakrishna, S. Surface engineering strategies to enhance the in situ performance of medical devices including atomic scale engineering. *Int. J. Mol. Sci.* 2021, *22*, doi:10.3390/ijms222111788.
15. Shibe, V.; Chawla, V. An overview of research work in hardfacing. *Mech. Confab* 2013, *2*, 105–110.
16. Mozetič, M. Surface modification to improve properties of materials. *Materials (Basel)* 2019, *12*, doi:10.3390/MA12030441.
17. Technology, T.S.; Tucker, R.C. Introduction to coating design and processing. *Therm. Spray Technol.* 2018, *5*, 76–88, doi:10.31399/asm.hb.v05a.a0005725.
18. Fotovvati, B.; Namdari, N.; Dehghanghadikolaei, A. On coating techniques for surface protection: A review. *J. Manuf. Mater. Process.* 2019, *3*, doi:10.3390/jmmp3010028.
19. Uusitalo, M.A.; Vouristo, P.M.J.; Mäntylä, T.A. High temperature corrosion of coatings and boiler steels in oxidizing chlorine-containing atmosphere. *Mater. Sci. Eng. A* 2003, *346*, 168–177, doi:10.1016/S0921-5093(02)00537-3.
20. Khun, N.W.; Tan, A.W.Y.; Liu, E. Mechanical and tribological properties of cold-sprayed Ti coatings on Ti-6Al-4V substrates. *J. Therm. Spray Technol.* 2016, *25*, 715–724, doi:10.1007/s11666-016-0396-6.
21. Malayoglu, U.; Tekin, K.C.; Malayoglu, U.; Shrestha, S. An investigation into the mechanical and tribological properties of plasma electrolytic oxidation and hard-anodized coatings on 6082 aluminum alloy. *Mater. Sci. Eng. A* 2011, *528*, 7451–7460, doi:10.1016/j.msea.2011.06.032.
22. Delaunois, F.; Lienard, P. Heat treatments for electroless nickel-boron plating on aluminium alloys. *Surf. Coatings Technol.* 2002, *160*, 239–248, doi:10.1016/S0257-8972(02)00415-2.
23. Krishnaveni, K.; Sankara Narayanan, T.S.N.; Seshadri, S.K. Electroless Ni-B coatings: Preparation and evaluation of hardness and wear resistance. *Surf. Coatings Technol.* 2005, *190*, 115–121, doi:10.1016/j.surfcoat.2004.01.038.
24. Dong, D.; Chen, X.H.; Xiao, W.T.; Yang, G.B.; Zhang, P.Y. Preparation and properties of electroless Ni-P-SiO_2 composite coatings. *Appl. Surf. Sci.* 2009, *255*, 7051–7055, doi:10.1016/j.apsusc.2009.03.039.
25. Chandra Yadaw, R.; Kumar Singh, S.; Chattopadhyaya, S.; Kumar, S.; Singh, C.R. Tribological behavior of thin film coating: A review. *Int. J. Eng. Technol.* 2018, *7*, 1656, doi:10.14419/ijet.v7i3.11788.
26. Stoyanov, P.; Boyne, A.; Ignatov, A. Tribological characteristics of Co-based plasma sprayed coating in extreme conditions. *Resul. Surf. Interf.* 2021, *3*, 100007, doi:10.1016/j.rsurfi.2021.100007.
27. Cavaliere, P. *Cold-Spray Coatings: Recent Trends and Future perspectives*; 2017; ISBN 9783319671833.
28. Aliofkhazraei, M.; Macdonald, D.D.; Matykina, E.; Parfenov, E.V.; Egorkin, V.S.; Curran, J.A.; Troughton, S.C.; Sinebryukhov, S.L.; Gnedenkov, S.V.; Lampke, T.; et al. Review of plasma electrolytic oxidation of titanium substrates: Mechanism, properties, applications and limitations. *Appl. Surf. Sci. Adv.* 2021, *5*, 100121, doi:10.1016/j.apsadv.2021.100121.
29. Sahoo, P.; Das, S.K. Tribology of electroless nickel coatings – A review. *Mater. Des.* 2011, *32*, 1760–1775, doi:10.1016/j.matdes.2010.11.013.
30. Sivakumaran, I.; Alankaram, V. The wear characteristics of heat treated manganese phosphate coating applied to AlSi D2 steel with oil lubricant. *Tribol. Ind.* 2012, *34*, 247–254.

31. Chauhan, K.V.; Rawal, S.K. A review paper on tribological and mechanical properties of ternary nitride based coatings. *Procedia Technol.* 2014, *14*, 430–437, doi:10.1016/j.protcy.2014.08.055.
32. von Fieandt, L.; Fallqvist, M.; Larsson, T.; Lindahl, E.; Boman, M. Tribological properties of highly oriented Ti(C,N) deposited by chemical vapor deposition. *Tribol. Int.* 2018, *119*, 593–599, doi:10.1016/j.triboint.2017.11.040.
33. Din, S.H.; Shah, M.A.; Sheikh, N.A. Effect of CVD-diamond on the tribological and mechanical performance of titanium alloy (Ti6Al4V). *Tribol. Ind.* 2016, *38*, 530–542.
34. Dobrzański, L.A.; Pakuła, D.; Křiž, A.; Soković, M.; Kopač, J. Tribological properties of the PVD and CVD coatings deposited onto the nitride tool ceramics. *J. Mater. Process. Technol.* 2006, *175*, 179–185, doi:10.1016/j.jmatprotec.2005.04.032.
35. Jafari, M.; Han, J.C.; Seol, J.B.; Park, C.G. Tribological properties of HVOF-sprayed WC-Co coatings deposited from Ni-plated powders at elevated temperature. *Surf. Coatings Technol.* 2017, *327*, 48–58, doi:10.1016/j.surfcoat.2017.08.026.
36. Kim, J.M.; Zhang, X.; Zhang, J.G.; Manthiram, A.; Meng, Y.S.; Xu, W. A review on the stability and surface modification of layered transition-metal oxide cathodes. *Mater. Today* 2021, *46*, 155–182, doi:10.1016/j.mattod.2020.12.017.
37. Muthaiah, V.M.S.; Indrakumar, S.; Suwas, S.; Chatterjee, K. Surface engineering of additively manufactured titanium alloys for enhanced clinical performance of biomedical implants: A review of recent developments. *Bioprinting* 2022, *25*, e00180, doi:10.1016/j.bprint.2021.e00180.
38. Maleki, E.; Bagherifard, S.; Bandini, M.; Guagliano, M. Surface post-treatments for metal additive manufacturing: Progress, challenges, and opportunities. *Addit. Manuf.* 2021, *37*, 101619, doi:10.1016/j.addma.2020.101619.
39. Matthews, A.; Franklin, S.; Holmberg, K. Tribological coatings: Contact mechanisms and selection. *J. Phys. D. Appl. Phys.* 2007, *40*, 5463–5475, doi:10.1088/0022-3727/40/18/S07.
40. Avcu, Y.Y.; Yetik, O.; Guney, M.; Iakovakis, E.; Sınmazçelik, T.; Avcu, E. Surface, subsurface and tribological properties of ti6al4v alloy shot peened under different parameters. *Materials (Basel).* 2020, *13*, 1–22, doi:10.3390/ma13194363.
41. Manuscript, A. ce Ac us pt. 2019, 5.
42. Chang, Y.P.; Wang, H.Y.; Chu, L.M.; Wang, J.C. Effects of the cryogenic treatments on tribological properties of the high carbon chromium alloy steel after induction hardening. *Proc. 4th IEEE Int. Conf. Appl. Syst. Innov. 2018, ICASI 2018* 2018, 1214–1217, doi:10.1109/ICASI.2018.8394507.
43. Rahel, M.; Tadepalli, L.D.; Balubai, M.; Sunil Kumar Reddy, K.; Subbiah, R. Tribological properties on p91 alloy steel treated with normalizing & carburizing process. *Int. J. Eng. Adv. Technol.* 2019, *9*, 2078–2082, doi:10.35940/ijeat.A9573.109119.
44. Amanov, A.; Choi, J.H.; Pyun, Y.S. Effects of pre-and post-carburizing surface modification on the tribological and adhesion properties of heat-resistant khr 45a steel for cracking tubes. *Materials (Basel)* 2021, *14*, doi:10.3390/ma14133658.
45. Bhaskar, S.V.; Kudal, H.N. Tribology of nitrided-coated steel-a review. *Arch. Mech. Technol. Mater.* 2017, *37*, 50–57, doi:10.1515/amtm-2017-0008.
46. Flodström, I. Nitrocarburizing and high temperature nitriding of steels in bearing applications. Thesis, 2012.
47. Vaxevanidis, N.M. The effect of shot peening on surface integrity and tribological behaviour of tool steels. *AITC-AIT 2006 Int. Conf. Tribol.* 2006, 1–8.
48. Khalaj, O.; Saebnoori, E.; Jirková, H.; Chocholatý, O.;Kučerová, L.; Hajšman, J.; Svoboda, J. The effect of heat treatment on the tribological properties and room temperature corrosion behavior of Fe–Cr–Al-based OPH alloy. *Materials* 2020, *13*(23), 5465.

49. Nissan, A.B.; Findley, K.O. *Microstructures and Mechanical Performance of Induction-Hardened Medium-Carbon Steels*; Elsevier, 2014; vol. 12; ISBN 9780080965338.
50. Babu, P.D.; Balasubramanian, K.R.; Buvanashekaran, G. Laser surface hardening: A review. *Int. J. Surf. Sci. Eng*. 2011, *5*, 131–151, doi:10.1504/IJSURFSE.2011.041398.
51. Jin, J.; Chen, Y.; Gao, K.; Huang, X. The effect of ion implantation on tribology and hot rolling contact fatigue of Cr4Mo4Ni4V bearing steel. *Appl. Surf. Sci.* 2014, *305*, 93–100, doi:10.1016/j.apsusc.2014.02.174.
52. Aktarer, S.M.; Sekban, D.M.; Yanar, H.; Purcek, G. Effect of friction stir processing on tribological properties of Al-Si alloys. *IOP Conf. Ser. Mater. Sci. Eng*. 2017, *174*, doi:10.1088/1757-899X/174/1/012061.
53. Liu, W.; Man, Q.; Li, J.; Liu, L.; Zhang, W.; Wang, Z.; Pan, H. Microstructural evolution and vibration fatigue properties of 7075-T651 aluminum alloy treated by nitrogen ion implantation. *Vacuum* 2022, *199*, 110931, doi:10.1016/j.vacuum.2022.110931.
54. Kennedy, D.; Xue, Y.; Mihaylova, E.; Kennedy, D.M.; Xue, Y.; Mihaylova, E. Current and future applications of surface engineering. *Eng. J.* 2005, *59*, 287–292.
55. Mishra, R.S.; Ma, Z.Y. Friction stir welding and processing. *Mater. Sci. Eng. R Reports* 2005, *50*, 1–78, doi:10.1016/j.mser.2005.07.001.
56. De, P.S.; Mishra, R.S.; Smith, C.B. Effect of microstructure on fatigue life and fracture morphology in an aluminum alloy. *Scr. Mater*. 2009, *60*, 500–503, doi:10.1016/j.scriptamat.2008.11.032.
57. Lorenzo-Martin, C.; Ajayi, O.O. Rapid surface hardening and enhanced tribological performance of 4140 steel by friction stir processing. *Wear* 2015, *332–333*, 962–970, doi:10.1016/j.wear.2015.01.052.

11 A Review on Tribological Investigations for Automotive Applications

Vipin Goyal, Pankaj Kumar, Pradyumn Kumar Arya, Dan Sathiaraj and Girish Verma
Indian Institute of Technology, Indore, India

CONTENTS

DOI: 10.1201/9781003243205-11

11.1 INTRODUCTION

Tribology analyzes interacting exteriors at the user interface of two substances in their relative motion and produces friction at the rubbing user interface. As a result, heat energy is made between two meting sides which triggers wear on either one or both substances [1]. It illustrates the consequences of friction, wear, lubrication, and circumstances that influence an automotive system's energy losses. The word 'tribology' derives from a Greek word, which means 'the study of objects that rub' [2]. Automotive manufacturing is rapidly expanding, with a global market revenue of almost US$1,700 billion during 2016 [3]. Modernization in automotive manufacturing is driven primarily by growing concerns regarding environmental impact, energy efficiency, and safety.

Consequently, automobile manufacturers, research associations, and component providers are consistently engaged in broad research and development, with the expectation that the latest innovations will make them distinctive in an ever more competitive market [4]. In the USA, the Environmental Protection Agency (EPA) has placed a benchmark for new automobiles to achieve 36 miles per gallon by 2025 [5]. Minimizing energy shortfalls from automobiles is an essential factor in enhancing gasoline efficacy and reducing the quantity of gasoline needed to drive an automobile. Almost ~65% of the energy wasted in an automobile is dropped from the engine systems, meaning that an enhancement of the operation of engine systems appears to be the standard position to begin when seeking to improve fuel productivity. One specific approach to strengthening engine systems' energy efficiency is to increase their tribological functioning [6]. Tribological components of engine systems include pistons, clutches, bearings, gears, and transmissions systems. Enhancing the tribological functioning of engine systems can offer reduced gasoline and oil consumption, a fall in exhaust emanations and automobile maintenance, increased engine power output, and superior sturdiness. The tribological operation of engine systems can be enhanced in three distinct ways: coatings of body surfaces; the use of lightweight material for mechanical systems; and proper lubrication in automotive systems.

11.1.1 Automotive Friction

Lateral resistance toward the movement between dual contacting objects is called friction. The significance of its resistance is a function of the objects' external characteristics, resources, and geometries, including their ecosystem and operational conditions [7].

Minimizing the friction of automotive components is advantageous in boosting the effectiveness of automobile performances. Friction rises along with an increase in load and component's surface roughness, which can be reduced using a suitable lubricants [8]. When the two body comes in contact generally two laws of friction are

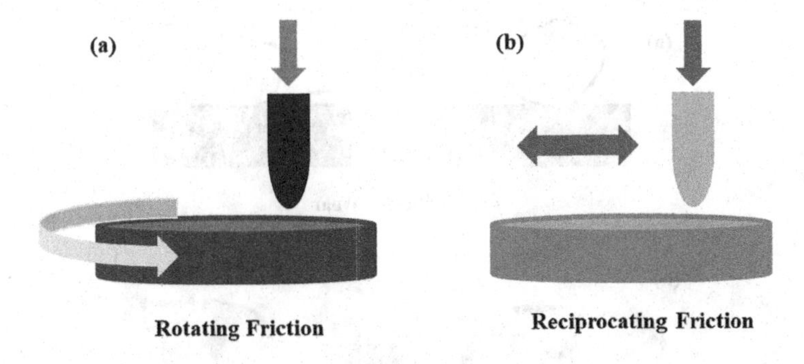

FIGURE 11.1 Schematic depiction of rotating and reciprocating friction.

applied. According to the first law, a frictional force is directly proportional to the normal applied load independent of the nominal contact area. The second law states frictional force is directly proportional to the normal area, which provides the following equation as $F = \mu N$, where F is the frictional force, N is the normal load, and μ is the coefficient of friction (COF). The frictional force is recorded by utilizing an apparatus called a tribometer. The coefficient of friction is measured by multiplying the normal applied. Rotating and reciprocating friction are shown in Figure 11.1.

11.1.2 Tribological Wear

The deficiency of materials arises due to skidding over the two-contact surface is called wear. Usually, wear headed to intensify friction and eventually to the failure of a subsystem, which is undesirable. Wear is typically curtailed by employing an appropriate lubricant to split the two rubbing contact bodies. The popular forms of wear are abrasive wear and adhesive wear [9].

A more rigid substance strip off a substance from a weaker one is called abrasive wear. When two contact substances hold on to other substances locally, the aim of substance is transported from one surface to another surface is called adhesive wear. The process of removing unevenness on the more rigid material by ploughing action is known as two-body abrasive wear, whereas removing asperity through ploughing action with the help of wear debris is referred to as three-body type abrasive wear. These two popular forms of wear are illustrated in Figure 11.2. Wear quantity can be evaluated either with regard to the amount of difference or the dimension of a worn region of the wear surface using optical microscopy or a profilometer.

11.1.3 Automotive Lubricants

Lubricants are employed to split up two sliding planes in order to curtail friction and wear. It also transfers heat and impurities away from the surface interface. Lubricants are frequently deployed in liquid form, comprising oils and additional chemicals for the improved functioning of particular purpose in automobiles [10]. The performance of lubricated surfaces is mainly defined through viscosity. Due to this parameter, it provides resistance to shear inside the fluid so that the resistance offered by

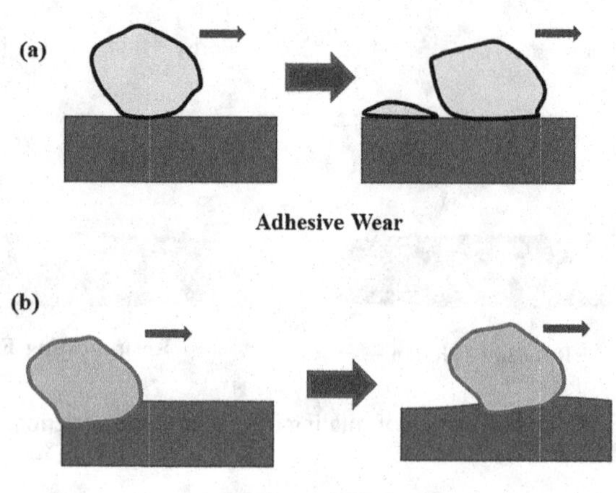

FIGURE 11.2 Illustration of abrasive and adhesive wear.

viscosity is less than the resistance that occurred due to friction during the dry sliding of surfaces [11]. The magnitude of viscous friction is a function of fluid viscosity and working situations. Figure 11.3 illustrates the Stribeck curve of lubrication regarding the effect of various loads, relative speed, and viscosity on surface friction. It recognizes three primary schemes: boundary lubrication, mixed lubrication, and full-film lubrication. When fluid cannot sustain the applied load at low viscosity and speed or higher load, direct surface contact is created called a boundary lubrication scheme.

When fluid surface splits into two surfaces at higher speed, viscosity or low applied load is known as a full film lubrication scheme. This is also referred to as a hydrodynamic lubrication scheme. A surface interface is separated through fluid

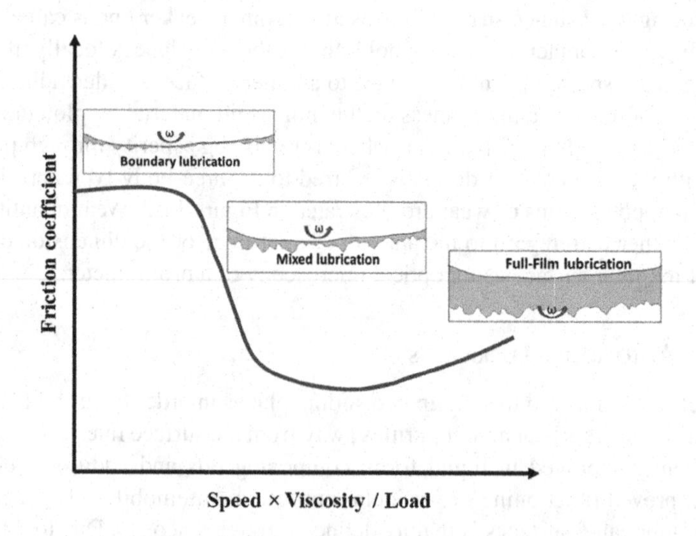

FIGURE 11.3 Schematic illustration of the Stribeck curve of lubrication scheme.

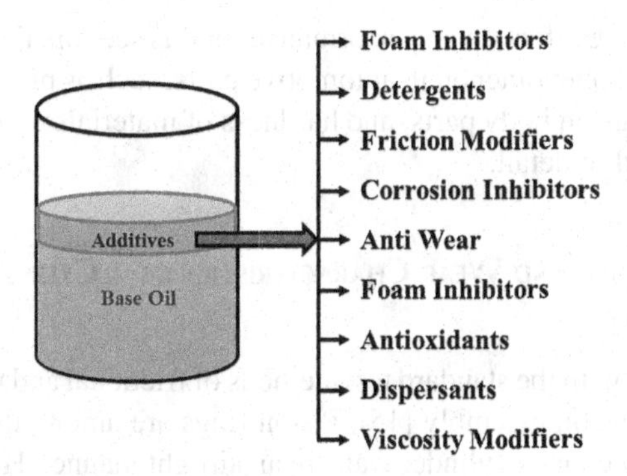

FIGURE 11.4 Formulation of lubricants.

between the boundary lubrication scheme, and a hydrodynamic lubrication scheme is called a mixed lubrication scheme [12]. Figure 11.4 describes the formulation of lubricants with amalgamations of synthetic base oils or minerals. Lubricants are prepared to meet the requirements of a broad range of applications whose preparations start with amalgamations of synthetic base oils or minerals. Synthetic base oil is derived from plants, animals, natural gas, and crude oil, reinforced through chemical-based additives to enhance their optimal functioning [13].

11.2 TRIBOLOGY PHENOMENON IN AUTOMOTIVE APPLICATIONS

Tribology in automotive applications is about contact surfaces, component materials, lubrication, and the interaction which governs their wear and frictional properties. The working condition of an internal combustion (IC) engine relies on tribological aspects of 100–300 moving parts, including engine systems working in the range of 2000–60,000 hours per year in applications, ranging from a commuter vehicle to more extensive commercial, automotive vehicles [14]. Friction remains a ubiquitous phenomenon in the automotive industry. For example, many issues that have a crucial impact on fuel consumption optimization are related to frictional phenomena, such as energy dissipation due to aerodynamic drag and the heating of moving engine parts. Furthermore, the durability of engine components can be severely compromised if effective lubrication treatments do not minimize the effects of friction. On the other hand, braking without friction would be hardly conceivable. Good adhesion properties between the road and the tires are necessary to guarantee adequate safety standards under all possible environmental conditions.

11.3 APPLICATION EXAMPLES OF TRIBOLOGY IN AUTOMOTIVE

Automotive vehicles consist of engines, chassis, bodies, electrical equipment. The engine contains fuel supply, lubrication, cooling, ignition, starting system. Tribology behavior plays a significant role in the performance of automotive bodies under motion. Some of the tribological outcomes are beneficial for the automobile, whereas other have an adverse effect on its performance. In this section, the application example of a tribological phenomenon in automotive parts, such as piston, cylinder, clutch, adhesion of coatings on body parts, and hardness of materials used in the automotive parts, are discussed in detail.

11.3.1 Frictional and Wear Characteristics of a Cylinder and Piston Assembly

ASTM G181 deals with the standard test methods of frictional and wear characteristics of a cylinder and piston assembly [15]. Piston rings are among the essential parts of an engine, and they contact cylinder walls in an airtight manner. High mechanical and tribological performance is needed for piston rings and cylinder surfaces at elevated temperatures. The fuel-saving and the efficiency of compression ignition (CI) and spark ignition (SI)-based engines are impacted by friction due to the presence of several moving components. However, no standard measurement exists for IC engines.

In contrast, one-half of the frictional losses in IC engines is estimated to be occurring due to friction between liner and ring interface [16]. Due to variations of engines system and their working environment, standard test methods for calculating the energy losses in piston assembly permit the operators to select applied loads, different speeds, suitable lubricants, and the testing duration as per their convenience. Variables parameters, such as normal force, oscillation speed, testing temperature as well as duration, length of stroke, sample preparation of specific method, and suitable lubricants, can be adapted for the assessment of energy losses [17].

11.3.2 Friction of Different Clutches

Measurement of coefficient of friction and their reporting are done in accordance with ASTM G115-10 standard, which also provides a manual for performing an applicable test for the frictional coefficient measurement of tribo-systems [18]. Apart from the negative effect of friction, some components require very high friction values for efficient working. Clutch is an essential component for coupling shafts transmitting torque in various mechanical structures. In the automotive industry, different clutch plates are used to transmit motion without unnecessary slipping [19]. The friction generated between the transmission and the engine delivers the required force to move the automotive vehicles. During the engagement of clutch plates, the frictional plate is inserted between the engine flywheel and a steel pressure plate through a fastener mechanism. In the event of there being insufficient pressure among pressure plate and flywheel, a frictional plate may slip; accordingly, the automobile will not operate appropriately. Wear of friction lining materials occurs due to friction at the running surface of the clutch plate. Commonly, it is believed that the frictional wear of the operating clutch surface is directly proportional to the frictional workforce, which can be understood through Equations (11.1) and (11.2).

$$S_L = \frac{V_w}{S_f f_w m_w} \tag{11.1}$$

$$V_w = A_s t_f \tag{11.2}$$

Where, S_L is a strength of frictional clutch surface, V_w is the wear volume, S_f is the specific wear constraint of frictional lining material, f_w is a frictional workforce of a clutch in a single coalition, m_w is the number of starts per hour of the clutch plate, A_s is operating friction of clutch plate and t_f is the thickness of a frictional lining [20].

11.4 FACTORS AFFECTING TRIBOLOGICAL PERFORMANCE IN THE AUTOMOBILE

The tribological performance is affected by a combination of variables, including temperature, pressure, lubricants, roughness, hardness, wear mechanism, relative movements, and materials, etc. [21]. Enhanced tribological performance of an engine part can result in lower fuel consumption, a higher output power of the engine, low oil consumption, lower emissions, better durability, and lower car maintenance. There are three methods for improving tribological performance: (i) improving the tribological features of vehicle component materials; (ii) employing surface coatings; and (iii) using energy-efficient lubricants.

11.4.1 Energy Losses in Vehicles

IC engines are essential parts of vehicles because they transform the chemical energy of fuels into mechanical energy and allow pistons to move from top to bottom. Friction arises at the cylinder–piston interaction, resulting in frictional heat and approximately 60 to 65% of fuel energy loss in the form of heat [22]. Several forms of accessories, including air conditioning, wipers, and power steering, consume around 3 to 5% of the energy generated by the engine. Approximately 4 to 6% of energy is lost primarily on transmissions and other drive shaft components. However, some newly developed technologies in transmission systems, such as automatic manual transmissions and continuously variable transmission techniques, have reduced these losses significantly [23]. Aerodynamic drag in automobiles wastes about 2 to 4% of energy which can be substantially reduced by utilizing a smoother exterior surface [24]. Some energy losses are due to vehicle inertia when traveling forward, and lightweight materials with ceramic constituents could be employed to minimize friction loss and wear.

11.4.2 Energy Losses in the Engine

The engine system is accountable for more than 60% of total losses in an automobile [25]. Those losses comprise cylinders' energy loss of 20 to 25%, exhaust system losses of 25%, cooling losses of 10%, and friction losses of 10 to 12%, depicted in Figure 11.5 [26]. Because of such losses, approximately 30 to 40% of an engine's total output power might potentially be utilized. The total output power within engines could be increased by 10 to 15% by minimizing cylinders' energy loss, exhaust system losses, and friction losses [27]. Friction losses account for about 5 to 10% of

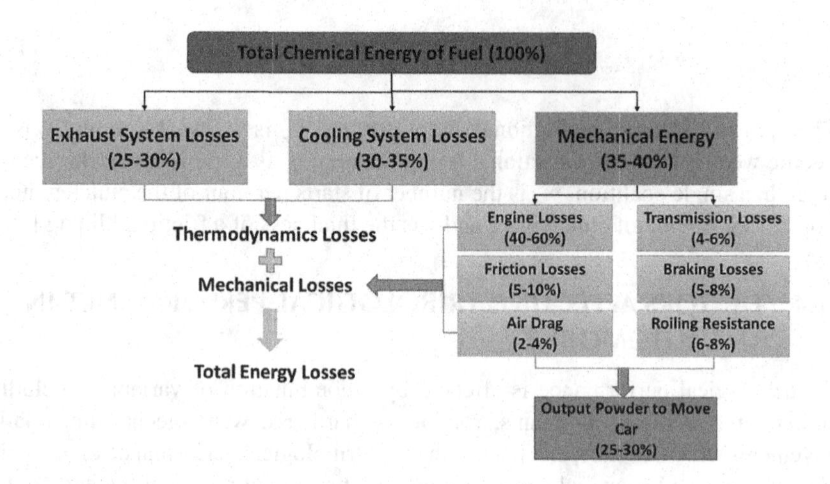

FIGURE 11.5 Energy losses in an engine.

overall power losses in an engine [28]. The critical components of a vehicle engine that contribute to mechanical friction are piston assembly, bearings and crankshaft mechanism, and auxiliary valve drive.

11.4.3 Energy Losses in Lubricants

Lubricants may significantly improve the energy efficiency of automotive engines. Utilizing an energy-efficient commercial lubricant can save 4 to 8% of energy, resulting in lower carbon emissions. Lubricants produce a thin coating between two meeting surfaces of a vehicle, preventing them from rubbing against each other and wearing [28]. After an extended usage period, the lubrication tends to degrade or become inefficient and needs to be replaced. The depreciation of lubricants is caused by several variables, including a thermal breakdown of lubrication, oxidation, micro-dieseling, contamination, and additive depletion. The oxidation of lubricants promotes corrosion and rust in automobile parts. The thermal stability of lubricants is related to their temperature; therefore, as the lubricant gets heated, its thermal stability decreases.

11.5 AUTOMOTIVE TRIBOLOGY

11.5.1 Engine

The enhancement of engine performance, as well as their efficiency, are primary requirements for automotive industries to manage greenhouse emissions and combat climate change. Therefore, the performance and efficiency of an engine may be increased by improving the fuel efficiency of automobiles while minimizing the various energy losses from the engine. Lower mechanical and thermal efficiency is the primary disadvantage of IC engines since up to 60% of the fuel energy is lost through friction and heat energy. According to various research, it is found that IC engines operated by motor vehicles are responsible for between 60 and 70% of worldwide

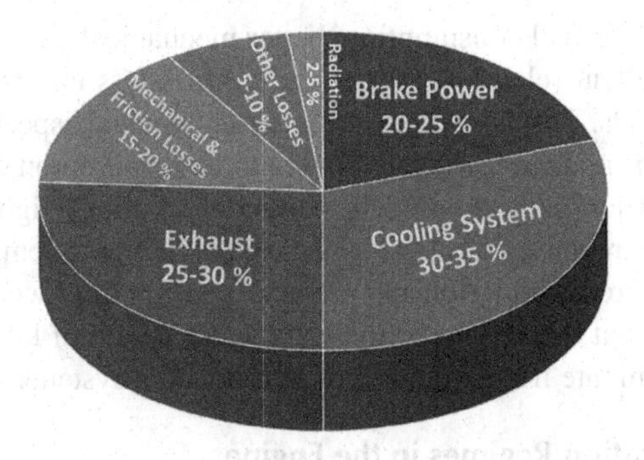

FIGURE 11.6 Percentage fuel energy distribution in an IC engine.

air pollution [29]. Figure 11.6 depicts the distribution of the fuel's energy of the IC engine in the most extensive context.

The enhancement of tribological performance of the engine may result in reduced exhaust emission and fuel consumption by employing suitable materials with excellent tribological properties as well as an improved lubrication system.

11.5.1.1 Importance of Engine Tribology

In the last 10 to 15 years, engine tribology research has progressed rapidly, with piston assemblies, valve train parts, and engine bearings receiving particular attention as the primary frictional components in vehicle engines. Figure 11.7 exhibits percentage frictional and mechanical losses distribution in an IC engine.

Engine tribology plays an essential part in developing, researching, and quality standards of several components for IC engines. The piston assembly is the most significant source of frictional losses in IC engines; however, IC engines' energy efficiency can be increased by supplying sufficient lubrication between the cylinder and piston. Engine frictional losses due to a piston sliding against the cylinder wall is approximately 60 to 70% of overall frictional losses. In contrast, engine bearings, valve trains, gears, and transmission lines are 30 to 40% of overall frictional losses. Tribological studies are needed to diminish wear and friction for all engine rotating parts by providing proper lubrication. Improved tribological effectiveness of IC

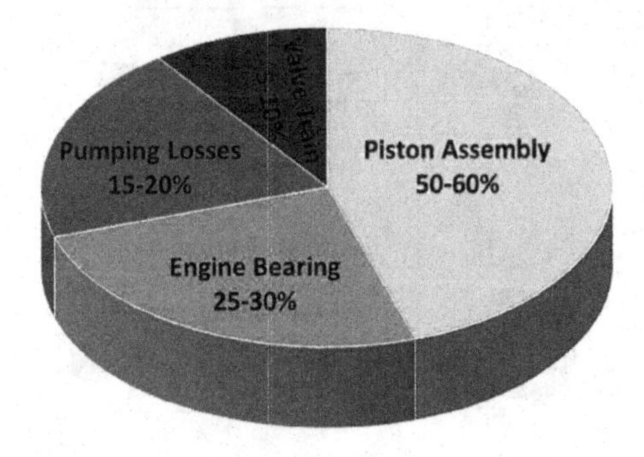

FIGURE 11.7 Percentage frictional and mechanical losses distribution in an IC engine.

engines results in less fuel consumption, better mechanical power, less harmful pollutants, more excellent reliability, better engine life, less engine maintenance, and better service periods. Tribology has multiple fundamental aspects, including Tribo-system, Tribo-film, and the Stribeck Curve, to evaluate automobiles' friction and wear behavior [20]. The tribological coatings of various lightweight and high-strength materials are another critical aspect of the Tribo-system. Tribological coatings can be used to reduce friction and protect elements from wear. Tribo-films are a solid surface film that occurs due to the chemical reaction of lubricant surfaces and help reduce or eliminate friction and wear in lubricated systems.

11.5.1.2 Lubrication Regimes in the Engine

Lubrication is the process of minimizing wear and friction between movable contact areas by providing a friction-reducing layer between them. This friction-reducing layer, often referred to as a lubricant, might be a solid, liquid, or plastic substance, the most prominent of which are oil and grease. Lubricants perform several functions: minimizing wear and friction; reducing corrosion; maintaining temperature and heat; transmitting power; and establishing a liquid sealing. Lubrication regimes specify the nature of lubrication film and are classified into three categories, boundary, mixed, and full film, as shown in Figure 11.8. Boundary lubrication is related to metal-to-metal interaction and occurs when the equipment is started and stopped frequently, as well as under heavily loaded conditions. There are two types of full-film lubrication known as hydrodynamic and elastohydrodynamic lubrication. Several parameters, such as load, speed, motion, and temperature, affect the engine's tribological performance, which must be analyzed before the lubricant selection. The first parameter of the tribological system is a type of motion such as sliding or rotary motion. When motion between two contact surfaces is sliding, hydrodynamic lubrication theory would be used for analysis, whereas elasto-hydrodynamic lubrication theory would give better results for rolling motion.

The tribological system's second parameter is speed, classified into three categories: slow, moderate, and fast. The third tribological variable is temperature, which is extremely important because the work efficiency of the lubricants varies with the temperature changes. A tribo-engineer can precisely identify a lubricant that will provide the best possible operating performance by determining the temperature of

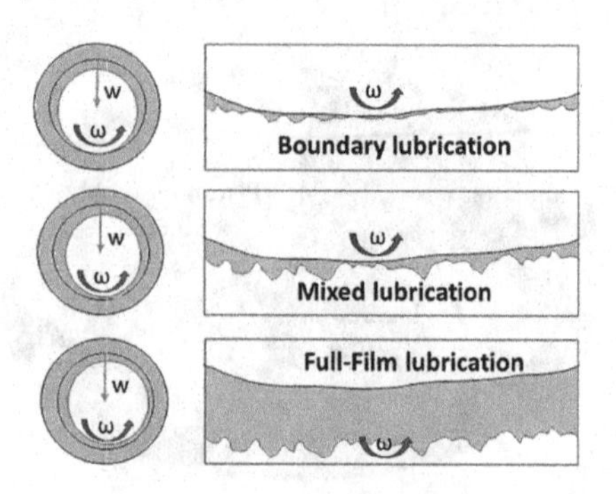

FIGURE 11.8 Lubrication regime.

the tribological system. Load is the fourth tribological parameter, and it has a significant impact on the lubrication requirements. The working environment is the last tribological variable, and it also has a substantial effect on lubrication requirements. The tribo-engineer researched various lubricant chemistries to determine the system and its parameters so that the optimal lubricant could be chosen for multiple applications.

11.5.1.3 Engine Bearing

The critical function of engine bearing is to assist and allow the rotating motion of the crankshaft connecting rod and camshaft during the engine function. Incorrect bearing selection and design can cause various issues, ranging from low oil consumption to complete engine failure. One of the most excellent essential features of a bearing would be its higher rate of embeddability, which means it is more likely to catch up foreign particles, preventing them from scoring the crankshaft or floating in the oil film. Plain bearings, often known as sleeve bearings, are the most common and widely used nowadays. Bearing material selection could influence various characteristics such as engine lifespan, corrosion resistance, fatigue resistance, and detonation due to higher load capacities. In recent years, Babbitt, copper-lead alloy, and aluminum have been among the most extensively employed as bearing surface-coating materials to improve automobile engine performance further. The thickness of such a coating applied to the surface of bearing materials is between 0.01 and 0.02 inches. Babbitt is one of the oldest bearing surface coatings, but it is also one of the least efficient and long-lasting nitro and alcohol engines today [30]. On the other hand, copper-lead alloy bearings are particularly resistant to fatigue and highly durable for both race and street engines.

11.5.1.4 Piston Assembly

The piston assembly is the center of the engine system. The piston's primary function is to convert the energy produced by the combustion of gasoline into valuable mechanical energy. The most frequently used piston assembly materials are aluminum alloys and cast iron. Aluminum alloy pistons are commonly used because they have high conductivity, allowing more excellent heat dissipation and heat transfer. The piston assembly system consists of piston ring, cylinder liner, piston, and connecting rod. Piston rings are critical to ensuring that the piston and cylinder valve is adequately sealed. There are two types of piston rings: compressor ring and oil regulator ring. The tribological characteristics of piston rings seem highly complicated due to the broad load, speed, temperature, and lubricant accessibility. All the lubrication regimes (boundary, mixed, and full film) exist at the cylinder wall surface, and piston rings for one operating cycle. The lubrication regime varies from full film lubrication to boundary lubrication as the piston system moves from bottom dead center to top dead center; however, in the middle position, mixed lubrication is sustained. To reduce friction and wear, the piston rings, piston, and cylinder liner are coated, further improving engine performance. Thus, lubrication analysis is needed to understand the complicated relationships between wear and lubrication. Wear occurs due to mechanical interactions of surfaces, and corrosion occurs in the top portion of cylinders, causing engine failure. Wear is also influenced by the surface smoothness of the piston rings and cylinder liner. During engine running, the surface finish of the piston rings, cylinder liner, and piston skirt influences oil retention and decreases scuffing.

11.5.1.5 Valve Train

The primary purpose of the valve train is to transform rotational camshaft movement into linear valve movement so that airflow into the combustion chamber can be controlled. Friction is an essential factor when selecting a valve train for an IC engine. Typically, two types of followers are used: roller follower and sliding follower. Roller follower is the optimum arrangement for valve train compared to sliding follower because roller contact gives a smaller coefficient of friction. Friction losses in the valve system are approximately 5–10% of mechanical losses, which is less than piston assembly. The inclusion of friction modifier chemicals like molybdenum dithiocarbonate into the lubricating oil reduces the friction coefficient in the valve train [31]. The wear of valve seats and their grooves seems to be a significant issue in the valve train that influences engine performance. So, materials for valve seats are chosen that are resistant to wear and corrosion and have high-temperature strength. The materials used for the inlet valves are hardened low alloy steel. Still, the materials used for the exhaust valves are hot hardened stainless steel because they require more excellent corrosion resistance and high-temperature strength.

11.5.2 Transmission and Driveline

11.5.2.1 Transmission

The transmission system consists of clutches and transmission bands that transmit power from the engine. The tribological study is essential for optimizing the vehicle's performance and durability. The primary function of an automatic transmission system is to transfer engine power to the driveshaft and in an actual wheel. It comprises two essential tribological components: steel reaction plates and friction-lined clutch plates. Various clutch plates such as compound clutch plate, band clutch, and torque converter clutch are utilized in automatic transmission systems. Dynamic and static friction torque seem to be essential clutch performance factors because they depend highly on transmission fluid attributes. The transmission fluid significantly affects chatter, stationary holding strength, and the friction connection between clutch materials and transmission fluid. Automobile manufacturers suggest the use of automatic transmission fluids that contain friction modifying compounds to ensure the best transmission performance [32]. The quantity of clutch energy wasted during clutch engagement operation could impact the life of the transmission fluid and the clutch friction materials [33]. Higher substrate temperatures of clutch begin to reduce clutch durability due to excessive power dissipation, but selecting the suitable transmission fluid can extend clutch life. In conclusion, clutch and band friction properties, the viscosity of transmission fluid, fluid shear consistency, fluid oxidation, and wear characteristics of friction material all have a significant impact on the complete performance and endurance of the transmission system.

11.5.2.2 Universal and Steady Velocity Joints

Universal joints have been necessary to deliver energy from the significant driveshaft to the vehicle's tyres. The whole assembly is encased inside a rubber boot and packed with grease, whereas rolling elements, raceway grooves, and cage initiate connection inside the joint. Where grease plays an essential role in reducing the friction of the mating components [34]. When shafts are aligned with the rolling element, no more friction occurs while they are misaligned with its axis during the oscillating motion. This leads to high friction, wear, and the breakdown of common elements due to severe damage. Long-term use of these joints may cause the balls and raceways to wear out, resulting in reduced geometrical quality.

11.5.2.3 Wheel Bearing

Wheel bearings guide the wheel and support shafts and axles by absorbing axial and radial forces. It is designed to resist high loads and high levels of environmental stress by reducing the friction of the wheel. Axial forces are produced during cornering, which acts out on the wheel bearing in the direction of the longitudinal axis. At the same time, radial forces are produced at a right angle to the longitudinal axis because of the rotation of wheels. The depreciation of wheel bearings occurs due to robust axial and radial forces as well as extreme cornering, which affects the service life of wheel bearings. The service life of bearings can be improved by implementing proper lubrication of the rolling element of bearing with lubricating oil and grease. Wheel bearings are responsible for stable driving performance, and it is a significant safety component for automobiles.

11.5.2.4 Drive Chains

Chain drives are used in automotive engines to control the operation of additional components and connect with many other parts such as chain guides and sprockets, as well as tensioners. During the operation of the drive chain, high noise and friction occur, which can be reduced by lubrication. Generally, two techniques are used in lubrication: splash lubrication and jet lubrication. The chain is immersed in the sump, where splash lubrication techniques lubricate it. However, when lubricant impinges on the chain, it becomes lubricated by jet lubrication techniques. Wear on guide surfaces occurs due to chain tension which influences stress between guides and chains. As a result, the wear mechanism escalates on bush and pin contact surfaces, resulting in chain elongation after some time and loss in automobile efficiency.

11.6 ADVANCE DEVELOPMENT TRENDS

11.6.1 Automotive Tribology

Forthcoming aspects of the automotive system tribology in expansion and revolution include technical advancement, low fuel efficiency, eco-friendly system, the better utilization of lubrication, high durability, and profitability. There is a need to address problems such as friction, wear, and fracture. This is very important to improve the life of the automotive component either by modifying surface technology or changing the material, or by using the appropriate lubrication in different operating conditions.

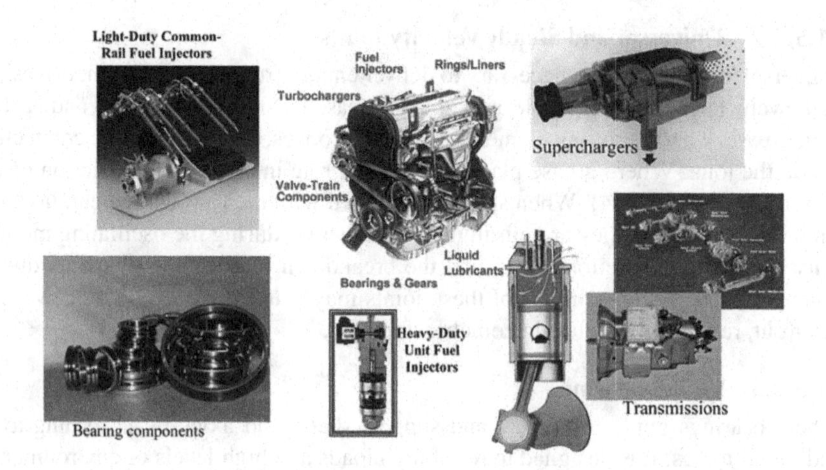

FIGURE 11.9 Various approaches for enhancing the fuel efficiency [1].

Develop a lubricant and additives added to reduce the effect of friction between the two rubbing surfaces, improving the life of automotive components [34]. The additive will enhance the life of the lubricant and make it environment-friendly. It will also be helpful to reduce emissions and improve fuel economy. Additive mixed lubricant is highly compatible with the new ferrous material [35] and does not create any toxic element during the operation. In addition to the numerous methodologies for enhancing fuel efficiency is demonstrated in Figure 11.9. Lubricating oil is the way to maximize an engine's energy efficiency.

11.6.2 Nanotribology Developments and Industrial Needs

Surface adhesion forces dominate the automobile components in lubricated connections as the size of the components varies from macroscale to nanoscale. Nanolubrication plays a vital role in improving automobile performance [36]. It is defined as a lubrication system essential to control the friction, adhesion, wear, and stiction of tribo-surfaces coming into contact at the nanoscale. As a result, nanolubrication must be considered to identify specific mechanisms distinct from those seen in traditional lubrication [38]. When the size of components decreases, the necessity to lubricate and secure the interfaces of these components rises. The authors did a literature survey to discover the industrial requirements for nanotribology or nanolubrication technology innovation. The following criteria have been observed:

Design of nanostructured materials:

a. Develop nanostructured materials with a strength of 5 to 10 times the conventional steel.
b. Use multifunctional materials to minimize wear, friction, and corrosion.
c. Inclusion of nanoparticles or nanomaterials coating to achieve excellent wear resistance and low friction.
d. Materials reinforced with nanoparticles can replace metallic components in automobiles.

11.7 FUTURE RESEARCH WORK

The future growth of the automotive industry will look for an enhancement in the performance of automotive vehicles by improving their tribological phenomenon whose primary focus will be on replacing heavy automotive subparts, large-scale surface design, and mineral-based oil lubricants. A tribological prodigy in an automotive system can be reduced by using lightweight automotive subparts; nanoscale surface designed parts; the use of water gases; reproducible oil-based lubricants with minimum supply; and on-purpose control wear particles and running in surfaces.

11.8 SUMMARY AND CONCLUSIONS

This chapter primarily focused on a comprehensive summary of automotive tribological components, effects of friction, wear, energy losses in automotive systems, and various lubrication characteristics in an engine and drive train, as well as transmission system. The main conclusions derived from this chapter are as follows:

a. In the area of automotive tribology and lubrication systems, along with various lubricants, a lot of special provisions have been made to enhance the performance as well as the efficiency of automobiles. In the upcoming decade, the tribological issues in automobiles will be more complex in response to improving fuel economy and reducing hazardous emissions through automobiles. Thus, the durability and wear resistance of tribological components need to be addressed for efficient engines systems with higher specific outputs.
b. Manufacturing investigators are acquiring lightweight materials for transmission systems and engines. Currently used heavyweight cast iron blocks will be replaced by non-ferrous alloys such as Al- and Mg-based alloy to meet their tribological constraints.
c. This chapter reviewed the recent automotive lubricant improvement involving the analysis of present and projected upcoming requirements of automotive engine oils.
d. In automotive industries, the technologies involved in systems and subsystems are moving from macro- to nanoscale, overshadowed by apparent adhesion forces. Hence, nanolubrication establishes distinctive structures that diverge from conventional lubrication. In this chapter, existing progress and upcoming developments in nanotribology have also been discussed.

REFERENCES

1. Tung SC, McMillan ML. Automotive tribology overview of current advances and challenges for the future. *Tribology International*. 2004;37(7):517–536. doi: 10.1016/j.triboint.2004.01.013
2. Williams JA. Wear and wear particles—some fundamentals. *Tribology International*. 2005;38(10):863–870. doi: 10.1016/j.triboint.2005.03.007
3. Spikes H. Tribology research in the twenty-first century. *Tribology International*. 2001;34(12):789–799. doi:10.1016/S0301-679X(01)00079-2

4. Taylor CM. Automobile engine tribology—design considerations for efficiency and durability. *Wear*. 1998;221(1):1–8. doi:10.1016/S0043-1648(98)00253-1
5. Andersson BS. Paper XVIII (iii) Company Perspectives in Vehicle Tribology – Volvo. In: Dowson D, Taylor CM, Godet M, eds. *Tribology Series. Vol 18. Vehicle Tribology*. Elsevier; 1991:503–506. doi:10.1016/S0167-8922(08)70168-8
6. Wilson B. *History of Tribology* (2nd ed.). Industrial Lubrication and Tribology. 1998;50(6). doi: 10.1108/ilt.1998.01850fae.001
7. *Tribology of Reciprocating Engines* (1st ed.). Accessed November 6, 2021. https://www.elsevier.com/books/tribology-of-reciprocating-engines/dowson/978-0-408-22161-0
8. Akiyama K, Masunaga K, Kado K, Yoshioka T. Cylinder wear mechanism in an EGR-equipped diesel engine and wear protection by the engine oil. *SAE Transactions* 1987; 96:1010–1016.
9. Chen G, ed. 3 – Fundamentals of Contact Mechanics and Friction. In: *Handbook of Friction-Vibration Interactions*. Woodhead Publishing; 2014:71–152. doi:10.1533/9780857094599.71
10. Komvopoulos K. Adhesion and friction forces in microelectromechanical systems: Mechanisms, measurement, surface modification techniques, and adhesion theory. *Journal of Adhesion Science and Technology*. 2003;17(4):477–517. doi:10.1163/156856103605543 84
11. Bhushan B, Davis RE. Surface analysis study of electrical-arc-induced wear. *Thin Solid Films*. 1983;108(2):135–156. doi:10.1016/0040-6090(83)90499-6
12. Bhushan BE, Davis R, Gordon M. Metallurgical re-examination of wear modes I: Erosive, electrical arcing and fretting. *Thin Solid Films*. 1985;123(2):93–112. doi:10.1016/0040-6090(85)90012-4
13. Maboudian R, Howe RT. Critical review: Adhesion in surface micromechanical structures. *Journal of Vacuum Science & Technology B: Microelectronics and Nanometer Structures Processing, Measurement, and Phenomena*. 1997;15(1):1–20. doi:10.1116/1.589247
14. Bhushan B. Adhesion and stiction: Mechanisms, measurement techniques, and methods for reduction. *Journal of Vacuum Science & Technology B: Microelectronics and Nanometer Structures Processing, Measurement, and Phenomena*. 2003;21(6):2262–2296. doi:10.1116/1.1627336
15. Adhesives in Automotive Assembly. Accessed November 6, 2021. https://www.adhesivesmag.com/articles/98072-adhesives-in-automotive-assembly
16. Deepika S. Nanotechnology implications for high performance lubricants. *SN Applied Science* 2020;2(6):1128. doi:10.1007/s42452-020-2916-8
17. Anand A, Haq MIU, Vohra K, Raina A, Wani MF. Role of green tribology in sustainability of mechanical systems: A state-of-the-art survey. *Materials Today: Proceedings*. 2017;4(2, Part A):3659–3665. doi: 10.1016/j.matpr.2017.02.259
18. Korcek S, Jensen RK, Johnson MD, Sorab J. Fuel Efficient Engine Oils, Additive Interactions, Boundary Friction, and Wear. In: Dowson D, Priest M, Taylor CM, et al., eds. *Tribology Series. Vol 36. Lubrication at the Frontier*. Elsevier; 1999:13–24. doi:10.1016/S0167-8922(99)80024-8
19. Taylor CM. Fluid film lubrication in automobile valve trains. *Proceedings of the Institution of Mechanical Engineers, Part J: Journal of Engineering Tribology*. 1994;208(4):221–234. doi:10.1243/PIME_PROC_1994_208_377_02
20. Lu X, Khonsari MM, Gelinck ERM. The Stribeck curve: Experimental results and theoretical prediction. *Journal of Tribology*. 2006;128(4):789–794. doi:10.1115/1.2345406
21. Wang Y, Wang QJ, Lin C, Shi F. Development of a set of Stribeck curves for conformal contacts of rough surfaces. *Tribology Transactions*. 2006;49(4):526–535. doi:10.1080/10402000600846110

22. Gale WF, Totemeier TC, eds. 24 - Lubricants. In: *Smithells Metals Reference Book* (8th ed.). Butterworth-Heinemann; 2004:24. doi:10.1016/B978-075067509-3/50027-0
23. Holmberg K, Erdemir A. Influence of tribology on global energy consumption, costs and emissions. *Friction*. 2017;5(3):263–284. doi:10.1007/s40544-017-0183-5
24. Zabala B, Igartua A, Fernández X, et al. Friction and wear of a piston ring/cylinder liner at the top dead centre: Experimental study and modelling. *Tribology International*. 2017; 106:23–33. doi: 10.1016/j.triboint.2016.10.005
25. Bao H, Zhang C, Hou X, Lu F. Wear characteristics of different groove-shaped friction pairs of a friction clutch. *Applied Sciences*. 2021;11(1):284. doi:10.3390/app11010284
26. Ost W, De Baets P, Degrieck J. The tribological behaviour of paper friction plates for wet clutch application investigated on SAE#II and pin-on-disk test rigs. *Wear*. 2001;249(5):361–371. doi:10.1016/S0043-1648(01)00540-3
27. Doerre M, Hibbitts L, Patrick G, Akafuah NK. Advances in automotive conversion coatings during pretreatment of the body structure: A review. *Coatings*. 2018;8(11):405. doi:10.3390/coatings8110405
28. Grujicic M, Sellappan V, Omar MA, et al. An overview of the polymer-to-metal direct-adhesion hybrid technologies for load-bearing automotive components. *Journal of Materials Processing Technology*. 2008;197(1):363–373. doi: 10.1016/j.jmatprotec.2007.06.058
29. Xu X, Dong P, Liu Y, Zhang H. Progress in automotive transmission technology. *Automotive Innovation* 2018;1(3):187–210. doi:10.1007/s42154-018-0031-y
30. Goyal V, Sharma SK, Kumar BVM. Effect of lubrication on tribological behaviour of martensitic stainless steel. *Materials Today: Proceedings*. 2015;2(4):1082–1091. doi: 10.1016/j.matpr.2015.07.013
31. Broitman E. Indentation hardness measurements at macro-, micro-, and nanoscale: A critical overview. *Tribology Letters* 2016;65(1):23. doi:10.1007/s11249-016-0805-5
32. Giampieri A, Ling-Chin J, Ma Z, Smallbone A, Roskilly AP. A review of the current automotive manufacturing practice from an energy perspective. *Applied Energy*. 2020; 261:114074. doi: 10.1016/j.apenergy.2019.114074
33. Taylor RI, Dixon RT, Wayne FD, Gunsel S. Lubricants & Energy Efficiency: Life-Cycle Analysis. In: Dowson D, Priest M, Dalmaz G, Lubrecht AA, eds. *Tribology and Interface Engineering Series. Vol 48. Life Cycle Tribology*. Elsevier; 2005:565–572. doi:10.1016/S0167-8922(05)80058-6
34. Nagendramma P, Kaul S. Development of ecofriendly/biodegradable lubricants: An overview. *Renewable and Sustainable Energy Reviews*. 2012;16(1):764–774. doi: 10.1016/j.rser.2011.09.002
35. Arya PK, Jain NK, Murugesan J, Patel VK. Developments in friction stir welding of aluminium to magnesium alloy. *Journal of Adhesion Science and Technology*. 2021;1–38. doi:10.1080/01694243.2021.1975614
36. Kishore K, Kumar P, Mukhopadhyay G. Resistance spot weldability of galvannealed and bare DP600 steel. *Journal of Materials Processing Technology*. 2019; 271:237–248. doi: 10.1016/j.jmatprotec.2019.04.005
37. Goyal V, Sharma, SK, Kumar BVM. Effect of lubrication on tribological behaviour of martensitic stainless steel. *Materials Today: Proceedings*. 2015; 2:1082–1091
38. Kumar A, Bijwe J, Sharma S. Hard metal nitrides: Role in enhancing the abrasive wear resistance of UHMWPE. *Wear*. 2017; 378:35–42.

12 Tribology in the Automotive Sector

Sudheer Reddy Beyanagari, P. Kumaravelu, Dhiraj Kumar Reddy Gongati, Yashwanth Maddini, S. Arulvel and Jayakrishna Kandasamy

Vellore Institute of Technology University, Vellore, India

CONTENTS

12.1 INTRODUCTION

Automotive vehicles are classified as follows: petroleum, gaseous, electric, and hybrid electric vehicles. Based on their usage, the automotive sector is categorized into four segments: two-wheelers, three-wheelers, passenger, and commercial vehicles.

DOI: 10.1201/9781003243205-12

The usage of automobiles is slowly increasing to meet the day-to-day requirements of human necessities [1–3]. According to the records of the International Organization of Motor Vehicle Manufacturers (OICA), nearly 4,4401,850 units of vehicles were sold or registered globally in Q1–Q2 2021 with 29% growth as compared to Q1–Q2 2020 [4]. Passenger vehicles and two-wheelers currently dominate the national and international market needs. In India, 231,633 units of passenger vehicles, 9,397 units of three-wheeler, and 1,055,777 units of two-wheelers are sold as per the records of the Society of Indian Automobile Manufacturers (SIAM) in 2021 [5]. The strong growth in sales is attributable to the rise in population and increased buying power.

The combustion of fuels inside the petrol engine leads to pollutants being emitted into the atmosphere. The huge number of petroleum-based vehicles, leading to the emission of CO, CO_2, nitrogen, etc., had led to an increase in atmospheric temperature and thereby resulted in global warming [6, 7]. The reciprocating internal combustion (IC) engine is prone to significant emissions of carbon and releases heat into the environment. Generally, the energy generated within the engine cylinder is transformed into heat, heat for power transfer, and/or energy to resist friction and wear losses, resulting in the tribological interaction of an automobile's moving components. It has been observed that the energy distribution will vary according to engine type and operating conditions [6, 8]. Nearly 60% of the energy is lost as heat, either in the engine or in the exhaust system, depending on the combustion situation (Figure 12.1). Mechanical activities may incur an extra frictional loss of up to 15%, leaving just 25% of the original energy accessible in the form of braking force [3, 9].

To overcome the emission of pollutants and heat energy from the combustion of engines, green energy regulations have been implemented by the governments with the aim of reducing the tribological issues in the automobile industry [10, 11]. One major change is found in the locomotives, where almost 80% of the diesel-run locomotives are being converted into electric motor run locomotives. The US Environmental Protection Agency (USEPA) defines green engineering as the "design commercialisation and use of processes and products those are technically and

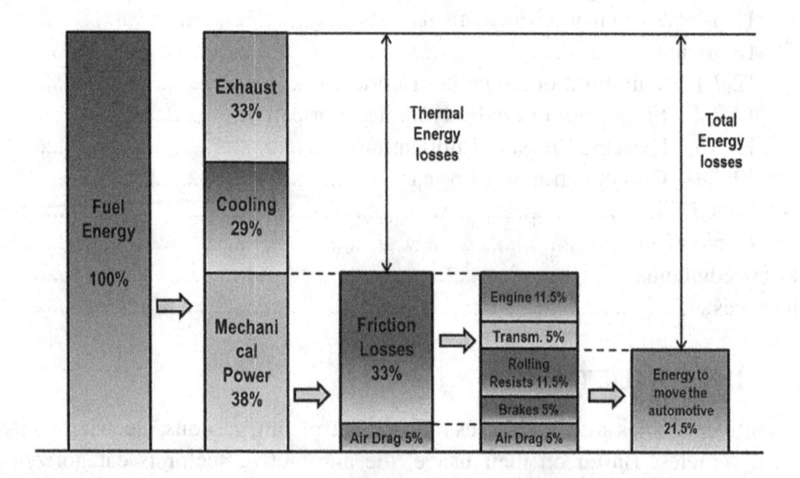

FIGURE 12.1 Fuel energy distribution in an IC engine based automobile.

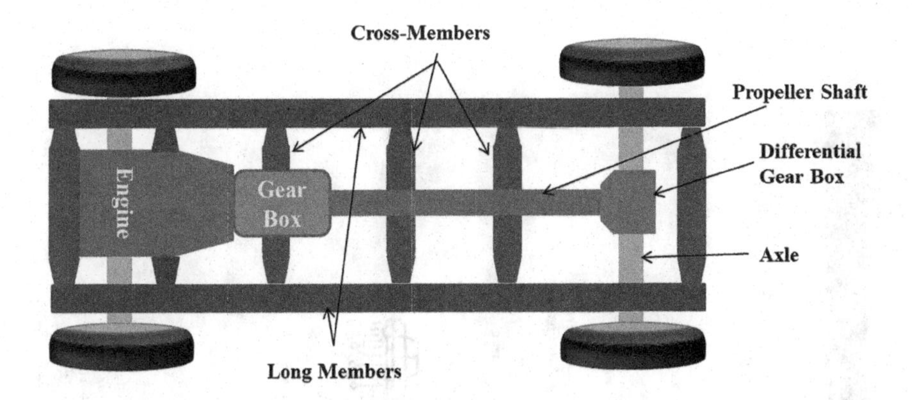

FIGURE 12.2 Bottom view of an automobile layout.

economically feasible by reducing the cause of pollution and risk associated with human health and the environment" [10, 12, 13]. The automobile is built on the concept of vehicle dynamics, durability (warranty), life cycle, damage ability, repairs, and crashworthiness, in spite of the energy used to operate the automobiles or automotive vehicles, whether driven by IC engines or by electrical motors. The layout of an automobile with different subsystems involved in transmission power is shown in Figure 12.2. By contrast, the critical components involved in automobile vehicles driven by IC engines as well as electric motors are presented in Figure 12.3.

Tribology deals with the friction, wear, and lubrication of contacting surfaces while they are moving relative to one another [14, 15]. Accordingly, tribology testing is intended to simulate the conditions encountered in real-world situations. Tribology plays a vital role in the automotive sector because friction, wear and lubrication are commonly dissipated in automotive vehicle [16–19]. Various components of an automotive vehicle will interact with each other, and the life of such components has to be extended by improving tribological characteristics. Automotive tribology starts from the drive connection of prime mover/motor to the road wheels through: (i) the power train (transmission system); (ii) the suspension and braking system; (iii) the steering system; (iv) the electric power and lighting system; and (v) auxiliaries.

Tribology has become an important subject for both the researchers and the manufacturers in the modern manufacturing arena as it is a topic which deals with the moving/running components/bearings regardless of the driving mechanism or prime mover [16]. Almost every moving element must overcome a degree of friction in order to perform its function. It is mandatory to maintain the friction and wear at their optimum levels, and thus bearings have to function with low friction and wear, whereas brakes have to have high friction and low wear. Accordingly, automobiles have varying frictional requirements and low wear. For instance, IC engines should have low friction and wear in order to improve the efficiency. Considering the recent developments in the field of automotive industries, the objective of this chapter is to present the fundamentals of automotive tribology. A detailed discussion is provided on current and past innovations in the motor vehicle industry to improve the tribology characteristics. The primary objective of this chapter is to concentrate on the

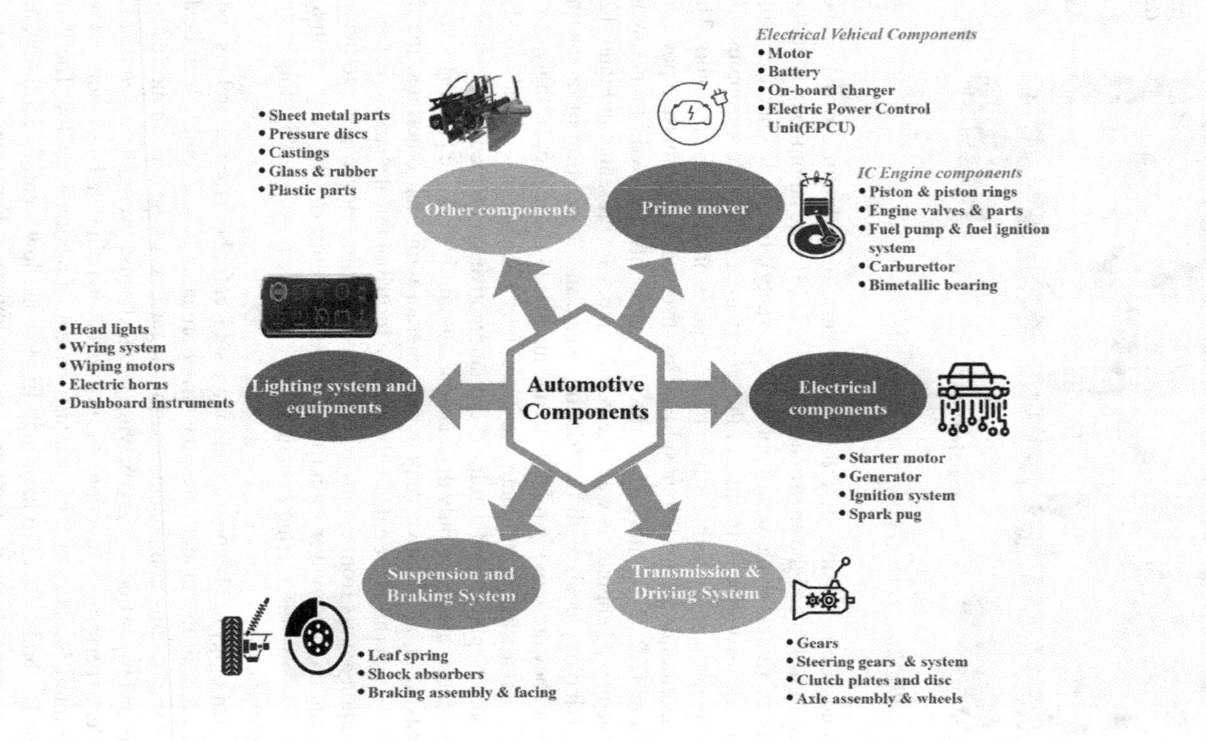

FIGURE 12.3 Components of the automobile.

tribological aspects of the automobile or related vehicles run by IC engines, electrical motor or batteries and hybrid electrical vehicles.

12.2 AUTOMOBILE TRIBOLOGY PHENOMENA

An automobile contains many components, including an engine, chassis, body, and electrical equipment. The engine is equipped with a gasoline delivery system, lubrication system, cooling system, ignition system, and starting system. Today, however, the trend is changed to electric vehicles with combustion inside the engine cylinder is being replaced by electric power in terms of electric motors or batteries to control or reduce environmental pollution [19, 20]. The electrical automobile vehicles (EAV) are run either by Direct Current (DC) motor or battery or fuel cell or fuel cell hybrid systems, something which has attracted researchers and manufacturers over the past decade. This has also led to the advancement of an alternative solution to decrease not only greenhouse gas emissions but also the extent of global warming. However, tribological issues (friction and wear characteristics) can be observed on the reciprocating/sliding components. For example, the contact between the vehicle wheels and the road could cause friction, which is needed to maintain stable movement of the vehicle. Here, in addition to the load and speed, the environment plays a vital role on the degree of friction. This can be seen clearly in the motion of vehicles during rainy seasons. The water molecules between the road and the tyres could act as a lubricant, which reduces the friction and may cause the vehicle to slip.

12.3 TRIBOLOGY AND THE VEHICLE BODY

The vehicle body is built on the chassis according to its size, which is determined by the seating capacity (passenger car) or transport products (a truck) or other requirements of the cabin, such as a 4-wheeler, a 6-wheeler, or other variations. There are many types of vehicle bodies that vary in terms of their function (passengers, cargo-passenger, trucks, and special type), as well as their structure (non-skeleton, skeleton, and half-skeleton) [1, 2]. In addition, it is built to reduce and absorb all of the shocks and bumps that an automobile encounters while being driven on the road. A vehicle's body must be sturdy enough to resist all sorts of forces, including the weight of the automobile, inertia, baggage, and the cornering, and braking forces occurring on it. Stresses that are produced in the body should be dispersed uniformly across the whole body. The body should be as light as feasible in order to maximise performance with a decent amount of fatigue.

When an automotive vehicle travels at high speeds, the air exerts a force on the vehicle according to its height and breadth, thereby resisting aerodynamic forces. During severe weather with strong airflow (floods), an arbitrarily-shaped body has high air resistance, which implies that there is a greater loss of engine power and the body subjected to various environmental conditions. Hence, the body must be designed and built to withstand forces acting in the opposite direction of the drag force as well as the other perpendicular direction of motion (i.e., lift force). The magnitude of friction drag and drag pressure depends upon the shape of the body. For example, a flat plate portion experiences only friction drag, resulting in the bending of the body surface undergoing corrosion or corrosive wear, whereas the flat portion perpendicular to the direction of flow exerts pressure drag which may lead to breakage or cracks in the body. In such situations, the drag and lift force depends on the co-efficient of drag or lift [1, 2].

Various materials, such as nickel, steel, and aluminium alloy sheets, are used in the manufacture of the vehicle body. The prime reason for using steel sheets for the body structure is its inherent capability to absorb impact energy in a crash situation. At present, Al usage in the automotive industry has grown due to its low density and high specific energy absorption performance and good specific strength. The use of Al can potentially reduce the weight of the vehicle body, thereby it helps to enhance the fuel efficiency. Recent developments have shown that up to 50% weight saving for the body is achieved by the substitution of steel by Al alloys (AA) [21]. AA are used for body structures, chassis applications, closures and exterior attachments such as crossbeams, doors or bonnets.

In addition, the vehicle is subjected to shear force in one direction and pressure force in the other. As such, the multidirectional forces acting on the vehicle's body lead to aerodynamic friction, which further resulting in erosive and corrosive wears. Hence, the body is built by considering such technical issues and suitable materials to overcome the wear, corrosion and other environmental conditions, which will improve the tribological properties. Therefore, the automobile body has to withstand the extreme environment operating conditions and often demand highly corrosion- and wear-resistant coatings to ensure the long-term functionality of technical systems. In addition, the thermal spraying of Al_2O_3 via atmospheric plasma spraying (APS) or flame spraying are among the commercially used coatings on the vehicle body. Therefore, coatings and painting are a good way to mitigate such drawbacks due to environmental conditions [22].

12.4 TRIBOLOGY IN THE PRIME MOVER

The prime mover is responsible for delivering the power for all of the numerous operations that the vehicle or any component of it may be required to execute on its own. The prime mover is typically composed of an internal combustion engine, which may be either spark ignition (petrol engine) or compression ignition (diesel engine). Power is generated inside the engine cylinder through the combustion of hot gases and their expansion, which push a piston within the cylinder, converting the piston's reciprocating movement into the rotational motion of a crankshaft connected by connecting rod and thus it is known as an IC engine. The piston is the most critical component of an engine because it makes airtight contact with the cylinder. Thermal conductivity, mechanical performance at high temperatures, excellent running-in, and wear resistance are all important characteristics of pistons and cylinders to develop the required output power. The sectional view of the piston inside the cylinder is shown in Figure 12.4.

Engine tribology is an important factor to minimise fuel consumption, lube oil consumption, vehicle hazardous emissions, service requirements, and longer service lives, while simultaneously improving engine power production, durability, dependability, and engine life, among other things [19]. The motion and the movement of

FIGURE 12.4 Piston assembly and piston ring function of an IC engine.

the piston and its related piston assembly, bearings, valve train etc., are the main components of the engine. In the process of the development of power, the said components or parts do undergo wear and tear; they develop friction over the contact surface, which affect tribological behaviour. Therefore, the study of automotive tribology is important.

Generally, around two-thirds of the friction loss in an engine is caused by piston skirt, piston rings, and bearings (66%); the valve train, crankshaft, gear-box and related gears make up the rest (34%). The movement of the piston surrounded by rings and the skirt of the piston against the cylinder wall contribute a major portion to friction loss in power generation and power train [19]. Good engine performance depends greatly on the piston assembly's ability to reduce friction and wear behaviour. In the area of piston ring or cylinder bore contact, many research possibilities may be explored by developing low friction coatings, wear-resistant coatings, surface texture, and some other surface treatments. Though friction and wear cannot be eliminated, they can be reduced. At present, there is a continued emphasis on improving tribological characteristics by reducing the friction so as to increase the IC engine power output, which is developed inside the cylinder around 700 °C and above temperature. At this junction, the material science is an option available for designers and manufacturers to produce IC engines based on the requirements as per SAE guidelines. In the production process, cast-iron or cast AA is used for engine blocks, whereas grey cast-iron for cylinder head if required power is more, else cast AA is preferred. The pistons are manufactured by forged AA, piston rings, and gudgeon pins by nickel steel, connecting rod by alloy steel or AA and crankshafts is manufactured by forging of medium carbon alloy steels [16, 23, 24].

The cylinder liner surfaces are widely investigated for the better understanding of frictional losses, emissions, and oil consumption. In such a process, surface texture can be used to monitor and manage the control of the production process. Engine cylinder liners with the optimal surface texture help to provide efficient running in, minimal friction, low oil consumption, and good motor engine operating characteristics for effective power and minimal fuel consumption. Linear surface texturing help to improve the tribological properties of a piston ring–cylinder liner frictional pair [25]. In order to reduce the wear on the piston rings, dual-layer coatings in a combination of hard (diamond-like-carbon (DLC) or tungsten carbide (WC)) and soft (epoxy composite) are carried out on the piston rings via physical vapour deposition followed by epoxy/graphene/base oil SN150. During the operation of the engine, the top ring (uncoated) has the lowest wear rate compared to the middle and lower rings of the piston in all the conditions of being coated (DLC/EGNSN & EC/EGNSN) or uncoated. By contrast, the uncoated rings (bulk material) lose more mass than dual-coated rings, resulting in smaller wear volume and hence a lower wear rate. Notably, the dual-coated top rings solely experienced soft layer wear, with no intermediate hard layer or ring bulk material wear [26].

Accordingly, the impact of tribology on energy usage leading to the emission of CO, CO_2 and such toxic pollutants from IC engines worldwide has resulted in the replacement of IC engine runs with petroleum-based fuels or gases. Gear change is a difficult procedure in IC engines. Indeed, the environmental issue and scarcity of petroleum fuels, which resulted in an increase of gasoline prices, led to development of electrically powered automobiles [19, 20]. Whatever type of vehicle runs on the road, tribology plays an important role. The tribology relating to EAVS is explained in the following section.

12.4.1 Triblogy in Electrical Automotive Vehicles (EAVs)

EAVs as an alternative solution not only decrease greenhouse gas emissions but also combat global warming. EAVs use electrical motors for propulsion. The efficiency of electrical motors (EMs) is almost three times that of IC engines, which can be observed from the direct current motor (DCM), and also it is possible to achieve the efficiency with DCM as maximum as possible around 78% from between 40 and 60 KW motors [27]. The other advantages of EMs over IC engines are that the EMs do not produce soot during their working and also the non-contamination of lubricating oils leading to not only environmental care but also extended oil services. The power generated in an automobile has to be utilised by the transmission system from part to part or section to section, to reach road wheels which produces friction [28].

Although EAVs develop substantial higher efficiency in terms of energy consumption, and at the same time more challenges have to be faced. Such challenges can be overtaken by the way of reduction of energy losses developed in EMs and power electronic devices, consisting of battery functioning, cabin heating/cooling and ventilation, air dragging, and friction. It is seen that about 57% of the total electrical energy supplied to an EAV is utilized only to overcome the friction losses [7]. Hence, the friction is to be reduced by improving the tribological properties or development of suitable tribological working conditions.

The power developed is distributed as follows: 1% in the electrical motor (EM), 3% in mechanical transmission loss, 41% in the rolling resistance, 12% in brakes and the rest in friction losses [2, 27]. Hence, OLA Manufacturers have planned to produce two lakhs electric driven two-wheelers per day to replace petrol/gas-run two-wheelers. Subsequently, General Motors has announced that it would stop producing IC engine light motor vehicles by 2035 as part of its plan to become carbon-neutral by 2040. Further, the friction is predominated in either the roller bearings or reciprocating parts and it plays a major role in the transmission of power from the motor to the road wheels, via transmission mechanism. Once the friction is generated, the next effect is the wear between two or more moving contact surfaces. Hence, the reduction of wear is an important aspect since it leads to not only wearing out of the contact surfaces in motion but also generates heat. With suitable lubrication, the reduction of friction and wear may happen and also attain better tribological conditions in any automobile.

12.5 TRIBOLOGY IN THE POWER TRAIN OR TRANSMISSION SYSTEM AND DRIVING SYSTEM

Power is transferred from the engine to the driving wheels, via the transmission system consisting of clutch, then gearbox, propeller shaft, differential gear, rear axle (or front axle as the case may be) and road wheels with braking subject to wear.

12.5.1 Propeller Shaft

A propeller shaft, also known as a drive shaft, is a component that transfers torque from the transmission to the transfer case and driving axles. The drive shaft is stressed by torsion or shear as it carries torque; as a result, it must be able to withstand the load while avoiding the addition of too much weight, which will simply increase their inertia. While the vehicle travels on the road, it has to experience rough road conditions containing ups and downs. Thus, the propeller shaft has to withstand shear force, transfer layer torque, higher dynamic and vibrating forces to the rear axles. Otherwise, the vehicles cannot run at this junction. While doing so, high friction takes place and this, in turn, leads to more wear and tear. Moreover, the drive shaft is employed differently in front-wheel drive, four-wheel drive, and front-engine, i.e., rear-wheel drive in automobiles. It is also seen in motorbikes, locomotives, and ships. Hence, a hollow and a solid shaft move relative to each other. A rubber element is placed between the sliding tube and sliding shaft to absorb shocks. In addition, the engine and the gearbox are linked to the vehicle's frame, via flexible mountings or bearings, which help to improve tribological conditions and smooth operating motion. Thus causes a change in the length of the propeller shaft, which is adjusted by the slip joint so as to reduce vibration. Generally, the propeller shaft is manufactured with tabular hardened alloy steel and sometimes it is comprised of steel, spring steel and Al/SiCp composites [24].

12.5.2 Clutches

A clutch is a device that sits between the flywheel and the gearbox and acts as aa regulator of the movement of electrical power to the flywheel. In normal situations, the clutch engages and transfers power from the motor to the gearbox. When the power flow takes place, the clutch engages and the power is transmitted depending on the speed requirements. When it is in neutral, the motor runs but the vehicle remains stationary. Foot pedals control the clutches in four- and six-wheelers; by contrast, hand levers control clutches in two- and three-wheelers. In the process of transmission of power from the flywheel to the gearbox, friction is generated between the clutch plates and the spring due to unexpected loads when operating at higher speeds, which also causes wear on the clutch plates and increased heat generation. In addition, when the pedal is just partially pressed/only half-pressed, the clutch is subjected to maximum wear, tear and also noise is generated. Thus, a good clutch design with suitable material to resist wear and co-efficient of friction (CoF) is found necessary. Since the friction is developed during the working conditions, it is necessary to reduce the friction leading to lower wear and tear of clutch plates, to attain improved tribological properties.

Hence, different clutch plates are used. Among them, the usually used clutch plates are shown in the following:

i. Friction clutch plates are usually produced by cast-iron and high carbon steel and possess high compressive strength, with higher resistance to friction without ductility.
ii. In the case of a pressure plate clutch, grey cast iron is used, since it possesses high hardness and can be easily machined.
iii. In the process, wet and dry clutches are employed, depending on the lubrication type.
iv. The electromagnetic clutch eliminates the need for lubrication with reduction in friction and wear by embedding solid lubricants into the base material.

As a result, the research is focused on varying parameters like temperature, rotation speed, and load. The wear and frictional characteristics of Cu-based friction pairs in wet clutches are investigated through a pin-on-disc tribometer. The CoF of Cu-based friction pairs remain steady at 120 °C. When the temperature reaches 420 °C, the CoF starts to vibrate. The CoF increases from 0.28 μ to 0.35 μ from 120 °C to 270 °C and falls to 0.30 μ from 270 °C to 420 °C. Similarly, as the temperature rises from 120 °C to 420 °C the wear factor increases drastically from 7.9×10^{-8} g/Nm to 112.2×10^{-8} g/Nm followed by increased abrasive and ploughing wear mechanisms at a lower temperature of 345°C; and it drastically increased adhesive and delamination wear mechanisms above 345°C [29]. Later, the wear behavior of paper-based friction materials with and without carbon fibers is investigated using a pin-on-reciprocating plate tribometer with changing sliding speed and normal force under boundary lubrication. The paper-based friction materials containing carbon fibres had higher friction and wear properties than those without carbon fibres because the sliding contact fractured the carbon fibres, perhaps contributing to the wear growth [30].

12.5.3 Gearbox

A gear is a device with two-toothed circular wheels fitted on shafts for power transmission. The gear mechanism is essential in transferring power from the engine to the driving wheels. Components of the gear mechanism include a hand lever for changing the gear speed; clutch; an idler; lay shaft; spur gears; lock washer, locking ball and spring; bearing; bearing rollers; pinion driver; and gear retaining ring. When the engine is running, the gear arrangement will be in the normal position, i.e., allowing direct power transmission from the clutch to the gear input shaft, via the gear shaft, and output splined shaft without any change in speed; this is also known as the neutral position. Depending on the requirement of speeds, the road wheels have to be controlled/operated through a gear lever by the driver. Thus, while the vehicle is running and changing speed, the gear wheels are prone to friction, wear, and tear, resulting in tribological difficulties. To minimise such tribological issues, the gears are designed to ensure high durability and resistant to corrosion, friction and wear. The gears are often made of cast iron or alloy steel, although they are hefty. As a result, researchers and modern manufacturing industries are working to develop gears with low weight and high specific strength. At this junction, light weight MMCs (AA MMCs) are found to be more useful/advantages, since hybrid MMCs exhibit high specific strength, improved wear resistance and tribological characteristics to withstand high thermal working condition.

12.5.4 Axle Assembly and Wheels

The axle shafts are located at the front and back of the vehicle, with a drive to connect the road wheels at front and back; they are termed as Front Wheel Drive (FWD) and All-Wheel Drive (AWD) automobiles, respectively. An automobile cannot move without an axle component. They just transfer power from the engine motor (as the case may be) to the wheels, and sustain vehicle self-weight and dynamic force while running. In a FWD vehicle, power is delivered from the engine to the transaxle, where transmission and differential are fused/merge into one in the transaxle when shear stresses are produced. In contrast, newer AWD cars feature a more sophisticated transaxle. The casing holds the crown wheel and differential gears (box) as well as two half-shafts on a rear axle, which is mounted either on a leaf spring or on a coil spring. The springs and components of the braking systems are also supported by this structure. The differential gearbox is connected to the propeller shaft by a knuckle joint with a splined end, so that the torque from the gearbox is transmitted to the road wheels. As a consequence, a significant amount of friction is generated, which not only reduces mechanical efficiency but also leads to excessive wear and tear, resulting in tribological issues. Therefore, the selection, design and manufacture of the components play an increased role. In addition, the front axle has to steady the dynamic motion of the vehicle whereas the rear axle bears the forces and loads while stationary as well as running. As such, the axles are manufactured with high-strength materials, namely carbon steel or nickel alloy steels.

12.5.5 Road Wheels/Tyres

Wheels are vital components of a vehicle because they carry the whole weight of the vehicle (passengers or cargo) and protect it from road shocks, even at high speeds while running with passengers or loads. In addition, the rear wheels must transfer power/torque to the front wheels and steer the vehicle. As a result, all of the wheels have to withstand and resist the barking pressures while balancing and enduring side thrust so as to maintain perfect balance. Hence wheel alignment is vital. The wheel alignment has to facilitate easy steering effort, differential stability and minimal tyre wear. Friction is generated throughout the running process of wheels, which is a significant factor since each wheel develops around 2% to be considered the COF of the rolling tyre. The wheel is an assembly of three main components, consisting of the wheel drum or disc, the brake drum and rubber tyre housing a tube (filled with air).

The wheel drum is often made of alloy steel, whereas the brake drum is made of cast iron having high strength and ductility, which can absorb shocks and resists wear and tear. The wheels are made up of steels are longevity, high strength and toughness. The steel wheels are quite cheaper than the alloy wheels because the manufacturing process consists of stamping the wheels into the desired shape. However, the steel wheels possess many disadvantages because the weight of the wheels is heavy and thus this will increase the unsprung weight of the vehicle; such conditions force to trash out the suspension system after a certain period. Moreover, steels are easily oxidised and corroded due to environmental changes leading to the formation of corrosion and pitting wears. With the growth of automobile industries, alloy wheels have come into existence, which are made up of Al and Mg alloys. This is because they are lightweight and thus help to reduce the overall weight of the vehicle and improves millage, braking system and acceleration. Furthermore, it distributes the minimal load to the coil spring, resulting in improved grip and traction. In comparison to steel wheels, the alloy wheels transfer heat faster; thereby, it helps to improve the life of the tyre. Moreover, these wheels are corrosion-proof.

Natural rubbers is used to manufacture tyres and tubes, although synthetic rubber is also used. The braking mechanism on the wheels depend entirely upon the resiliency of the tyres from road shocks and hence the tyre body plies are composed of cotton cord fabric in the production of rubber tyres. The inside of the wheel is an endless tube of rubber fitted with a valve through which air is pumped and retained under pressure.

12.5.6 Steering Mechanism

When a driver rotates the steering wheel in their hands, the steering mechanism converts that rotary action into an angular motion of the front wheels, which subsequently turns the automobile in a desired direction. So, the steering system has to satisfy the leverage of the driver's efforts to make the vehicle turn effortlessly. The front wheels are attached to knuckle spindles that are visible in tapered roller setup and assembled in a steering gearbox. The steering gearbox and steering mechanism of an automobile are prone to load variations, fluctuation, and vibration at high speeds and shocks when travelling over uneven surfaces, which may cause damage to the steering system. So, in order to avoid wear and tear as well as friction, the gear bearings and sealed washers must be capable of withstanding jolts and torsional vibrations. Furthermore, proper gear mechanism and lubrication are required to solve such challenges.

12.6 TRIBOLOGY IN THE BRAKING AND SUSPENSION SYSTEMS

12.6.1 Braking System

Brakes are the devices used to slow down or stop a vehicle. Braking is required to reverse the vehicle's acceleration and lead it rearward. During the braking operation, the vehicle's kinetic energy is converted to heat and dispersed into the air; thus, the brakes are more important in vehicle control. The braking system acts on the wheels; it is controlled by a foot pedal/hand. Friction resists the motion of the vehicle, when a moving vehicle is abruptly stopped by applying brakes. The motion of the brake shoes/pads creates friction between the brakes and the braking drum/disc and then tyre-road friction slows or stops vehicle motion. In the present scenario, the majority of low-weight vehicles employ disc brakes rather than drum brakes [31]. Thus, the tribological characteristics of discs and pads have a considerable impact on the braking process. The discs are often made of grey cast iron due to its excellent wear and frictional resistance, high thermal efficiency, and anti-vibration characteristics, but they are heavy in weight. So, discs made with composite materials have developed, particularly for sports cars. Brake friction components may wear out in many ways. The major prevalent kind of wear is abrasion, which occurs when friction parts rub against each other. Scratching, micro-grinding, and groove development cause the material loss of the pads.

Different forces are applied on the vehicle during braking operation; for example, an overall force is applied on the moving vehicle that causes it to slowdown is referred to as the braking force. When the pressure applied on the brakes, the friction between the breaks and tyres produces work to retardation of vehicle's kinetic energy thereby slowing it down. While the vehicle runs on the wet roads, high braking effect at the front would lead to the skidding of the front wheels, due to a decrease in weight transfer. Manual, servo, or power brakes are used in general in automobiles. When the brake pedal is pressed, cam turns through linkages. The cam turns the shoes outwards, causing them to come into contact with the retarding drums, which eventually stops the wheel and the vehicle. On the release of the brake pedal, the retracting spring helps the brake shoes return to their original position and release the brakes, allowing the vehicle to move after a halt. Hence, many forces act on the components of the braking system, causing wear, tear, and increased friction, resulting in heat and sometimes developing vibrations and noise, resulting in tribological issues [12, 32]. Therefore, proper lubrication is needed as per the SAE standards. Further, it is suggested to produce the parts with 2-D solid lubricants reinforced metal matrix composites so as to reduce tribological issues and enhance the mechanical efficiency of the vehicle.

12.6.2 Suspension Mechanism

Suspension is the mechanism that links a vehicle to its wheels and permits relative motion. There are a number of components in a vehicle's suspension system, including springs, linkages, dampers, and shock absorbers, as well as wheels and tyres. Such components are subjected to undergo shocks due to self-weight, and passenger loads while running at various speeds, followed by the variation in torque while transferring thrust or twist to the wheels during the travel on even surfaces. Therefore, laminated leaf springs or coil-type springs are used to join the frame over the axle and the front axle. In addition to friction, the frame, axles and suspension systems are exposed to wear and tear during function, causing severe tribological difficulties due to high dampening forces, tensile and shock loads. Accordingly, leaf springs and coil springs with shock absorbers are undergoing compression during the heavy loads/thrust force and release to restore its original position after the release of surfaces. Hence, flexible materials such as stainless steel and spring steels are used to manufacture springs. In addition, springs need to be lubricated since they oscillate and then return to their previous position, causing friction and wear [12, 21, 33]. Therefore, suitable and appropriate lubricants are necessary for applications.

12.7 LUBRICANTS

Friction is developed during the contact surfaces of two or more components in action. Once the friction is developed, it immediately leads to wear and tear on the contact surfaces and also generates heat resulting in tribological impact. Therefore, a proper lubricant has to be applied between the contact/moving parts such as gears, bearings, rollers, clutches, steering system, brakes, etc., in automotive vehicles. Hence, in order to minimize the frictional resistance between the contact or sliding surfaces, a proper lubricant is applied and the resulting phenomenon is called lubrication [28, 34].

Whatever the type of lubricant, it has to possess the following characteristics so as to (i) minimise frictional resistance, (ii) reduce the developed heat, (iii) control deformation or prevent deformation, (iv) inhibit the passage of moisture, (v) allow the moving components to move more smoothly, (vi) keep the cost-effective maintenance of machine components, (vii) reduce the corrosion, (viii) prevent the asperities on the surface by interlocking or interjoint welding and (ix) mitigate the amount of noise produced by bearings and other components. The lubrication regime varies with the approach or separation of moving contacts. The asperity locations in various lubrication regimes and the asperity state of the lubrication regime are shown in Figure 12.5. The sealing action of the piston ring and cylinder liner are lubricated with a partial contact.

12.7.1 Thin Film or Boundary Lubrication

A small film of lubricant is deposited onto the metal surface of the two contact parts. This helps in the prevention of direct metal-to-metal contact. A thin layer of oil is supplemented to the shaft when it first begins to rotate, or when the speed seems to be very minimal or the load is extremely high, or when the viscosity of the lubricant is extremely low. This phenomenon separates two contacting surfaces by forming a thin film layer of film [1, 14, 35]. In such boundary lubrication, the

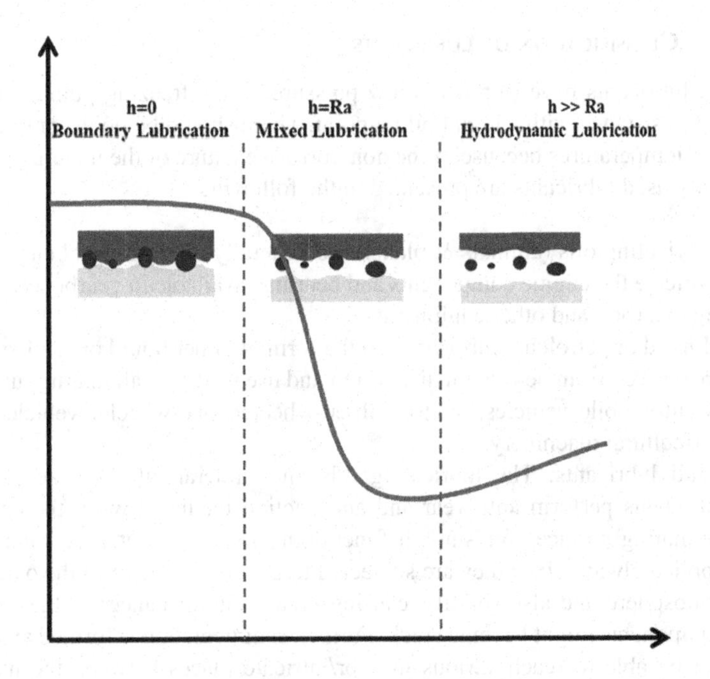

FIGURE 12.5 Stribeck curve and various lubrication regimes.

grease is applied to journal bearings, slider bearings to avoid wear or to minimise friction.

12.7.2 Fluid Film or Hydrodynamic Lubrication

A thick layer of fluid lubricant with a thickness of 1000 A° is applied between the sliding surfaces, for the prevention of direct surface-to-surface contact and thereby minimising friction as well as wear and tear [1, 14, 36]. Such lubricants as SAE 10W30, SAE 10W60, SAE 0W5, SAE 20W 40 and so on are used in the operation of high contact pressure surfaces such as roller bearings, gear teeth, cam and followers. In addition, they are specifically used in sensitive machinery equipment such as sewing machines, watches, clocks, sensitive and scientific and measuring instruments, speedometers, and gyroscopes.

12.7.3 Extreme Pressure Lubrication

When the rotating interfaces are subjected to high pressure as well as speed, frictional heat is generated, resulting in a significant increase in surface temperature [37]. This occurs specifically at crankshaft and main bearings, crankshaft and camshaft bearings, universal joints, bearings and distributors, locker arm, safety values and so on. In such circumstances, special lubricants such as chlorinated esters, sulphurized oils, and tricresyl phosphate are used.

12.7.4 Classification of Lubricants

Superior lubricants have to possess low pressure, a low freezing point, and strong oxidation resistance with a high boiling point. There should be no decomposition at operating temperatures because of the non-corrosive nature of the material [38]. The commonly used lubricants are presented in the following:

i. Lubricating oils (animals & plants) are generally used in diesel engines to lubricate the engine components and bearings available in gearboxes, steering, gearbox and other equipment.
ii. Mineral or petroleum lubricants (in the form of either liquid or semi-solids) are derived from petroleum distillation and use in almost all mating surfaces in automobile vehicles, tractors, three-wheeler, two-wheeler vehicles and agricultural machinery.
iii. Solid lubricants: The lubricating oils and mineral oil-based semi-solid lubricants perform anti-wear and anti-friction media between the contact or mating surfaces for smooth functioning [39, 40]. But they cannot be applied always since they are subjected to undergo reaction in the oxidized atmosphere and also solidify, causing greasy circumstances in the mating components might become stuck. Further, sedimentation is formed and will not be able to reach various interior/intricate places between the mating components. Hence, the solid lubricants have come into existence in the modern nanotechnological world. The solid lubricants such as Gr, h-BN, MoS_2, WS_2, CNTs and so on are used due to their characterisation and properties to withstand the extreme environmental and surrounding working conditions, pressure, temperature and space applications. Therefore, SL nanoparticles are blended with engine lubricants to reduce wear and friction between engine components. They can also be used as reinforcements in the fabrication of light-weight-to-high-strength automobile components. Such MMCs will reduce the adhesive wear and CoF between the parts and will thereby enhance the tribological properties.

During the application of any kind of lubricant, the quality as well as the most essential characteristics, such as viscosity, flash and fire points, cloud and pour points, aniline point, neutralisation number, and mechanical strength, have to be taken into consideration based on SAE standards. Further, with the growth of the research on lightweight material and MMCs and pollutants due to emission from the automotive vehicles, a concept of green tribology has come into existence which is presented in the following section.

12.8 GREEN TRIBOLOGY

In recent years, there has been a growth of investigations into sustainable lubrication in the tribological environment in automotive vehicles. It is observed that the lubricants disposed into the atmosphere after their usage in machine parts poses a serious threat to the flora and fauna. The disposed wastes not only contain toxic harmful particles, but also metallic wear debris. Recent research studies have revealed that there

FIGURE 12.6 Aspects of green tribology.

is a serious concern about the use of mineral oil lubricants. As a result, bio-lubricants and their additives are recommended to prevent such negative consequences [10]. Such activities paved the way for the identification of means and methods for the effective usage of biofuels as lubrication media leading to green tribology. The green tribology deals mainly with the science and technology of tribological aspects of environmental and biological impacts associated with various tribological systems. It is more compatible to that of normal tribology characteristics, which is one of the important factors from the energy consumption point of view.

Further, green energy and green chemistry are the two closely linked areas of green tribology. Green tribology mainly deals with interdisciplinary areas attributed to various concepts relating to energy, material science and technology, process monitoring, green bio-lubrication, surface topology, environmental studies including waste disposal. The concept of green tribology is outlined in Figure 12.6.

12.9 CONCLUSIONS

The present chapter addressed the tribology-related issues in various automobile components. It is evident that tribology played a vital role in emission characteristics. Design modifications, the use of composites, the lubrication system, and lubricants were effectively used for controlling the friction and wear of the automobile components. The influence of different types of lubricants on tribology performance was dominant in automobile industries; however, it is quite difficult to maintain the lubrication system because of the contamination generated during the operation. Hence, a solid lubricant-reinforced composite comes into existence for tribology applications. Though solid lubricants controlled the friction and wear, it is very challenging to improve the interface strength of reinforcement particles and matrix. Hence, a today's research was heading towards enhance the interface strength between the solid lubricants and matrix, which is also capable of reducing the emissions, as well as being environmentally friendly.

ACKNOWLEDGEMENTS

The authors would like to convey sincere thankfulness to the administration of VIT University, Vellore, India, for their continuous support and encouragement during the research.

REFERENCES

1. Holmberg, K.; Andersson, P.; Erdemir, A. Global Energy Consumption Due to Friction in Passenger Cars. *Tribol. Int.*, 2012, *47*, 221–234. https://doi.org/10.1016/j.triboint.2011.11.022.
2. Holmberg, K.; Andersson, P.; Nylund, N. O.; Mäkelä, K.; Erdemir, A. Global Energy Consumption Due to Friction in Trucks and Buses. *Tribol. Int.*, 2014, *78*, 94–114. https://doi.org/10.1016/j.triboint.2014.05.004.
3. Taylor, C. Automobile Engine Tribology—Design Considerations for Efficiency and Durability. *Wear*, 1998, *221* (1), 1–8. https://doi.org/10.1016/S0043-1648(98)00253-1.
4. OICA. Global Sales Statistics 2019–2020. *International Organization of Motor Vehicle Manufacturers OICA is the Voice Speaking on Automotive Issues in World Forums*; 2021.
5. SIAM. Auto Industry Sales Performance of June 2021 & April–June 2021. *Press Releases*; 2021.
6. Holmberg, K.; Kivikytö-Reponen, P.; Härkisaari, P.; Valtonen, K.; Erdemir, A. Global Energy Consumption Due to Friction and Wear in the Mining Industry. *Tribol. Int.*, 2017, *115* (May), 116–139. https://doi.org/10.1016/j.triboint.2017.05.010.
7. Holmberg, K.; Erdemir, A. The Impact of Tribology on Energy Use and CO2 Emission Globally and in Combustion Engine and Electric Cars. *Tribol. Int.*, 2019, *135*, 389–396. https://doi.org/10.1016/j.triboint.2019.03.024.
8. Holmberg, K.; Erdemir, A. Influence of Tribology on Global Energy Consumption, Costs and Emissions. *Friction*, 2017, *5* (3), 263–284. https://doi.org/10.1007/s40544-017-0183-5.
9. Priest, M.; Taylor, C. Automobile Engine Tribology—Approaching the Surface. *Wear*, 2000, *241* (2), 193–203. https://doi.org/10.1016/S0043-1648(00)00375-6.
10. Anand, A.; Irfan Ul Haq, M.; Vohra, K.; Raina, A.; Wani, M. F. Role of Green Tribology in Sustainability of Mechanical Systems: A State of the Art Survey. *Mater. Today Proc.*, 2017, *4* (2), 3659–3665. https://doi.org/10.1016/j.matpr.2017.02.259.
11. Shoeab, M.; Mishra, N. Study of Tribology Application and Its Impact on Indian Industries. *Int. J. Sci. Eng. Technol.*, 2014, *1312* (3), 1310–1312.

12. Aranke, O.; Algenaid, W.; Awe, S.; Joshi, S. Coatings for Automotive Gray Cast Iron Brake Discs: A Review. *Coatings*, 2019, *9* (9). https://doi.org/10.3390/coatings9090552.
13. Cha, S. C.; Erdemir, A. Coating Technology for Vehicle Applications. *Coat. Technol. Veh. Appl.*, 2015, 1–240. https://doi.org/10.1007/978-3-319-14771-0.
14. Erdemir, A. Review of Engineered Tribological Interfaces for Improved Boundary Lubrication. *Tribol. Int.*, 2005, *38* (3), 249–256. https://doi.org/10.1016/j.triboint.2004.08.008.
15. Meng, Y.; Xu, J.; Jin, Z.; Prakash, B.; Hu, Y. A Review of Recent Advances in Tribology. *Friction*, 2020, *8* (2), 221–300. https://doi.org/10.1007/s40544-020-0367-2.
16. Becker, E. P. Trends in Tribological Materials and Engine Technology. *Tribol. Int.*, 2004, *37* (7), 569–575. https://doi.org/10.1016/j.triboint.2003.12.006.
17. Cole, G. S.; Sherman, A. M. Light Weight Materials for Automotive Applications. *Mater. Charact.*, 1995, *35* (1), 3–9. https://doi.org/10.1016/1044-5803(95)00063-1.
18. Gangopadhyay, A. A Review of Automotive Engine Friction Reduction Opportunities Through Technologies Related to Tribology. *Trans. Indian Inst. Met.*, 2017, *70* (2), 527–535. https://doi.org/10.1007/s12666-016-1001-x.
19. Tung, S. C.; McMillan, M. L. Automotive Tribology Overview of Current Advances and Challenges for the Future. *Tribol. Int.*, 2004, *37* (7), 517–536. https://doi.org/10.1016/j.triboint.2004.01.013.
20. Ravishankar, B.; Nayak, S. K.; Kader, M. A. Hybrid Composites for Automotive Applications – A Review. *J. Reinf. Plast. Compos.*, 2019, *38* (18), 835–845. https://doi.org/10.1177/0731684419849708.
21. Fentahun, M. A.; Savas, M. A. Materials Used in Automotive Manufacture and Material Selection Using Ashby Charts. *Int. J. Mater. Eng.*, 2018, *8* (3), 40–54. https://doi.org/10.5923/j.ijme.20180803.02.
22. Lampke, T.; Meyer, D.; Alisch, G.; Wielage, B.; Pokhmurska, H.; Klapkiv, M.; Student, M. Corrosion and Wear Behavior of Alumina Coatings Obtained by Various Methods. *Mater. Sci.*, 2011, *46* (5), 591–598. https://doi.org/10.1007/s11003-011-9328-2.
23. Aman, G.; Shubham, S.; Sunny, N. Materials for Engine 7.1. In *Combustion Engines: An Introduction to Their Design, Performance, and Selection*; John Wiley & Sons, 2016; pp 95–102.
24. Prasad, S. V.; Asthana, R. Aluminum Metal–Matrix Composites for Automotive Applications: Tribological Considerations. *Tribol. Lett.*, 2004, *17* (3), 445–453. https://doi.org/10.1023/B:TRIL.0000044492.91991.f3.
25. Grabon, W.; Koszela, W.; Pawlus, P.; Ochwat, S. Improving Tribological Behaviour of Piston Ring-Cylinder Liner Frictional Pair by Liner Surface Texturing. *Tribol. Int.*, 2013, *61*, 102–108. https://doi.org/10.1016/j.triboint.2012.11.027.
26. Kumar, V.; Sinha, S. K.; Agarwal, A. K. Wear Evaluation of Engine Piston Rings Coated with Dual Layer Hard and Soft Coatings. *J. Tribol.*, 2019, *141* (3), 3–13. https://doi.org/10.1115/1.4041762.
27. Farfan-Cabrera, L. I. Tribology of Electric Vehicles: A Review of Critical Components, Current State and Future Improvement Trends. *Tribol. Int.*, 2019, *138* (April), 473–486. https://doi.org/10.1016/j.triboint.2019.06.029.
28. Van Rensselar, J. Lubrication and Tribology Trends and Challenges in Electric Vehicles. *Tribol. Lubr. Technol.*, 2020, *76* (7), 26–33.
29. Zhao, E. H.; Ma, B.; Li, H. Y. Study on the High Temperature Friction and Wear Behaviors of Cu-Based Friction Pairs in Wet Clutches by Pin-on-Disc Tests. *Adv. Mater. Sci. Eng.*, 2017, *2017*. https://doi.org/10.1155/2017/6373190.
30. Cho, H. R.; Je, Y.; Chung, K. H. Assessment of Wear Characteristics of Paper-Based Wet Friction Materials. *Int. J. Precis. Eng. Manuf.*, 2018, *19* (5), 705–711. https://doi.org/10.1007/s12541-018-0084-1.

31. Borawski, A. Common Methods in Analysing the Tribological Properties of Brake Pads and Discs-A Review. *Acta Mech. Autom.*, 2019, *13* (3), 189–199. https://doi.org/10.2478/ama-2019-0025.
32. Stoica, N. A.; Petrescu, A. M.; Tudor, A.; Predescu, A. Tribological Properties of the Disc Brake Friction Couple Materials in the Range of Small and Very Small Speeds. *IOP Conf. Ser. Mater. Sci. Eng.*, 2017, *174* (1). https://doi.org/10.1088/1757-899X/174/1/012019.
33. Macke, A.; Schultz, B.; Rohatgi, P. Metal Matrix Composites. *Adv. Mater. Process.*, 2012, *170* (3), 19–23.
34. Priest, M.; Taylor, C. M. Automobile Engine Tribology—Approaching the Surface. *Wear*, 2000, *241* (2), 193–203. https://doi.org/10.1016/S0043-1648(00)00375-6.
35. Manu, B. R.; Gupta, A.; Jayatissa, A. H. Tribological Properties of 2D Materials and Composites—A Review of Recent Advances. *Materials (Basel)*, 2021, *14* (7). https://doi.org/10.3390/ma14071630.
36. Azmi, M. A. M.; Mahmood, W. M. F. W.; Ghani, J. A.; Rasani, M. R. M. A Review of Surface Texturing in Internal Combustion Engine Piston Assembly. *Int. J. Integr. Eng.*, 2020, *12* (5), 146–163. https://doi.org/10.30880/ijie.2020.12.05.018.
37. Reeves, C. J.; Menezes, P. L.; Lovell, M. R.; Jen, T.-C. Tribology of Solid Lubricants. In *Tribology for Scientists and Engineers*; Menezes, P. L., Nosonovsky, M., Ingole, S. P., Kailas, S. V., Lovell, M. R., Eds.; Springer: New York, 2013; pp. 447–494. https://doi.org/10.1007/978-1-4614-1945-7_13.
38. Sahoo, S. Self-Lubricating Composites with 2D Materials as Reinforcement: A New Perspective. *Reinf. Plast.*, 2021, *65* (2), 101–103. https://doi.org/10.1016/j.repl.2020.06.007.
39. Uzoma, P. C.; Hu, H.; Khadem, M.; Penkov, O. V. Tribology of 2D Nanomaterials: A Review. *Coatings*, 2020, *10* (9). https://doi.org/10.3390/COATINGS10090897.
40. Scharf, T. W.; Prasad, S. V. Solid Lubricants: A Review. *J. Mater. Sci.*, 2013, *48* (2), 511–531. https://doi.org/10.1007/s10853-012-7038-2.

13 Tribocorrosion in the Automotive Sector

Pravesh Ravi and Jitendra Kumar Katiyar
SRM Institute of Science & Technology, Chennai, India

CONTENTS

13.1 INTRODUCTION

Tribology and corrosion are the two major problems that exist in nature from the inception of the world. Tribological study reveals the phenomenon of friction, wear and lubrication in between the mating surface under applied load. was Although they have been mentioned briefly in previous chapters, this chapter gives a fuller discussion of both corrosion and tribocorrosion. It is very well known that friction consumes roughly one-third all energy used in transportation. The CO_2 emissions are several times higher for a combustion engine car than an electric car. The IC engine is used as the power source for major transportation. Only 21% of fuel is effectively converted fuel energy into mechanical energy of engine in the passenger vehicles, as shown in Figure 13.1.

It has been mentioned that approximately 200,000 million litres of fuel is used yearly to overcome the friction in passenger cars. Thus, the reduction of frictional losses in engine may lead to a threefold increase in fuel economy and also a reduction in both exhaust and cooling losses. It also decreases the remanufacture of worn parts and spare equipment due to corrosive wear. The total CO_2 emission due to

DOI: 10.1201/9781003243205-13

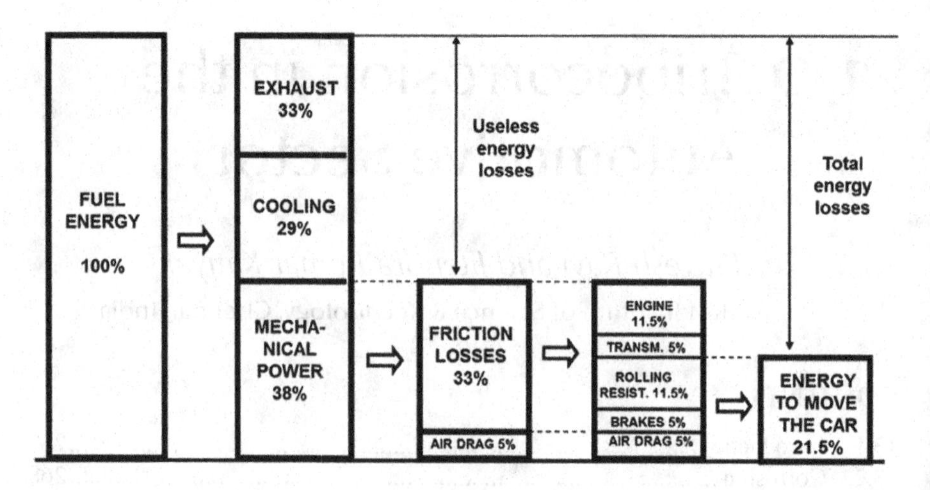

FIGURE 13.1 Breakdown of Internal combustion engine- passenger car energy consumption. ([1].)

friction and wear is estimated to be about 8120 Mt/year, including transportation, manufacturing, power generation, residential and others. Moreover, the frictional losses in electric cars are almost half of the frictional losses in IC Engine cars [2]. However, the development of new materials for use in manufacturing industries for the development of cutting tools, drill bits, mining, and high wear resistant parts, due to its superior qualities with regards to its thermal, chemical, and mechanical properties, is very much required to improve the wear resistance of the parts that are used in manufacturing industries [3]. Further, the engine oil that are used for transportation of vehicles also degrades and decomposes into corrosive sulfur-derived products, these degradations are the primary aspects, causing corrosive wear at the piston ring/cylinder contact. The corrosive media can be gone deeper into the fatigue cracks of piston, and these will accelerate the wear and fatigue on the surface of piston ring.

13.2 CORROSION

Corrosion is one of the major reasons for the failure of metal components. Although corrosion cannot be eliminated from metals, its intensity can be reduced through the selection of suitable materials, such as new alloys or protective coatings and films were deposited on the metal surface [4]. Further, to enhance the corrosion resistance on the metal surfaces, various types of protective coatings can be used. The most common eight forms of corrosion are briefly discussed here.

13.2.1 Major Forms of Corrosion

- Uniform (general) corrosion.
  Uniform corrosion is the most common form of corrosion; it takes place in ferrous materials and alloys where the surface is not protected through coating, as shown in Figure 13.2(1) [5]. It occurs on many different materials such as steel,

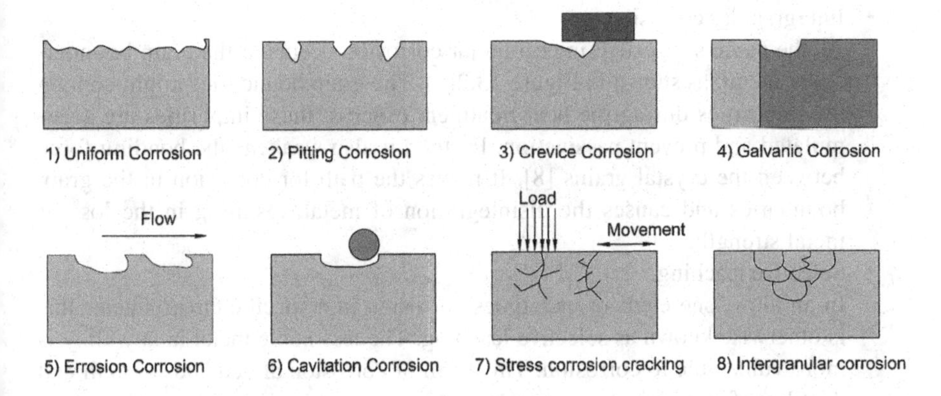

FIGURE 13.2 Form of corrosion.

aluminum, copper, which commonly become victims of uniform corrosion. This form of corrosion occurs more or less uniformly throughout the thickness of the material. These types of corrosion eat into the material and reduce the thickness. This is continued until the metal is completely vanished.

- Galvanic (two-metal) corrosion.

  Galvanic corrosion is occurred on non-homogeneity materials, such as metals and alloys, as shown in Figure 13.2(4). It is an electrochemical process where the metal corrodes when the potential is applied/flow between bimetallic substrate that are immersed in electrolyte. The rate of corrosion is dependent upon the potential applied and the distance between two metals. The low standard electrode potential decreases the corrosion resistance of the specimen [6].
- Crevice corrosion

  Crevice corrosion is a localized corrosion that occurs in crevices on the surface of a metal that contains water, mud, dirt, biofouling, etc. This form of corrosion is occurred only on a few areas of surface, where the other surface is unaffected by crevice corrosion, as shown in Figure 13.2(3). The stagnation of water inside the crevice leads to the development of crevice corrosion, which accelerates the electrochemical reacting that leads to the corrosion. [7] stated that crevice corrosion has taken place in the higher chloride concentration and temperature above ~35°C.
- Pitting corrosion.

  Pitting corrosion is another localized form of corrosion that occurs on the metal surface at various coordinates of same substrate, as shown in Figure 13.2(2). Pitting corrosion occurs when the passive layer of protective coating is damaged over a larger area when compared to the exposed metal [5]. This form of corrosion is initiated as small cavities and pin holes then begin to grow, where the other surface remains unaffected. It penetrates the metal and makes failure almost unpredictable, making pitting corrosion one of the most dangerous forms of corrosion. The damaged passive layer is the cathode, and the metal acts as an anode where the moisture present in the atmosphere is act as electrolyte that facilitate the corrosion.

- Intergranular corrosion.
  As the name suggested, intergranular corrosion occurs at the grain boundaries of metal, as shown in Figure 13.2(8). The grain boundaries might contain the impurities during the heat treatment process; these impurities are accumulated and prevent passivation. Further, it also weakens the bonding force between the crystal grains [8]. It makes the path for corrosion in the grain boundaries and causes the disintegration of metal, resulting in the loss of metal strength.
- Selective leaching
  In an alloy, one element undergoes corrosion in a specific environment; this is otherwise known as selective leaching. The less noble metal in any alloy is more vulnerable to corrosion. This form of corrosion affects the strength and ductility of materials. For example, the selective corrosion of zinc in brass is a process known as dezincification.
- Erosion corrosion
  Erosion corrosion is another form of corrosion, in which the metal deteriorates due to the relative motion between the corrosive fluid and the metal surface, as shown in Figure 13.2(5) [9]. The fluid may be multiphase flow i.e., liquid-gas phase or liquid-solid phase. The fluid flows at high velocity, which removes the passive layer of metal and accelerates the corrosion process. Erosion corrosion is one of the types of tribocorrosion, which is discussed elaborately in the next section.
- Stress corrosion cracking
  Stress corrosion cracking is developed by a crack formation due to static tensile stress in metals, as shown in Figure 13.2(7). It occurs in the grain boundaries where the intergranular corrosion has taken place. Further, the improper heating process leads to the development of internal residual stresses in metals, resulting in the corrosion process. This form of corrosion is generally unpredictable during the inspection and a sudden failure of the structure may occur.

13.3 TRIBOCORROSION

Tribocorrosion is the process that leads to the degradation of metallic surfaces by the simultaneous action of friction and corrosion [10]. It is the study of tribology under an electrochemical medium in which the material has started to degrade due to mechanical and electrochemical interaction. The degradation of materials is affected components in many industries, including mining, mineral processing, biomedical, automobile, food, nuclear, offshore, marine, oil, and gas production, etc. The materials that are coated for protection from corrosion and wear, when it is removed or damaged, the surface of the material is directly exposed to the corrosive environment, and it readily interacts with mating surfaces, resulting in tribocorrosion. Corrosion wear, erosion corrosion, fretting corrosion, and microabrasion corrosion are the most prevalent types of tribocorrosion, as shown in Figure 13.3 [11]. Corrosion wear is a type of wear that leads to the degradation of materials by mechanical and electrochemical reaction between the surface in a corrosive environment [12].

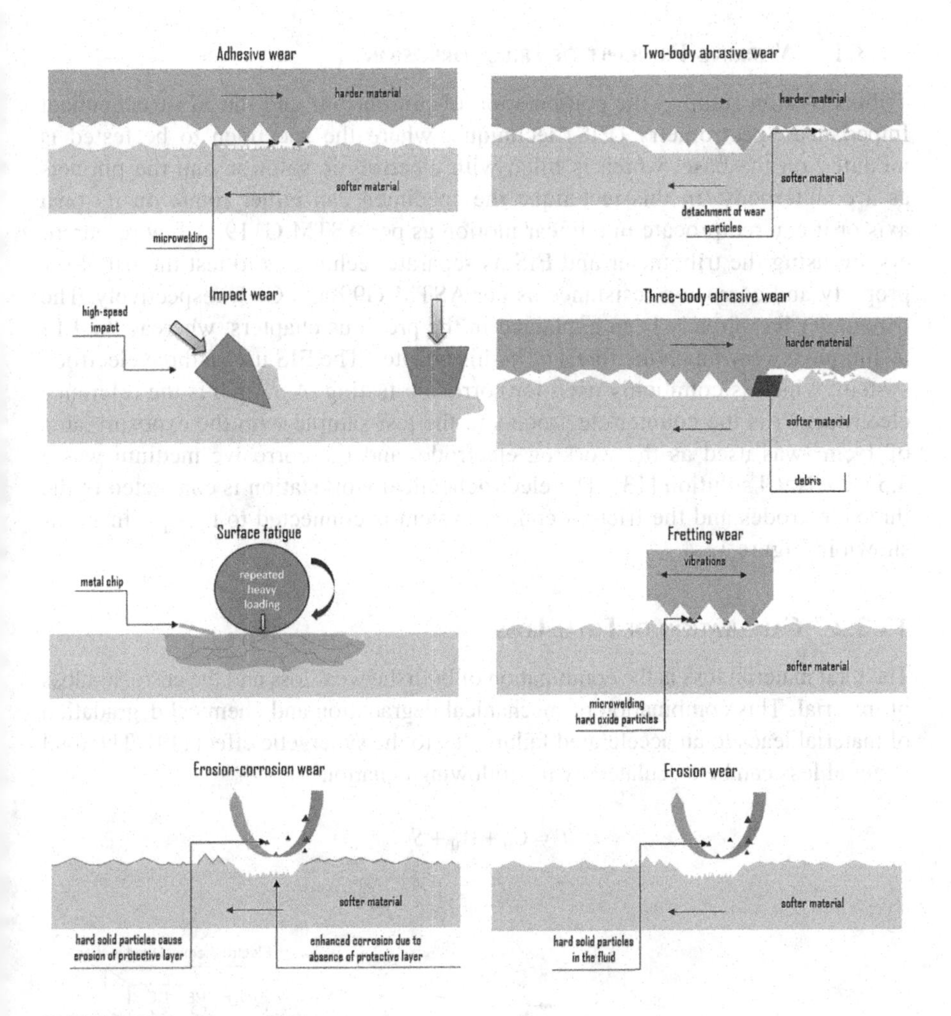

FIGURE 13.3 Form of tribocorrosion.

Erosion corrosion is another form of tribocorrosion. In this process the degradation of material occurs due to fluid flow coupled with the mechanical removal of protective layer, which increases the corrosion rate. It can be observed in pumps, impellers, propellers, valves, heat exchanger tubes and other fluid handling equipment. The morphology of erosion corrosion-affected surfaces can be seen in the form of shallow pits or horseshoes or other local phenomena related to the direction of flow. Fretting corrosion occurs due to the deterioration of material through corrosion at the asperities of mating surfaces in relative motion. The damage is induced under load with the presence of repeated relative surface motion. For example, due to vibration or oscillation.

Microabrasion corrosion is the result of the degradation of material by the combined action of particle abrasion, mechanical load and corrosivity of the medium. It is commonly observed in orthopedic implants, particularly on hip joints

13.3.1 Working Principle of Tribocorrosion

Tribocorrosion setup is the combination of pin-on-disc and the Electrochemical Impedance Spectrometry (EIS) technique, where the specimen to be tested is mounted on the base, which is filled with electrolytic solution and the pin acts as a counterbody. In this technique the specimen can either rotate on its own axis or it can reciprocate in a linear motion as per ASTM G119. A few researchers are using the tribometer and EIS as separate techniques to test the tribology property and corrosion resistance as per ASTM G99 and G03, respectively. The tribometer has already been explained in the previous chapters, whereas the EIS technique is very much the focus of in this chapter. The EIS uses a three electrode system, which is commonly used for corrosion testing. Ag/AgCl is the reference electrode, Pt is the counter electrode and the test sample with the exposure area of 1 cm^2 was used as the working electrode, and the corrosive medium was a 3.5 wt% NaCl solution [13]. The electrochemical workstation is connected to the three electrodes and the friction control system is connected to the specimen, as shown in Figure 13.4.

13.3.2 Calculation of Total Loss

The total material loss is the combination of both the wear loss and the corrosion loss of material. This combination of mechanical degradation and chemical degradation of material leads to an accelerated failure due to the synergetic effect [14]. The total material loss can be calculated by the following equation

$$T = C_0 + W_0 + S$$

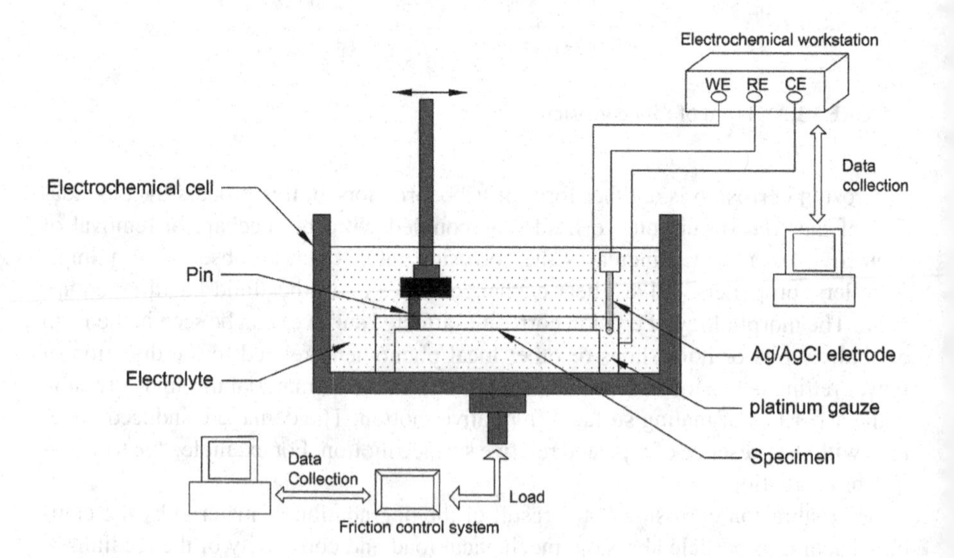

FIGURE 13.4 Schematic diagram of Tribocorrosion test rig.

where,

C_0 is the material loss due to corrosion alone,
W_0 is the material loss due to mechanical wear alone, and
S is the material loss due to wear-corrosion synergy.

13.3.3 Calculation of the Corrosion Rate

Electrochemical Impedance Spectroscopy (EIS) is one of the most reliable methods to measure the electrochemical characteristics of a system. The application of EIS is employed to determine the corrosion in semiconductors, batteries, supercapacitors, fuel cells, membranes etc. In this method a small potential or current is applied, and the response is measured. The magnitude of the impedance and phase shifts are then determined. When the material is exposed to the corrosive environment, the thickness of the material is reduced. The measurement of reduction in weight or thickness is known as the corrosion rate measurement, which can be expressed in mils per year (mpy), inches per year (ipy) and milligrams per decimeter per day (mdd).

$$\text{Mpy} = \frac{534W}{DAT}$$

Where,

W = weight loss, mg
D = Density of specimen, g/cm^3
A = Area of specimen, sq.in
T = Exposure time, hr

13.3.4 Bode and Nyquist Plot

Figure 13.5 presents the Bode plot and the Nyquist plot for the identification of the corrosive behavior of materials. In the Bode plot, the X-axis is denoted the logarithmic scale of frequency and the Y-axis is denoted the logarithm of impedance Z while the second ordinate is the phase shift Φ. In the Nyquist Plot, resistance, 'Z', is plotted on the X-axis while the reactive part, –"Z", is plotted on the Y-axis, as shown in Figure 13.5 [15]. When resistance is observed in parallel with a capacitor, the capacitive behaviour is presented at high frequency, while the resistive behaviour is shown at low frequency, which is used for the testing of corrosion. When resistance is parallel to capacitor, the Nyquist plot has shown a common semi-circumference movement in the graph. The size of the semicircle dots in Nyquist plots helps us to understand the material characteristic under low frequencies.

The epoxy and its composites are prepared and coated over the mild steel for the testing of EIS characteristics. The Bode plot and the Nyquist plot for epoxy and ts composites are shown in Figure 13.6.

The anticorrosion performance of epoxy coatings was examined using the value of impedance at a lowest frequency (f = 0.01 Hz) obtained from the Bode plot. The

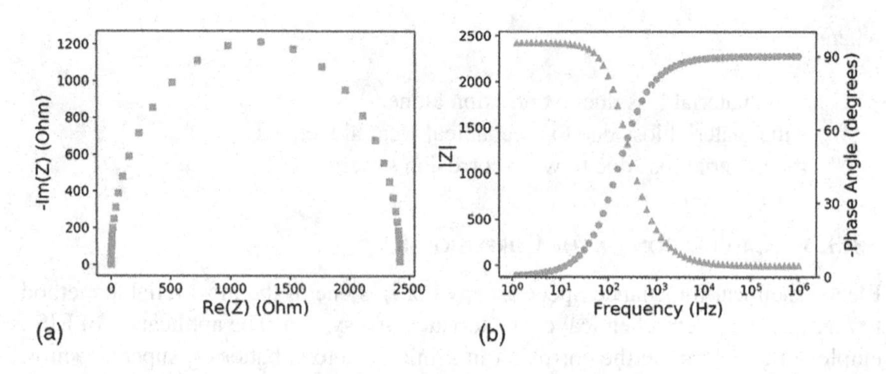

FIGURE 13.5 (a) Nyquist plot and (b) Bode plot. ([15].)

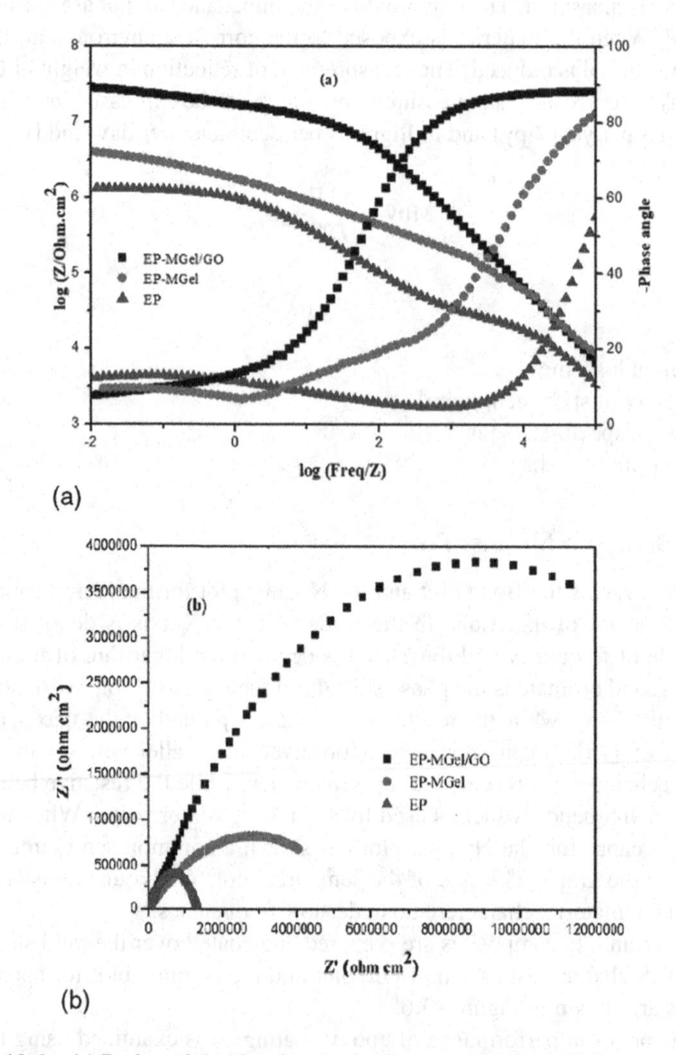

FIGURE 13.6 (a) Bode and (b) Nyquist plot for Epoxy composite. ([16].)

Bode plot reveals that the pure epoxy-coated specimens have exhibited least values of impedance. The MGel-EP and MGel/GO-Ep has shown significantly higher impedance magnitude that indicates the corrosive resistance of composite coating. As time passes, the electrolyte coating is penetrated and there is a reduction in performance. In general, larger-phase angle values are denoted the improved corrosion resistance. Thus, the MGel/GO-EP-coated sample has shown less degradation of coating during exposure to the NaCl solution. Similarly, different metal, metal oxide and polymers can be coated over any corrosive substrate and tested in the EIS for the Bode plot and the Nyquist plot.

13.3.5 Tafel Plot

Polarization technique consists of various methods, including Tafel extrapolation, potentio-dynamic measurements, and linear and cyclic polarization resistance. Tafel extrapolation is very important for the evaluation of corrosion property in a material [17]. Tafel extrapolation of polarization curves is plotted against log current density and electrode potential v/s SCE (saturated calomel electrode) that is used to determine the corrosion rate, Pi (mm year^{-1}), as shown in Figure 13.7. The imaginary

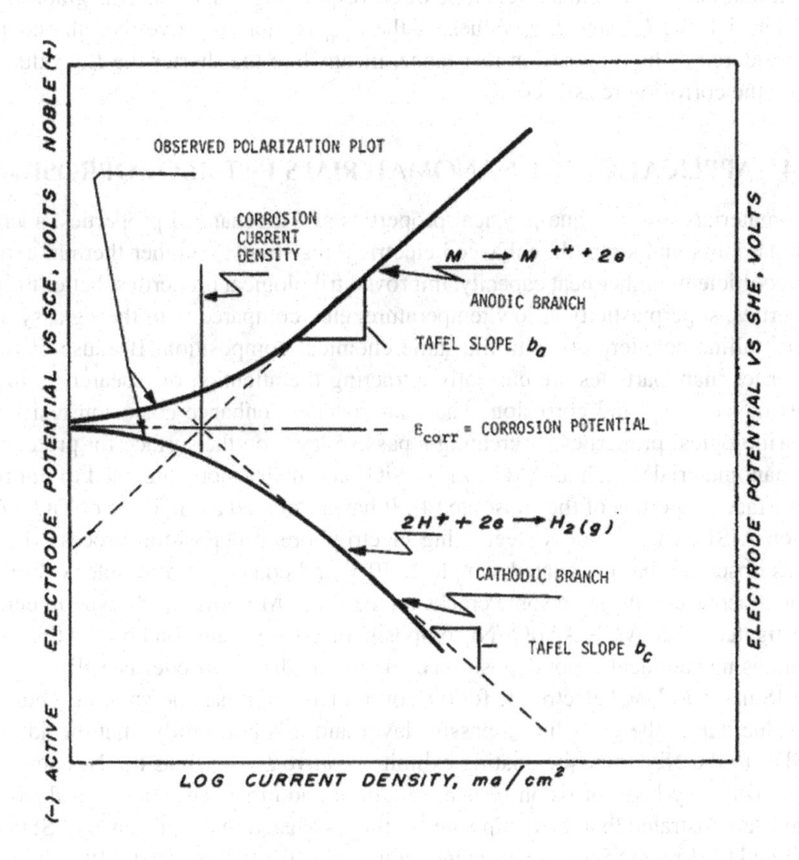

FIGURE 13.7 Cathodic and anodic polarization diagram.

lines are drawn in anodic and cathodic branches, and the corresponding slope on the anodic and cathodic branches are known as Tafel slopes. Further, the corrosion current density is an important parameter to determine the anticorrosion property of the coatings. The relationship between the polarization resistance, R_p (Ω cm^2) and corrosion current density, i_{corr} (mA cm^{-2}) is shown in the Stern-Geary equation below.

$$i_{corr} = \frac{\beta_a \beta_c}{2.3R_p(\beta_c - \beta_a)} = \frac{B}{R_p}$$

Where,

β_a and β_c are the anodic and cathodic Tafel slopes
B is the Tafel Constant.

The polarization diagram has two open-ended curves in which one is for the cathode branch and the other is for the anode branch. The anodic branch is known as the oxidation curve and cathodic branch is known as the reduction curve in which a slope is drawn that is known as the Tafel slope of corresponding branches. This graph helps to determine the I_{corr} and E_{corr} values. If the E_{corr} is more positive then the coating is considered as high corrosion resistance, meanwhile the shorter the I_{corr} value the higher the corrosion resistance.

13.4 APPLICATION OF NANOMATERIALS IN TRIBOCORROSION

Nanomaterials offer unique physical properties and mechanical properties, such as high hardness and strength, enhanced electrical resistivity, a higher thermal expansion coefficient, higher heat capacity, improved tribological properties, better fatigue properties, superplasticity at low temperature, etc., compared with their glassy and/or crystalline counterparts with the same chemical composition. Because of these properties, nanoparticles are currently attracting the attention of researchers in the fields of tribology and corrosion. The nanomaterials enhance corrosion resistance and tribological properties by creating a passive layer on the surface for protection. The nanomaterials, such as Al_2O_3, ZrO_2, SiO_2 and many more, are used to improve the surface properties of the substrate [4]. It has developed a thin layer of TiO_2 coating on AISI 316L stainless steel using electrophoretic deposition process (EPD). It was observed that the coated sample at 30 V and coated for 5 minutes exhibits a homogeneous coating with some cracks on the layer. Moreover, [18] experimentally investigated the MWCNT-Al_2O_3/Ni composite that were coated on brass of thickness 70 μm using chemical vapor deposition on a rotating disc electrode. The Ni are tested for EIS in 0.6 M NaCl electrolyte for 60 h of immersion. It has shown more a positive E_{oc} value due to the growth of a passive layer and it is noteworthy that the addition of SDS to the Ni composite coating exhibits positive E_{oc} whereas the Ni composite has shown very less corrosion resistance without additive. The tribological experiments demonstrated that the composite coating produced in the presence of SDS has exhibited the lowest friction coefficient (COF) value $\mu = 0.356 \pm 0.01$. [19] fabricated

TABLE 13.1
A Summary of Various Corrosion Studies

Author	Material	Thickness	Electrolyte	Condition	Ecorr (mV)	Icorr (uA)
[4]	TiO_2	11 microns	3.5% NaCl	316 L SS bare	−357.14	1.36
				TiO_2 coated at 30 V	−111.89	0.212
[13]	Zr/MoS2 1.5%	3.22 μm to 4.4 μm	3.5 wt% NaCl solution	Electrodeposited on Si wafer	−0.350 ± 0.002	4.869 ± 0.293e−8
[20]	Cobalt/graphene composite	50 μm	3.5 wt% NaCl	Electrodeposition on steel plate	−391	0.8

TABLE 13.2
A Summary of Various Tribology Studies

Material	Coating Thickness	Substrate	Method	Wear	COF	Microhardness (Hv)	Ref.
Zr/MoS_2 1.5%	3.22 μm to 4.4 μm	Si wafer	Electro deposition	~7.7×10^{-7} mm^3/Nm	0.05	9.8 GPa	[13]
Cr_3C_2-NiCr	300 μm	grey cast iron	High Velocity Oxygen Fuel	7.8×10^{-4} mm^3/Nm	—	766	[21]
Al_2O_3-TiO_2	550 μm		Plasma spray	10.2×10^{-4} mm^3/Nm		643	

a nano BN on the boronized steel using the dip coating process followed by drying and consolidation at higher temperature. The highest tribological performance was achieved for the B-rich BN-layer coated over the carbon steel. The lowest COF achieved ~0.18 and the lowest wear rate of 0.1×10^{-6} mm^3/Nm. Furthermore, the Zr doped MoS_2 was coated over the substrate through the sputter-depositing method. The 1.5% of Zr on MoS2 coating exhibited less COF of 0.05 at 78% relative humidity and better corrosive resistance due to the good quality of the transfer film [13]. Therefore, it is important to understand the corrosion and tribological phenomena of various materials due to which the published data on various materials by different researchers are presented in Tables 13.1 and 13.2, respectively, for their corrosion and tribological behaviour.

13.5 EFFECT OF TRIBOCORROSION IN THE AUTOMOTIVE SECTOR

Wear and corrosion are two important factors in the degradation of materials in the automotive industry. Automobile parts such as cylinders, pistons, cranks, crank shafts, inlet valves, exhaust valves, cams, followers, and brake pads are more susceptible to both wear and corrosion. Further, in the engine liquid, additives are added to reduce both friction and wear. Additives such as zinc thiophosphate are used as an additive in liquid lubricants for improving antiwear properties. Moreover, the brake pad is the other component, which wear off very frequently. Grey cast iron is used for the disk brake rotors in the automotive. As a material, it has poor wear and corrosion resistance under severe operating conditions. To improve the wear and corrosion resistance of grey cast iron, a coating was developed on the sliding surface [21]. This can be achieved by the spraying of powders such as Cr_3C_2-NiCr and Al_2O_3-TiO_2 using High Velocity Oxygen Fuel (HVOF) and plasma spray processes, respectively. Cr_3C_2-NiCr coatings have shown better corrosion and wear resistance than the plasma-sprayed Alumina-Titania composite. Further, Wan et al. [22] demonstrated the corrosion-fatigue wear on CrN-coated piston rings. Testing has been carried out by both experimentation and simulation methods. The CrN has coated over the piston ring by the arc ion plating system at −20 V in a N_2 medium. The thickness of the coating has lain at around 35 μm, and the intermediate Cr layer has been almost 3 μm. The piston ring was fixed to the 6-cylinder CRDI-Engine and tested for 10,000 h of run. The worn-out piston ring was undergone various inspections and it was found that the CrN coating has partly removed from the surface, although the intermediate layer of Cr has still adhered to the piston ring. The delaminated worn-out surface was likely around 25 microns in depth. Further, during the reciprocating action the fatigue wear has increased if some corrosive species are presented on the specimen. This study has revealed that the sulfur and calcium have played an important role that accelerates fatigue cracking. Furthermore, Wan et al. [23] fabricated the Al_2O_3 coating on the CrN layer using atomic layer deposition. This has increased the thickness of the coating and decreased the grain size and surface roughness, resulting in an increase in corrosion resistance. The Al_2O_3 interlayer has acted as a good sealing layer that inhibits charge transfer. Tung and Gao [24] developed a CrN using thermal spray and DLC using PVD coated on a nitrided stainless steel (NSS) piston rings and chrome-plated stainless steel rings and it has tested using a high-frequency reciprocating tribometer. The experimentation result has revealed that the DLC coating has produced the lowest wear on the cylinder liner segment and it has a similar ring wear to nitrided and CrN-coated piston rings. Similarly, Wan et al. [25] demonstrated CrN and CrN/GLC coatings in different lubricated conditions. A GLC surface reduced the friction by 67% under a low lubricating environment and it has decreased the wear by 70% in a lubricating environment compared with CrN.

13.6 A METHOD TO IMPROVE TRIBOCORROSION IN THE AUTOMOBILE SECTOR

The coating is the most common technique to improve the durability of material during a sliding motion. Materials such as chromium nitride, diamond like carbon (DLC), NiCr, Al_2O_3, TiO_2 and its composites are widely used as the coating materials in the piston ring for the improvement of materials' tribocorrosion properties. The suitable intermediate layer has always preferred to enhance the adhesion of coating towards the substrate. Chromium nitride is the commonly used intermate layer for improving the adhesion between the substrate and material to be coated. The WC/Co composite has shown the capability of fighting severe wear, as well as erosion and corrosion in service. There are few successful methods which are used to deposit

TABLE 13.3
Materials Used to Improve Tribocorrosion in Automobile

Author	Method	Coating Material	Remarks
(S. [22])	Arc ion plating system	chromium nitride (CrN)	after 10000 h of reciprocating testing of piston ring, fatigue wear is increased
[26]	PVD	Hydrogen free diamondlike carbon (DLC)	18% reduction of friction between cylinder bore and piston. Similarly, 10% reduction when coated to top ring
[3]	Atmospheric pressure chemical vapor deposition	Chromium nitride (CrN) with Graphene oxide	ZDDP with GO forms tribofilm that provides improved wear resistance to the ring
[21]	High Velocity Oxygen Fuel (HVOF) Plasma spray processes	Cr_3C_2-NiCr Al_2O_3-TiO_2	HVOF showed better tribocorrosion performance than Plasma spray
[27]	RF magnetron sputtering	Crx N	Piston ring coating exhibited no cracks or delamination.
[23]	Atomic layer deposition	CrN-Al_2O_3	Addition of Alumina on the CrN layer improves the corrosion resistant of the coating
[24]	thermal spray Physical vapor deposited (PVD)	CrN Diamondlike carbon (DLC)	DLC coating on piston ring showed lowest wear among other
[25]	Arc ion plating and magnetron sputtering	CrN, CrN/GLC	CrN/GLC surface reduces friction by 67% under low lubricating environment and decreases the wear by 70% in lubricating environment compared with CrN

materials on the substrate such as physical vapour deposition, arc ion plating, chemical vapour deposition, high velocity oxygen fuel, plasma spray process, atomic layer deposition, magnetron sputtering and others. The coating methods are adopted based on the material, which must be coated on the substrate. Table 13.3 shows the summarized data of various researchers adopted for the improvement in tribocorrosion properties in automobile.

13.7 THE EFFECT OF TRIBOCORROSION IN INDUSTRIAL APPLICATIONS

Wear and corrosion of metals are major problem for the oil and gas industry [28]. To overcome these issues, coatings are one of the most commonly used surface treatments to improve tribological applications. It provides lower friction coefficient (COF) and wear resistance for sliding tribo-pair in mechanical systems, along with improved adhesion. Further, the oil industries are faced severe corrosion- and wear-related problems because of water/oil contains the high salinity and sand particles. To prevent the machinery components from corrosion and wear, it could be coated with a high tribocorrosion-resistant materials. Presently, the additives are used to reduce the corrosion and wear of the coating, but they are affected by various factors such as velocity, pressure, density and type of crude oil and it should also be replaced within the appropriate time span. Moreover, other conventional methods such as polymeric coating have exhibited less hardness and degradation over time. As discussed above, materials such as DLC, CrN, nano Alumina-Titania, etc. can be used to improve the tribocorrosion properties. The coating techniques are very important to obtain the robust design of material on substrate. Techniques such as CVD, PVD, and other methods help to improve the adhesion of coating.

13.8 CONCLUSION

Corrosion and tribology is one of the major issues in automotive industry. Friction, wear and corrosion have impacted on energy consumption, economic and CO_2 emissions. Approximately 23% of world's total energy consumption due to the friction and wear. By improving the coating methods, materials and lubrication have reduced the friction and wear in industries. Further, through the application of coatings, it has been estimated that the automotive sector would improve the GDP by ~ 1.4% and ~8.7% of total energy consumption in the long term [29]. Advance research on nanomaterials and feasible coating techniques have been developed by the researchers at the laboratory scale to improve the properties of materials that would increase corrosive resistance and tribology properties. Furthermore, the oil and gas industries are facing a severe problem during the transportation of oil and gas, where increased corrosions is taking place, resulting in the loss of energy. Replacing those components has increased maintenance costs. These problems can be rectified by the use of advanced coating techniques with suitable parameters.

REFERENCES

1. Holmberg, K., Andersson, P., & Erdemir, A. (2012). Global energy consumption due to friction in passenger cars. *Tribology International*, *47*, 221–234. https://doi.org/10.1016/j.triboint.2011.11.022
2. Holmberg, K., & Erdemir, A. (2019). The impact of tribology on energy use and CO2 emission globally and in combustion engine and electric cars. *Tribology International*, *135*(January), 389–396. https://doi.org/10.1016/j.triboint.2019.03.024
3. Özkan, D., Erarslan, Y., Kıncal, C., Gürlü, O., & Yağcı, M. B. (2020). Wear and corrosion resistance enhancement of chromium surfaces through graphene oxide coating. *Surface and Coatings Technology*, *391*(December 2019), 125595. https://doi.org/10.1016/j.surfcoat.2020.125595
4. Dhiflaoui, H., Khlifi, K., Barhoumi, N., & Ben Cheikh Larbi, A. (2020). Effect of voltage on microstructure and its influence on corrosion and tribological properties of TiO2 coatings. *Journal of Materials Research and Technology*, *9*(3), 5293–5303. https://doi.org/10.1016/j.jmrt.2020.03.055
5. Guo, Z., Hui, X., Zhao, Q., Guo, N., Yin, Y., & Liu, T. (2021). Pigmented Pseudoalteromonas piscicida exhibited dual effects on steel corrosion: Inhibition of uniform corrosion and induction of pitting corrosion. *Corrosion Science*, *190*(July), 109687. https://doi.org/10.1016/j.corsci.2021.109687

6. Yang, W., Liu, Z., & Huang, H. (2021). Galvanic corrosion behavior between AZ91D magnesium alloy and copper in distilled water. *Corrosion Science*, *188*(December 2020), 109562. https://doi.org/10.1016/j.corsci.2021.109562
7. Yan, L., Song, G. L., Wang, Z., & Zheng, D. (2021). Crevice corrosion of steel rebar in chloride-contaminated concrete. *Construction and Building Materials*, *296*, 123587. https://doi.org/10.1016/j.conbuildmat.2021.123587
8. Zhou, Y., & Zuo, Y. (2015). The intergranular corrosion of mild steel in CO_2 + $NaNO_2$ solution. *Electrochimica Acta, 154*, 157–165. https://doi.org/10.1016/j.electacta.2014.12.053
9. Mostepaniuk, A. (2017). Tribocorrosion. https://doi.org/10.5772/63657
10. Kubecka, P., Tvrdy, M., Wenger, F., & Ponthiaux, P. (2000). Tribocorrosion Behaviour of 08Ch18N10T Steel. In *Proceedings of International Topical Meeting on VVER Technical Innovations for next Century*, 365–372. https://inis.iaea.org/search/search.aspx?orig_q=RN:32011708
11. Sankara Narayanan, T. S. N. (2012). Nanocoatings to Improve the Tribocorrosion Performance of Materials. In *Corrosion Protection and Control Using Nanomaterials*. Woodhead Publishing Limited. https://doi.org/10.1533/9780857095800.2.167
12. Alojali, H. M., & Benyounis, K. Y. (2016). Advances in Tool Wear in Turning Process. In *Reference Module in Materials Science and Materials Engineering* (pp. 1–15). Elsevier Ltd. https://doi.org/10.1016/b978-0-12-803581-8.04031-5
13. Li, L., Lu, Z., Pu, J., & Hou, B. (2021). Investigating the tribological and corrosive properties of MoS2/Zr coatings with the continuous evolution of structure for high-humidity application. *Applied Surface Science, 541* (October). https://doi.org/10.1016/j.apsusc.2020.148453
14. Chen, J., Mraied, H., & Cai, W. (2018). Determining tribocorrosion rate and wear-corrosion synergy of bulk and thin film aluminum alloys. *Journal of Visualized Experiments*, *2018*(139), 1–11. https://doi.org/10.3791/58235
15. Bardini, L. (2018). EIS 101, an introduction to electrochemical spectroscopy. What was a website is now available as a self-contained PDF. July 2015. https://doi.org/10.13140/RG.2.1.2248.5600
16. Rajitha, K., Mohana, K. N. S., Mohanan, A., & Madhusudhana, A. M. (2020). Evaluation of anti-corrosion performance of modified gelatin-graphene oxide nanocomposite dispersed in epoxy coating on mild steel in saline media. *Colloids and Surfaces A: Physicochemical and Engineering Aspects*, *587*(November 2019), 124341. https://doi.org/10.1016/j.colsurfa.2019.124341
17. Atrens, A., Song, G. L., Shi, Z., Soltan, A., Johnston, S., & Dargusch, M. S. (2018). Understanding the Corrosion of Mg and Mg Alloys. In *Encyclopedia of Interfacial Chemistry: Surface Science and Electrochemistry*. Elsevier. https://doi.org/10.1016/B978-0-12-409547-2.13426-2
18. Chronopoulou, N., Siranidi, E., Routsi, A. M., Zhao, H., Bai, J., Karantonis, A., & Pavlatou, E. A. (2018). Embedding of hybrid MWCNT-Al_2O_3 particles in Ni matrix: Structural, tribological and corrosion studies. *Surface and Coatings Technology*, *350*, 672–685. https://doi.org/10.1016/j.surfcoat.2018.07.034
19. Panda, J. N., Wong, B. C., Medvedovski, E., & Egberts, P. (2021). Enhancement of tribo-corrosion performance of carbon steel through boronizing and BN-based coatings. *Tribology International*, *153*(August 2020), 106666. https://doi.org/10.1016/j.triboint.2020.106666
20. Toosinezhad, A., Alinezhadfar, M., & Mahdavi, S. (2020). Cobalt/graphene electrodeposits: Characteristics, tribological behavior, and corrosion properties. *Surface and Coatings Technology*, *385*(January), 125418. https://doi.org/10.1016/j.surfcoat.2020.125418

21. Samur, R., & Demir, A. (2012). Wear and corrosion performances of new friction materials for automotive industry. *Metalurgija*, *51*(1), 94–96.
22. Wan, S., Wang, H., Xia, Y., Tieu, A. K., Tran, B. H., Zhu, H., Zhang, G., & Qiang zhu. (2019). Investigating the corrosion-fatigue wear on CrN coated piston rings from laboratory wear tests and field trial studies. *Wear*, *432–433*(June), 202940. https://doi.org/10.1016/j.wear.2019.202940
23. Wan, Z., Zhang, T. F., Lee, H. B. R., Yang, J. H., Choi, W. C., Han, B., Kim, K. H., & Kwon, S. H. (2015). Improved corrosion resistance and mechanical properties of CrN hard coatings with an atomic layer deposited Al_2O_3 interlayer. *ACS Applied Materials and Interfaces*, *7*(48), 26716–26725. https://doi.org/10.1021/acsami.5b08696
24. Tung, S. C., & Gao, H. (2003). Tribological characteristics and surface interaction between piston ring coatings and a blend of energy-conserving oils and ethanol fuels. *Wear*, *255*(7–12), 1276–1285. https://doi.org/10.1016/S0043-1648(03)00240-0
25. Wan, S., Pu, J., Li, D., Zhang, G., Zhang, B., & Tieu, A. K. (2017). Tribological performance of CrN and CrN/GLC coated components for automotive engine applications. *Journal of Alloys and Compounds*, *695*, 433–442. https://doi.org/10.1016/j.jallcom.2016.11.118
26. Higuchi, T., Mabuchi, Y., Ichihara, H., Murata, T., & Moronuki, M. (2017). Development of hydrogen-free diamond-like carbon coating for piston rings. *Tribology Online*, *12*(3), 117–122. https://doi.org/10.2474/trol.12.117
27. Friedrich, C., Berg, G., Broszeit, E., Rick, F., & Holland, J. (1997). PVD CrxN coatings for tribological application on piston rings. *Surface and Coatings Technology*, *97*(1–3), 661–668. https://doi.org/10.1016/S0257-8972(97)00335-6
28. Bueno, A. H. S., Solis, J., Zhao, H., Wang, C., Simões, T. A., Bryant, M., & Neville, A. (2018). Tribocorrosion evaluation of hydrogenated and silicon DLC coatings on carbon steel for use in valves, pistons and pumps in oil and gas industry. *Wear*, *394–395*, 60–70. https://doi.org/10.1016/j.wear.2017.09.026
29. Holmberg, K., & Erdemir, A. (2017). Influence of tribology on global energy consumption, costs and emissions. *Friction*, *5*(3), 263–284. https://doi.org/10.1007/s40544-017-0183-5

14 Numerical Solution of Reynolds Equation for a Compressible Fluid Using Finite Volume Upwind Schemes

M. Phani Kumar, Pranab Samanta and Naresh Chandra Murmu

CSIR-Central Mechanical Engineering Research Institute, Durgapur, India

CONTENTS

14.1 INTRODUCTION

Surfaces that are in relative motion are generally separated by a thin film of lubricant to minimize friction. If two mating surfaces during operation are completely separated by lubricant film, such a type of lubrication is known as fluid film lubrication. Thin viscous fluid films, which carry load due to the effect of hydrodynamic lubrication, can be found in journal bearings, one of the most widely used tribology components. Advances in engineering and technology demand bearings with high precision and optimum performance. The great majority of these demands can be found in specialized fields such as high-precision instruments and sensors, space technology, nuclear engineering, machine tools and computer applications. Journal bearings are lubricated by the hydrodynamic flow which is generated by relative surface motion and/or external pressurization. In order to meet the challenges of high-precision applications, it is necessary to operate with near frictionless motion and several methods are being explored to reduce the friction. This can be achieved with the use of gas lubrication as the major advantage of this material is its minimal frictional characteristics. The governing equation that describes the flow of a thin film

DOI: 10.1201/9781003243205-14

of lubricant separating two rigid surfaces in relative motion is given by the Reynolds equation, a partial differential equation, which can be derived from the Navier-Stokes equation. To design a hydrodynamic bearing, a few important characteristics, such as load carrying capacity, flow requirements and power loss due to viscous friction, are to be predicted accurately. It is essential to investigate these characteristics at the desired operating conditions during the design stage itself. These characteristics can be determined if the pressure distribution within the bearing is known. Although the pressure distribution can be obtained by experimental investigations by operating the bearing at operating conditions, however, such a method is not feasible on economic grounds. One can overcome such limitations by numerical analysis. The exact close-form solution of the Reynolds equation exists only for a few fortunate cases. However, for a complicated realistic system, the absence of close form solution to the full Reynolds equation renders it unsolvable by known analytical methods. Researchers have therefore resorted to numerical methods to solve the problem.

The numerical solution of some form of Reynolds equation is usually required in many fluid-film lubrication analyses or bearing designs. Therefore, the modelling and study of fluid flow in thin film is an important engineering issue since the pressure generated defines the performance of the device. The generalized Reynolds equation is presented in Eq. (14.1)

$$\begin{aligned}&\frac{\partial}{\partial x}\left(\frac{\rho h^3}{12\eta}\frac{\partial p}{\partial x}\right)+\frac{\partial}{\partial z}\left(\frac{\rho h^3}{12\eta}\frac{\partial p}{\partial z}\right)\\&=\frac{\partial}{\partial x}\left[\frac{\rho\left(U_1+U_2\right)h}{2}\right]+\frac{\partial}{\partial z}\left[\frac{\rho\left(W_1+W_2\right)h}{2}\right]+\rho\left(V_2-V_1\right)+\frac{\partial}{\partial t}\left(\rho h\right)\end{aligned} \tag{14.1}$$

Where:

p – is the pressure [Pa]
h – film thickness [m]
U, V, W – velocity components along x, y and z directions.
η – Viscosity of the lubricant [Pa s]
ρ – Fluid density [kg/m^3]

Equation (14.1) presents the generalized Reynolds equation and is applicable for both incompressible and compressible fluids. The two terms on the left-hand side of the equation describe the net flow rates due to pressure gradients. The first two terms on the right-hand side describes the flow rates due to surface velocities. These are known as Poiseulle and Couette terms, respectively.

The solution of the hydrodynamic lubrication problem requires obtaining an approximate numerical solution to the Reynolds equation. Prominent numerical methods such as finite difference (FDM), finite element (FEM) and finite volume (FVM) have been proposed to provide the solution of the fluid film lubrication problem. In brief, these methods follow the discretization of governing equations and obtain its solution by dividing the region of interest into finite subdivisions. Of the three numerical methods, the majority of researchers use the finite difference method (FDM) due to its relatively simplicity. In this method, the partial derivatives of the equation are approximated by finite difference formulas obtained from the truncation of the high-order terms of the Taylor expression. In order to obtain better approximations, second-order central differencing schemes are used. The finite difference representation of the second-order partial derivative is shown in Equation (14.2):

$$\left(\frac{\partial^2 u}{\partial x^2}\right)_{i,j} = \frac{u_{i+1,j} - 2u_{i,j} + u_{i-1,j}}{(\Delta x)^2} + O(\Delta x)^2 \tag{14.2}$$

From Equation (14.2), it is observed that information used in forming the finite difference quotient comes from both sides of the grid point located at(i,j); that is, it uses $u_{i+1,j}$ as well as $u_{i-1,j}$. Grid point (i,j) falls between two adjacent grid points. In order to calculate the finite difference quotient, information from the adjacent nodes is being used. It is expected that a numerical scheme, while obtaining solution for flow equations, should use information which is consistent with flow field. In simple terms it is like respecting the flow physics. To be precise, the central differencing schemes does not always follow the proper flow of information throughout the flow field. In most of the cases, they obtain information from outside the domain of dependence of a given grid point and this can compromise the solution.

The generalized Reynolds equation, as mentioned in Equation (14.1), comprises the terms which represent the net flow due to pressure gradient and relative surface velocities, i.e. Poiseulle flow and Couette flow terms respectively. Bearings which uses gas film lubrication to support the loads usually operate at higher speeds because of low viscosity and the compressibility nature of gas. To discretize the governing equation at the clearance region, finite difference methods, because of their simplicity in implementation have been the popular choice among the majority of researchers. Specifically, second-order central differencing schemes are used to obtain discretized relations. As bearings operating with gas film lubrication usually run at higher speeds, terms associated with velocity components, i.e. Couette flow terms, dominate the other flow terms. This results in non-physical oscillations, which initially appear at the regions where the second derivative of the pressure, i.e. Poiseulle terms, are very small and rapidly compromise the solution. Because of the strong dominance of Couette flow terms at higher-bearing numbers, the numerical method used for the analysis should accurately represent the sharp gradients caused by Couette terms. However, the inefficiency of the central difference schemes to capture sharp gradients makes the solution unstable [1]. A well-known remedy to suppress these unwanted oscillations of the Couette flow dominated problems, which are of purely numerical origin, is to employ the upwind schemes [2]. To discretize the Reynolds equation, finite volume discretization is used as it uses the governing equation in conservative form and ensures the conservation of all properties at control volume. This is not true with regard to the finite difference methods [3]. Therefore, in the present chapter, a numerical solution for the porous journal bearing with gas film lubrication is presented which is stable even at higher operating speeds.

14.2 NUMERICAL SCHEME

In order to show the effectiveness of the abovementioned solution procedure, steady-state characteristics of a porous journal bearing is obtained at higher-bearing numbers (Figure 14.1).

The governing differential equation for finite bearings using compressible lubricant in non-dimensional form is given as follows

In the porous bush

$$\bar{K}_x \frac{\partial^2 \bar{P}'^2}{\partial \theta^2} + \left(\frac{R}{H}\right)^2 \frac{\partial^2 \bar{P}'^2}{\partial \bar{y}^2} + \bar{K}_z \left(\frac{D}{L}\right)^2 \frac{\partial^2 \bar{P}'^2}{\partial \bar{z}^2} = 0 \quad (14.3)$$

In the clearance region

$$\frac{\partial}{\partial \theta}\left(\bar{h}^3 \frac{\partial \bar{P}^2}{\partial \theta}\right) + \left(\frac{D}{L}\right)^2 \frac{\partial}{\partial \bar{z}}\left(\bar{h}^3 \frac{\partial \bar{P}^2}{\partial \bar{z}}\right) = 2\Lambda \frac{\partial}{\partial \theta}\left(\bar{P}\bar{h}\right) + 4\Lambda\lambda \frac{\partial}{\partial \tau}\left(\bar{P}\bar{h}\right) + \beta \left|\frac{\partial \bar{P}'}{\partial \bar{y}}\right|_{\bar{y}=1} \quad (14.4)$$

The governing equation at the clearance region, i.e. the Reynolds equation, is discretized by using the finite volume discretization. In order to obtain a linear discretized relation, parameter values on the cell boundaries are expressed as a function of the parameter values at the surrounding nodes. The discretization procedure for a Poiseulle term and a Couette term is detailed below.

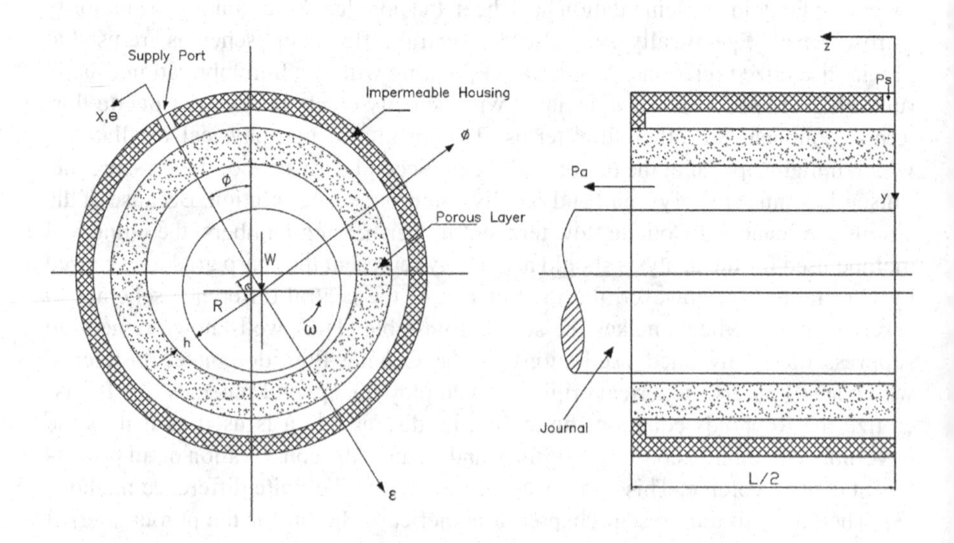

FIGURE 14.1 Schematic diagram of externally pressurized porous journal bearing.

Poiseuille term

$$\frac{\partial}{\partial\theta}\left(\bar{h}^3\frac{\partial\bar{P}^2}{\partial\theta}\right)=\frac{\left(\bar{h}^3_{i+\frac{1}{2}}\left(\frac{\partial\bar{P}^2}{\partial\theta}\right)_{i+\frac{1}{2}}-\bar{h}^3_{i-\frac{1}{2}}\left(\frac{\partial\bar{P}^2}{\partial\theta}\right)_{i-\frac{1}{2}}\right)}{\Delta\theta}$$

Where,

$$\bar{h}^3_{i\pm\frac{1}{2}}=\frac{1}{2}\left[\bar{h}_i^3+\bar{h}_{i\pm1}^3\right]$$

$$\left(\frac{\partial\bar{p}^2}{\partial\theta}\right)_{i+\frac{1}{2}}=\frac{\left(\bar{p}_{i+1}^2-\bar{p}_i^2\right)}{\Delta\theta}$$

Couette term

$$\frac{\partial}{\partial\theta}\left(\bar{P}\bar{h}\right)=\frac{1}{\Delta\theta}\left[\bar{P}_{i+\frac{1}{2}}\bar{h}_{i+\frac{1}{2}}-\bar{P}_{i-\frac{1}{2}}\bar{h}_{i-\frac{1}{2}}\right]$$

As discussed earlier, since the central differencing scheme evaluates the parameter at the grid point using the information from outside the domain of dependence, and this compromises the solution. For the cases where there is no dominance of Couette flow terms, i.e. when dealing with incompressible fluids, the use of a central differencing scheme doesn't cause convergence issues. However, in the case of compressible fluids, since the bearing operates at higher speeds, even for light loading conditions, there is a dominance of Couette flow terms and the solution fails to converge when central differencing schemes are used. Therefore, upwinding schemes are used to suppress these numerical oscillations to obtain the converged solution. Accordingly, the upwinding scheme is used to calculate the value of $\bar{P}$ at grid point center.

Figure 14.2 presents the schematic of the cell and its interface. The expression used to obtain $\bar{P}$ at the grid point center is taken from Kim and Kim [4] and is as follows

$$\bar{P}_{i+\frac{1}{2}}=\frac{\left(2\bar{P}_{i+1}+5\bar{P}_i-\bar{P}_{i-1}\right)}{6}$$

$$\bar{P}_{i-\frac{1}{2}}=\frac{\left(2\bar{P}_i+5\bar{P}_{i-1}-\bar{P}_{i-2}\right)}{6}$$

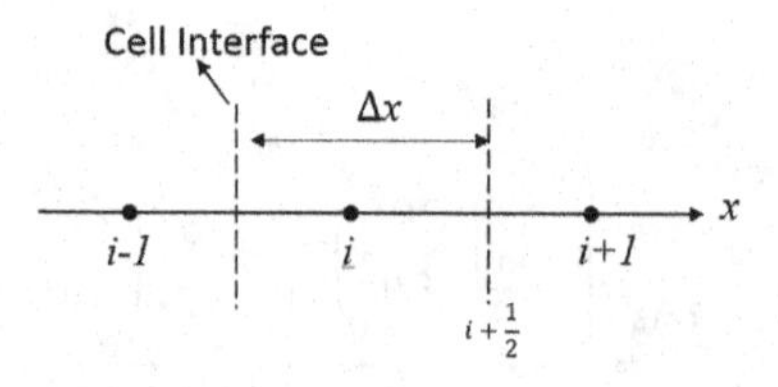

FIGURE 14.2 Cell-interface and cell center.

A steady-state solution was obtained for a porous journal bearing working with compressible fluid. A converged solution was obtained by considering the problem at hand as a pseudo-time problem. Discretized governing equation at the clearance region is solved by using the first-order three-stage Runge-Kutta scheme with optimal multistage coefficients as mentioned in [5]. The solution procedure begins with an initial guess values obtained by considering the non-rotating case. With the help of these initial guess values, residuals are obtained at all the grid points. Further, pressure values at the grid points are updated using these residual values as per the following relation

$$P_{(i,j)}^{\text{new}} = P_{(i,j)}^{\text{old}} + \Delta t.\alpha_r.\text{residual}_{(i,j)}$$

Where α_r is the optimal coefficients of the three-stage Runge-Kutta method.

Once the pressure values in the clearance region is updated, the governing equation in the porous region given by Darcy's law is solved by using a popularly known iterative solver biconjugate gradient stabilized method (BICGSTAB). This iterative method is applied until the residual at the grid points in the porous region reaches to 1e−12. With the time increment ((Δt) of 1e−4, the process is repeated until maximum residuals at the grid point in film region reaches to 1e−10. At such point, iterations are stopped and pressure values at the film region are used to obtain the steady-state characteristics. Figure 14.3 depicts the solution procedure followed. Figure 14.4 presents the residual values obtained using the FDM methods and the current solution method. It is observed that the residual values from the FDM method fluctuate during the iterative process, whereas the current solution method shows that the residual values are decreased gradually to a level of 1e−10. Therefore, it can be understood that, from Figure 14.4, the current solution procedure results in a stable numerical solution.

In order to validate the present numerical method, the load-carrying capacity of a gas lubricated porous journal bearing was obtained and compared with the published literature. Figure 14.5 presents the comparison plot of the present solution method results with the results from Sun [6].

14.3 RESULTS

According to the bearing geometry parameters provided, the steady-state characteristics, i.e. the load-carrying capacity, attitude angle, friction coefficient and flow rate are estimated from the pressure profile obtained for two supply pressure values, 2.0

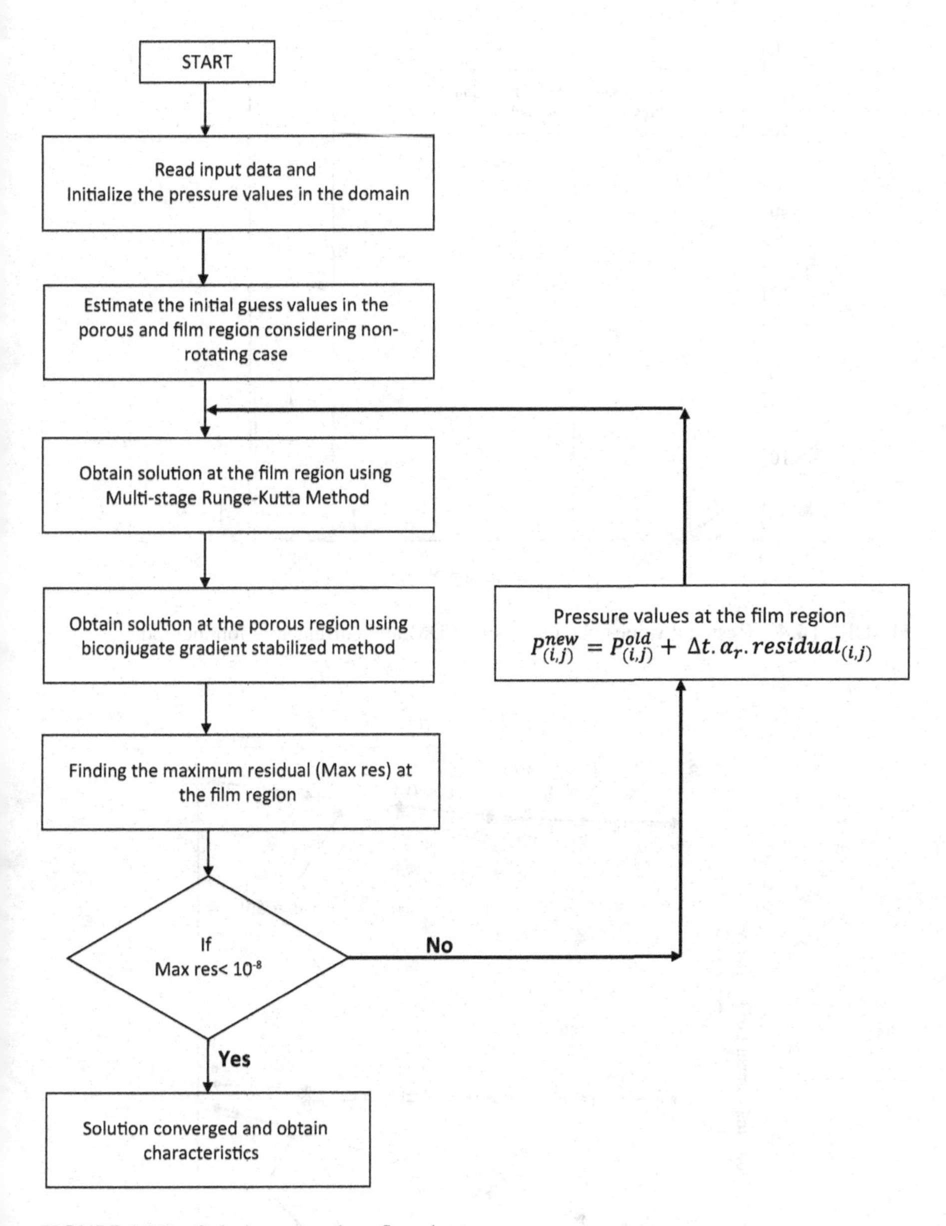

FIGURE 14.3 Solution procedure flowchart.

and 5.0. These steady-state performance characteristics were obtained up to a bearing number (Λ) of 150.

Figures 14.6 and 14.7 present the steady-state charcateristics of a porous journal bearing operating with a supply pressure of 2.0. Variation of load-carrying capacity ($\bar{W}_o$) and attitude angle (ϕ_o^0) with bearing number (Λ) for vairous values of ε_o is presented in Figure 14.6. It is observed that the load-carrying capacity increases with increase in the eccentricity ratio. Bearing operating at the high eccentricity ratio

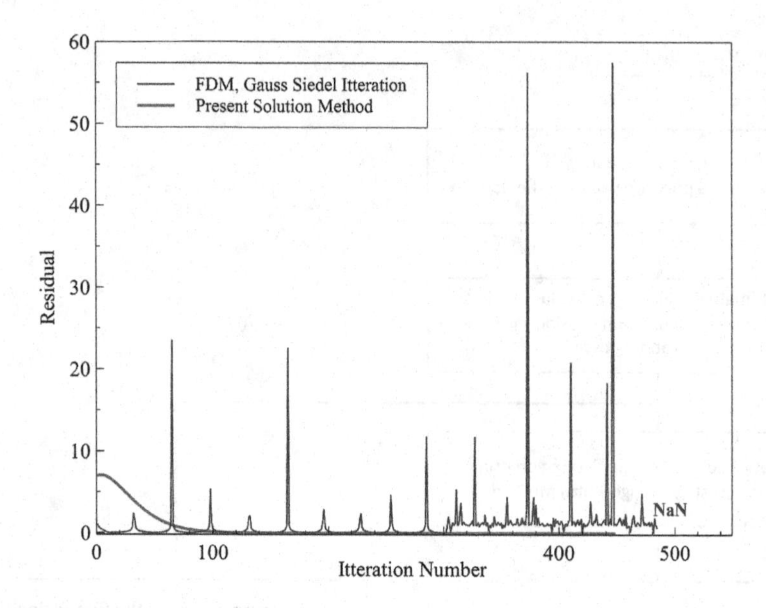

FIGURE 14.4 Residual values obtained using FDM and current solution method.

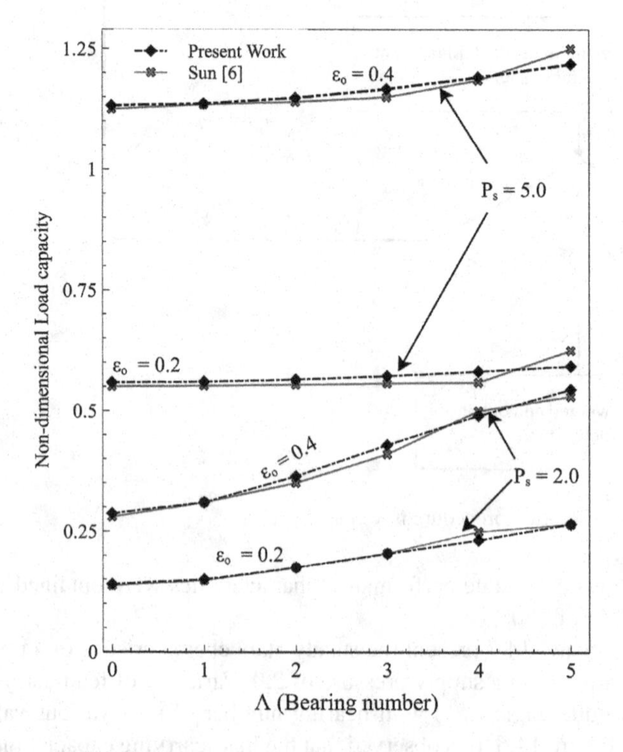

FIGURE 14.5 Validation of the present model toward the work of Sun [6].

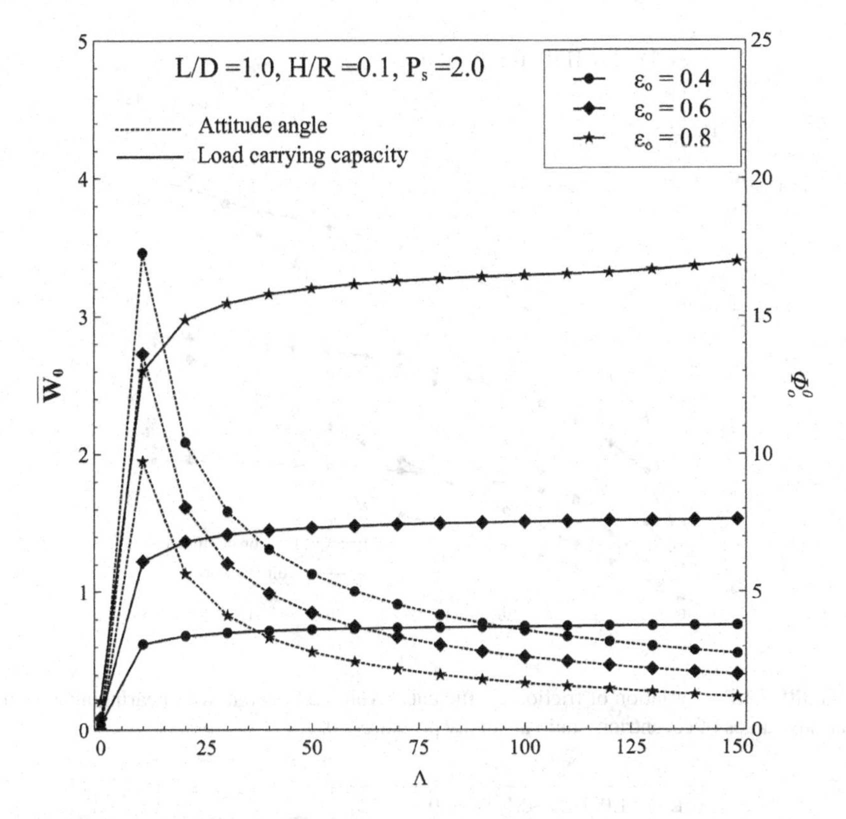

FIGURE 14.6 Variation of load & attitude angle with bearing number for various values of eccentricity ratio at supply pressure 2.0.

results in higher film pressure at the clearance and allows it to withstand higher loads. It is also observed that for an eccentricity ratio load-carrying capacity increases with increase in bearing number. The difference in the directions between the lines of applied load and corresponding reaction forces causes journal-bearing instability, measured by the attitude angle (ϕ_o^0). It is observed that the attitude angle initially increases with an increase in the bearing number and then decreases at higher bearing numbers. It can be understood from the figure that bearing operating at higher bearings numbers and eccentricity ratio are more stable.

Figure 14.7 presents the variation of volume flow rate ($\bar{Q}_o$) and friction coefficient ($\mu_f(R/C)$) with a bearing number for various ε_o values. It is observed that the volume flow rate increases and the friction coefficient value decreases with an increase in eccentricity ratio.

Figures 14.8 and 14.9 presents the steady-state characteristics of a porous journal bearing operating with a supply pressure of 5.0. The trend behaviour of the characteristics seems to be similar with the trends obtained with a supply pressure of 2.0. The load-carrying capacity increases with an increase in supply pressure. The attitude angle shows a similar trend as observed in Figure 14.4 and it shows that bearings that operate at higher speed and eccentricity ratio are more stable.

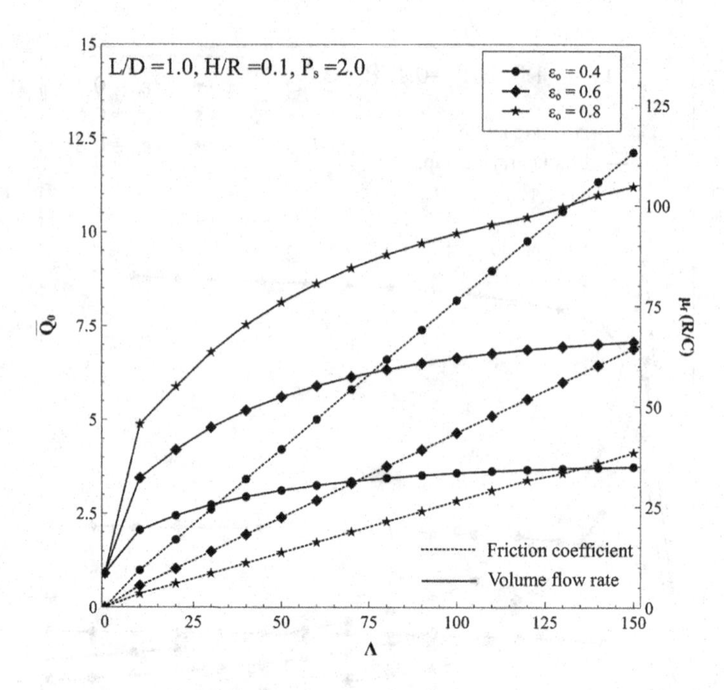

FIGURE 14.7 Variation of friction coefficient & volume flow rate with bearing number for various values of eccentricity ratio at supply pressure 2.0.

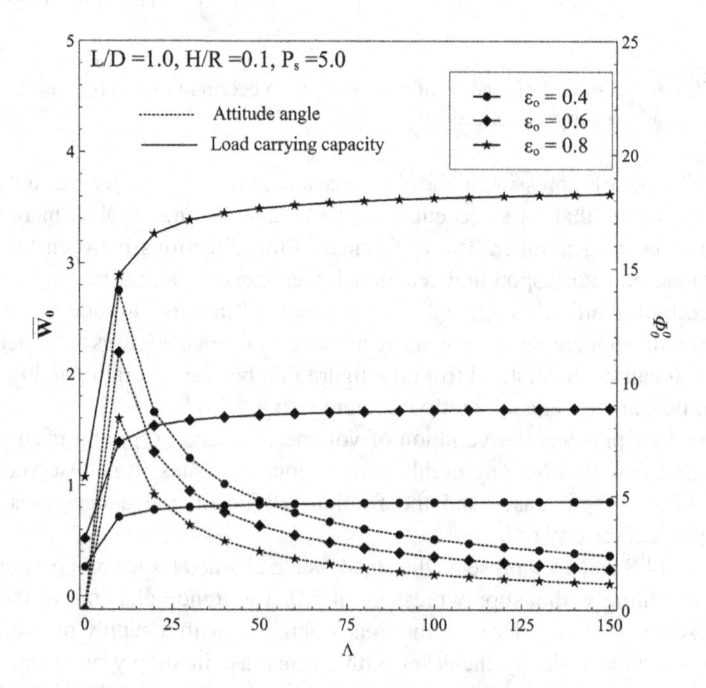

FIGURE 14.8 Variation of load & attitude angle with bearing number for various values of eccentricity ratio at supply pressure 5.0.

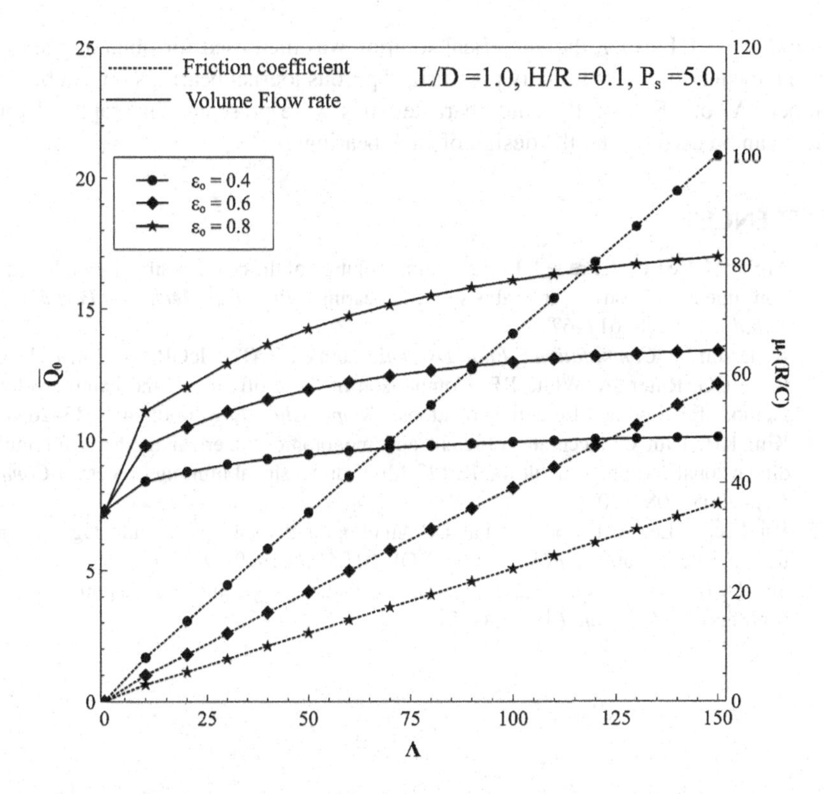

FIGURE 14.9 Variation of friction coefficient & flow rate with bearing number for various values of eccentricity ratio at supply pressure 5.0.

With the influence of supply pressure, the friction coefficient and the volume flow rate increases, as presented in Figure 14.9. However, operating at higher eccentricity ratio and higher speeds results in higher volume flow rates and lesser friction coefficient values.

14.4 SUMMARY

The present work, with inspiration from oscillation control schemes which are often used in the CFD analysis of compressible flows, uses an upwind scheme to obtain a stable numerical solution of the compressible Reynolds equation at higher bearings numbers. The governing equations are discretized using the finite volume method as it uses the governing equation in conservative form and ensures the conservation of all properties at control volume. After discretization, a solution was obtained using a first-order three-stage Runge-Kutta method and BiCGSTAB method to solve pressure values at clearance region and porous region, respectively. This solution method was validated by obtaining solution for a porous journal bearing and compared with published literature and the results seem to concur. It is shown that under the proposed method the maximum residual values decrease with iteration. However, the residual values fluctuate during the iterative process when FDM with a Gauss-Seidel method is used. Further, the numerical solution was then used for obtaining steady-state characteristics of externally pressurized porous journal bearings up to a bearing number (Λ) of 150. Steady-state characteristics were presented in graphical form which can be used during the design of such bearings.

REFERENCES

1. Arghir M, Le Lez S, Frene J. Finite-volume solution of the compressible Reynolds equation: linear and non-linear analysis of gas bearings. *Proc Inst Mech Eng Part J J Eng Tribol* 2006; 220: 617–627.
2. Anderson JD. *Computational Fluid Dynamics*. 6th ed. TATA McGRAW-Hill, 1995.
3. Botte GG, Ritter JA, White RE. Comparison of finite difference and control volume methods for solving differential equations. *Comput Chem Eng* 2000; 24: 2633–2654.
4. Kim KH, Kim C. Accurate, efficient and monotonic numerical methods for multi-dimensional compressible flows. Part II: Multi-dimensional limiting process. *J Comput Phys* 2005; 208: 570–615.
5. Van Leer B, Lee W-T, Roe PL, et al. Design of optimally smoothing multistage schemes for the Euler equations. *AIAA J* 1992. DOI: 10.2514/6.1989-1933.
6. Sun D. Analysis of the steady state characteristics of gas-lubricated, porous journal bearings. *J Lubr Technol* 1975; 44–51.

15 Tribological Performance of Optimal Compound-Shaped Texture Profiles for Machine Components

Nilesh D. Hingawe and Skylab P. Bhore

Motilal Nehru National Institute of Technology Allahabad, Prayagraj, India

CONTENTS

15.1 INTRODUCTION

At present, worldwide research is concentrated on energy savings. The leading source of energy loss is friction. Friction initiates wear that leads to the failure of machine components. It is thus essential to reduce the friction for improving tribological performance and to minimize the energy loss. Nevertheless, this issue can be addressed effectively using surface texturing. It is a well-established technique to improve the tribological performance of machine components, such as bearings, parallel slider, seals, piston rings, etc. [1].

Surface texture decreases the contact area of mating faces in boundary lubrication, acts as lubricant reservoir during startup, and increases the film thickness in hydrodynamic regime to generate an additional hydrodynamic effect that leads to the further separation of surfaces which decreases shear stress. These mechanisms in different lubrication regimes contribute to friction reduction [2, 3]. The surface texture fabricated using non-conventional machining processes are found to be more effective than conventional processes [4]. Non-conventional processes, such as laser surface texturing, micro-electric discharge machining, electrochemical machining,

DOI: 10.1201/9781003243205-15

chemical machining, and reactive ion etching, are widely used [5]. Among these, laser surface texturing is the most advance technique and it can generate accurate texture shape with controlled dimensions [6].

Widespread research on texture shape has been carried out. Siripuram and Stephens [7] investigated the hydrodynamic performance of simple texture shapes with constant depth. They observed that the friction coefficient is insensitive to texture shape, but is substantially sensitive to texture density. However, Qiu et al. [8, 9] found that the texture shape with round profile performs better than the straight profile. In addition, the non-standard texture shapes, such as short and long drop, egg-shape, star-like, heart-like, chevron, trapezoid and fusiform, have obtained better tribological characteristics than the standard/simple texture shapes [10–16]. In contrast, most of the research on non-standard texture shapes is based on numerical analysis. Experimentally, the generation of these non-standard textures on different machine components, such as bearings, piston rings, seals, etc., is difficult to achieve with controlled dimensions. Consequently, the fabrication of surface textures for different applications is still limited to standard texture shapes.

In the present chapter, the compound-shaped textures are introduced with the viewpoint of improvement in tribological performance and the feasibility of fabrication. Compound-shaped textures are modeled using straight and round edges, namely C1, C2, and C3. For each texture shape, the variable parameters: aspect ratio (Ar), texture density (Td), and slider speed (U) are first analysed using response surface methodology based central composite design (CCD) and then optimized using grey relational analysis (GRA) multi-objective optimization. Under optimal condition, the best texture shape in terms of maximum load capacity and minimum friction coefficient is evaluated.

15.2 NUMERICAL MODEL

Detailed analysis is performed on a single texture which can be applicable to a simplified case of machine components, such as journal bearing, thrust bearing, seals, parallel slider, etc. Schematic of a textured parallel slider are shown in Figure 15.1(a). A single texture is formed with an imaginary unit cell, and positioned at the center

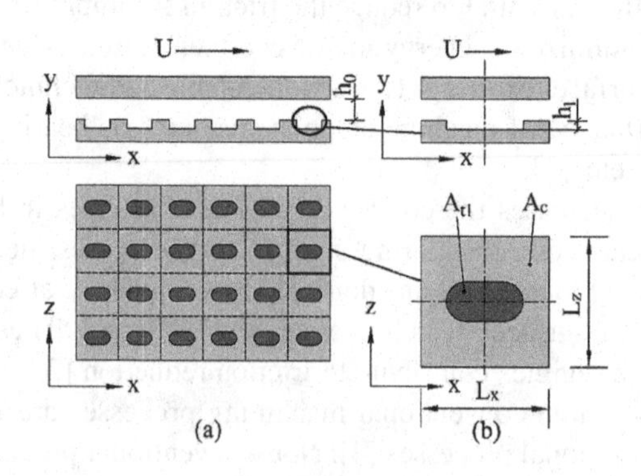

FIGURE 15.1 Surface textured (a) parallel slider, and (b) unit cell.

of a unit cell. This approach reduces the computation time and is relatively easier to model [7–9]. A typical texture and unit cell is shown in Figure 15.1(b).

The lubricant is assumed to be incompressible and follows Newton's law of viscosity. Neglecting thermal, inertia, and squeezing effect, the Reynolds equation is given as

$$\frac{\partial}{\partial x}\left(h^3\frac{\partial p}{\partial x}\right)+\frac{\partial}{\partial z}\left(h^3\frac{\partial p}{\partial z}\right)=6\mu U\frac{\partial h}{\partial x} \tag{15.1}$$

where, $h(x,z)=\begin{cases} h_0+h_1 & \text{(Textured surface)} \\ h_0 & \text{(Plane surface)} \end{cases}$

The boundary conditions to solve the Reynolds equation are:

Atmospheric boundary condition:

$$p\left(x,-\frac{L_z}{2}\right)=p\left(x,\frac{L_z}{2}\right)=p_{\text{atm}} \tag{15.2}$$

Periodic boundary condition:

$$\frac{\partial p}{\partial x}\left(-\frac{L_x}{2},z\right)=\frac{\partial p}{\partial x}\left(\frac{L_x}{2},z\right) \tag{15.3}$$

Reynolds cavitation condition:

$$p=p_{\text{cav}}=0 \text{ and } \frac{\partial p}{\partial x}=0 \quad \text{when } p<0 \tag{15.4}$$

Considering these boundary conditions, the Reynolds equation is discretized by the finite difference method with a Gauss–Seidel iterative scheme. To achieve an accurate pressure, the following convergence criterion has been satisfied.

$$\Sigma\Sigma\left|\frac{(p_{i,j})_{N+1}-(p_{i,j})_N}{(p_{i,j})_{N+1}}\right|\le 10^{-5} \tag{15.5}$$

The converged pressure is integrated by Simpson's 1/3 rule. Using this, the load capacity and frictional force are evaluated.

$$W=\int_{-L_Z/2}^{L_Z/2}\int_{-L_X/2}^{L_X/2}p(x,z)\,dx\,dz \tag{15.6}$$

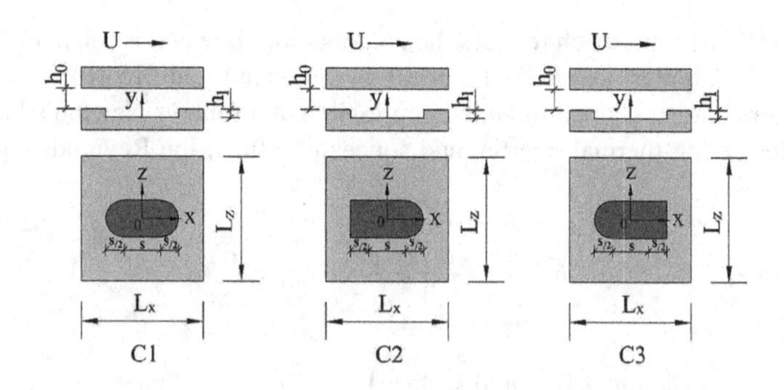

FIGURE 15.2 Composite shaped textures: (a) C1, (b) C2, and (c) C3.

$$F = \int_{-L_Z/2}^{L_Z/2} \int_{-L_Z/2}^{L_Z/2} \tau(x,z)\,dxdz \tag{15.7}$$

From these equations, friction coefficient is calculated as

$$f = \frac{F}{W} \tag{15.8}$$

For this, a MATLAB code was created and validated with Siripuram and Stephens [7]. Good agreement between the results was obtained.

In the present chapter, the compound-shaped textures, namely C1, C2, and C3, are introduced as shown in Figure 15.2. The textured geometry is specified by non-dimensional parameters, viz. Ar and Td. The maximum possible Td for composite textures C1, C2, and C3 are 0.446, 0.473, and 0.473, respectively. However, Ar remains unchanged for all the shapes.

15.3 DESIGN AND OPTIMIZATION

A. Texture design using response surface methodology
   In textured parallel sliding contact, Ar, Td, and U are the controllable parameters. These controllable parameters are thus considered to be variable parameters. However, μ, h, and T are uncontrolled parameters and thus considered to be fixed parameters (see Figure 15.3). To get an optimal variable parameter, the maximum possible range is considered. For Ar, Td, and U, the range is considered as 0.005–0.025, 0.2–0.4, and 0.66–2.66 m/s, respectively. Each range is split into five levels using response surface methodology based central composite design (CCD). The detailed formulation of the response surface model is given in [17].

B. Optimization of texture design parameters
   The objective is to improve the tribological characteristics (W and f) of textured parallel sliding contact. For this, the variable parameters (Ar, Td, and U) are optimized using GRA multi-objective optimization. The equal weightage

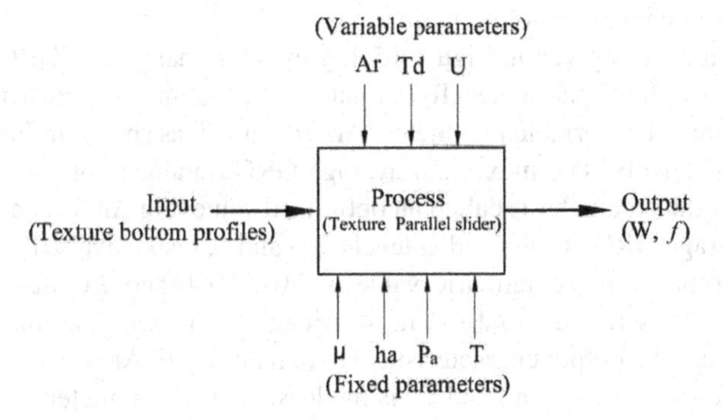

FIGURE 15.3 Texture design model.

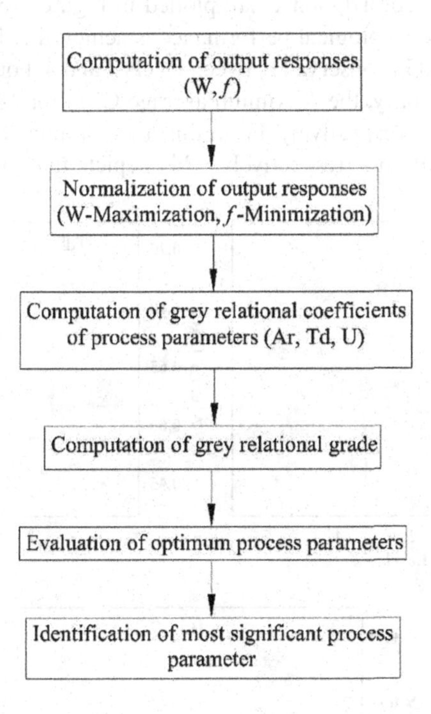

FIGURE 15.4 Steps in GRA multi-objective optimization.

is given to both the output responses [18]. Further, the most significant/critical variable parameter is evaluated using the steps given in Figure 15.4.

15.4 RESULTS AND DISCUSSION

For each compound-shaped texture the optimum variable parameters are evaluated. Further, the comparative analysis of these compound textures with each other is carried out.

A. Optimum parameters of compound texture C1
Using the steps given in Figure 15.4, grey relational grade (GRG) is obtained for each variable parameter. To evaluate the optimum value, an average GRG is determined for variable parameters Ar, Td, and U, as shown in Figures 15.5(a–c), respectively. The maximum average GRG obtained corresponding to that level of the parameter is called an optimized value. For Ar, Td, and U, the highest average GRG is achieved at levels 5, 3 and 5, respectively. For these levels, the corresponding parametric value is 0.015, 0.4, and 2.66m/s, respectively. Further, U is found to be the most critical parameter. This indicates that U influences the output characteristics more than that of Ar and Td. Furthermore, Ar is found to be second and Td is the least critical parameter.

B. Optimum parameters of compound texture C2
For compound texture C2, average GRGs for each level corresponding to variable parameters Ar, Td, and U are plotted in Figures 15.6(a–c), respectively. For Ar, the best tribological performance is achieved at level 4, while a slight deviation in GRG is observed between levels 3 and 4. For level 4, the optimum Ar is 0.02. Similarly, the maximum average GRG for Td, and U are obtained at levels 4 and 5, respectively. From this, the optimum Td and U are found to be 0.35 and 2.66 m/s, respectively. This depicts that higher Td and U gives

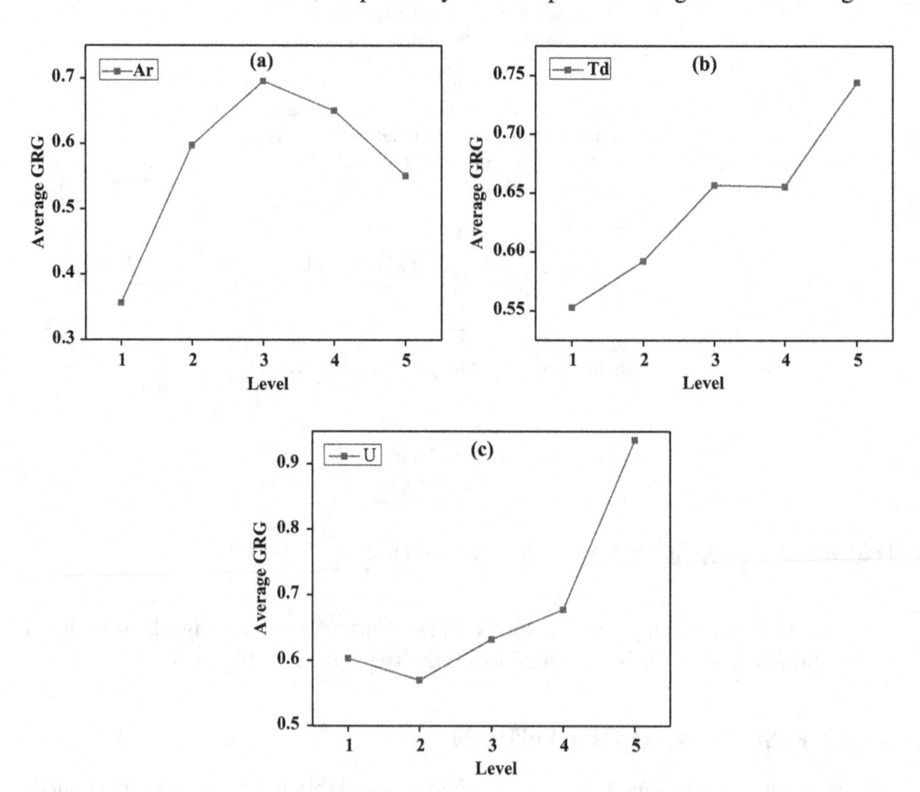

FIGURE 15.5 Influence variable parameters: (a) Ar, (b) Td, and (c) U on average GRG of compound texture C1.

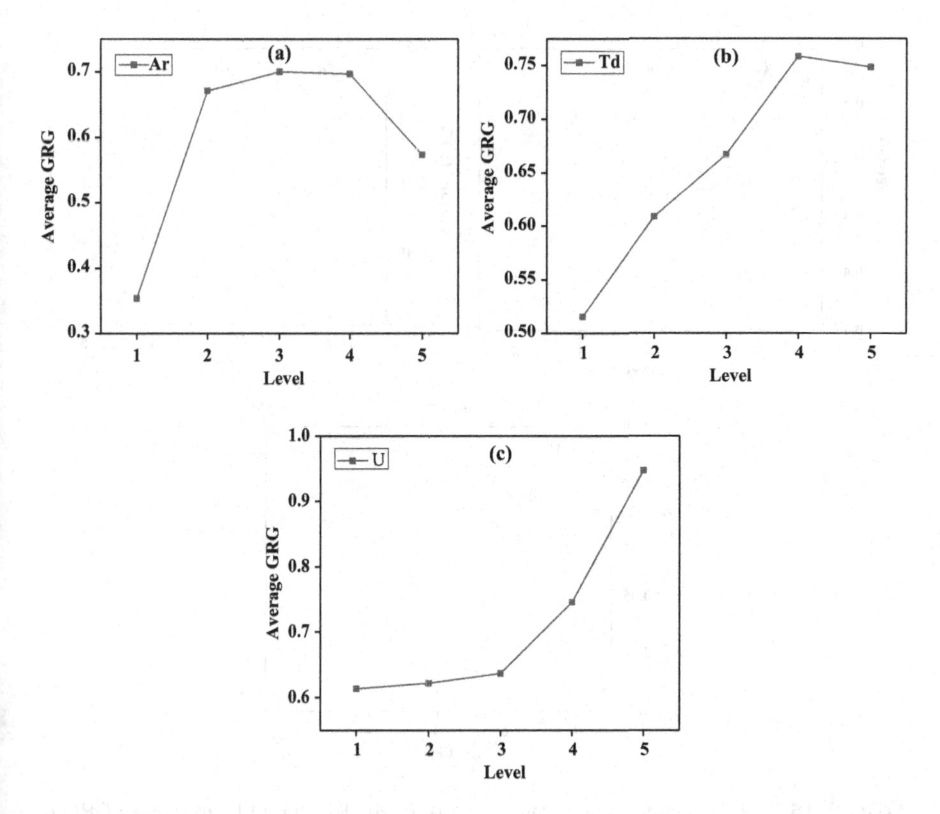

FIGURE 15.6 Influence variable parameters: (a) Ar, (b) Td, and (c) U on average GRG of compound texture C2.

better tribological performance. Furthermore, the slider speed is evaluated as being the most critical parameter and texture density is regarded as the least critical parameter in improving the tribological performance of parallel sliding contact.

C. Optimum parameters of compound texture C3

For compound texture C3, the maximum average GRG for Ar, Td, and U is obtained at levels 3, 4, and 5, as shown in Figures 15.7(a–c) respectively. For these levels, the parametric values are checked. It is observed that upon increasing Ar the tribological performance is improved up to level 3 and then decreases. Also, for Td, it increases up to level 4 and then diminishes. However, an increase in slider speed is observed to be beneficial and gives the best result at level 5. From these results, the optimum Ar, Td, and U are found to be 0.015, 0.3 and 2.66 m/s, respectively. Similar to compound textures C1 and C2, the slider speed is the most critical variable parameter for C3.

For each compound texture, the output responses, viz. the load capacity and the friction coefficient, are assessed at optimum parameters. It is observed that texture shape influences the output responses (see Figure 15.8). The compound texture C2 gives the best result. In contrast, C1 texture performs worst. The load

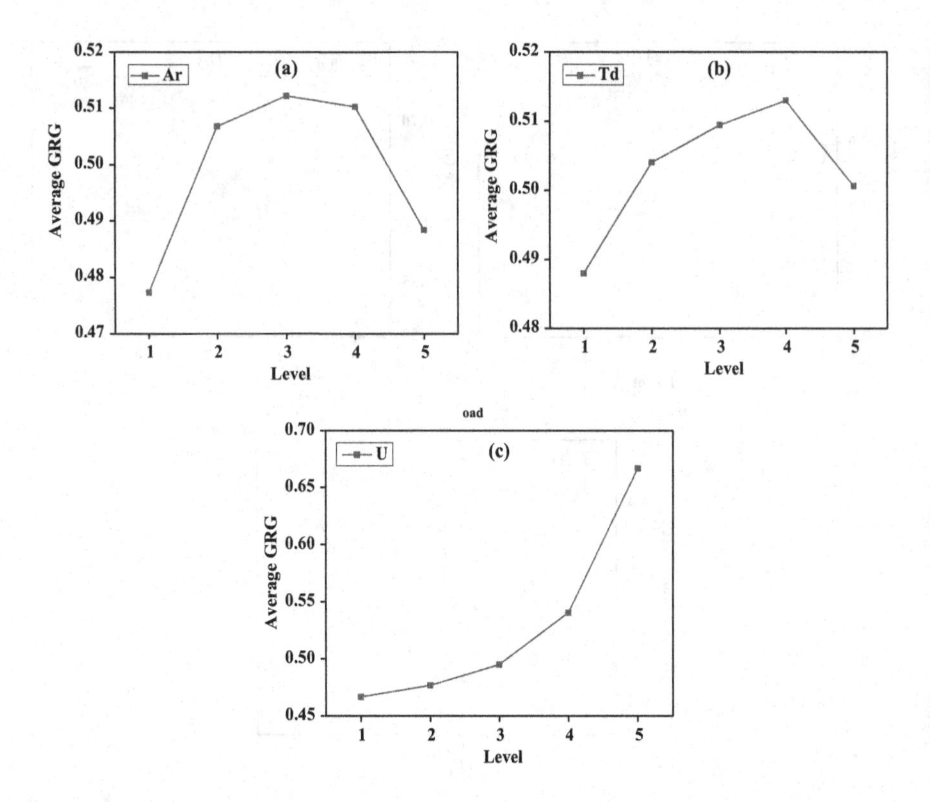

FIGURE 15.7 Influence variable parameters: (a) Ar, (b) Td, and (c) U on average GRG of compound texture C3.

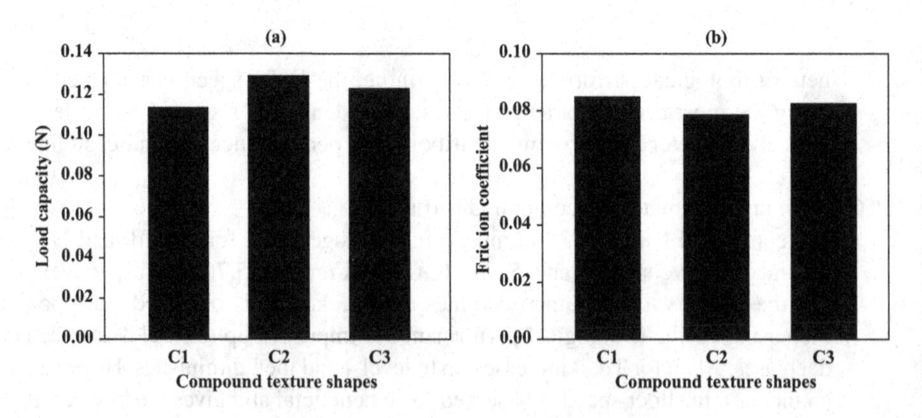

FIGURE 15.8 (a) Load capacity and (b) friction coefficient at optimum parameters.

capacity obtained by C2 texture is more than C1 and C3 shapes, by 13.56% and 4.96% respectively. Moreover, the friction coefficient obtained by compound texture C2 is lesser than C1 and C3 by 7.30% and 4.72%, respectively. This gain in better tribological characteristics by texture C2 justified that the volume of fluid accumulation in the texture has great significance. It is also observed that the round edge at the outlet of texture is more beneficial than a straight edge. In contrast, at the inlet of texture the straight profile should be preferred.

15.5 CONCLUSIONS

In the present work, the compound texture shapes, viz. C1, C2, and C3 are first analysed by response surface methodology-based CCD and then optimized using GRA multi-objective optimization. For improving the tribological performance, compound texture C2 performs better than C1 and C3. Under optimum condition, the load capacity obtained by compound texture C2 is more than C1 and C3 by 13.56% and 4.96% respectively. Moreover, the friction coefficient is lower by 7.30% and 4.72% than that of C1 and C3, respectively. To improve the tribological performance of compound texture C2, slider speed and texture density are, respectively, the most and least critical parameters.

REFERENCES

1. Etsion, I. Modeling of surface texturing in hydrodynamic lubrication. *Friction*, 2013, Vol. 1(3), pp. 195–209.
2. Etsion, I. State of the art in laser surface texturing. *Journal of Tribology*, 2005, Vol. 127(1), pp. 248–253.
3. Gachot, C., Rosenkranz, A., Hsu, S. M., and Costa, H. L. A critical assessment of surface texturing for friction and wear improvement. *Wear*, 2017, Vol. 372, pp. 21–41.
4. Arslan, A., Masjuki, H. H., Kalam, M. A., Varman, M., Mufti, R. A., Mosarof, M. H., Khuong, L. S., and Quazi, M. M. Surface texture manufacturing techniques and tribological effect of surface texturing on cutting tool performance: A review. *Critical Reviews in Solid State and Materials Sciences*, 2016, Vol. 41(6), pp. 447–481.
5. Coblas, D. G., Fatu, A., Maoui, A., and Hajjam, M. Manufacturing textured surfaces: State of art and recent developments. *Proceedings of the Institution of Mechanical Engineers, Part J: Journal of Engineering Tribology*, 2015, Vol. 229(1), pp. 3–29.
6. Shamsul Baharin, A. F., Ghazali, M. J., and Wahab, A.J. Laser surface texturing and its contribution to friction and wear reduction: A brief review. *Industrial Lubrication and Tribology*, 2016, Vol. 68(1), pp. 57–66.
7. Siripuram, R. B., and Stephens, L. S. Effect of deterministic asperity geometry on hydrodynamic lubrication. *Journal of Tribology*, 2004, Vol. 126(3), pp. 527–534.
8. Qiu, M., Delic, A., and Raeymaekers, B. The effect of texture shape on the load-carrying capacity of gas-lubricated parallel slider bearings. *Tribology Letters*, 2012, Vol. 48(3), pp. 315–327.
9. Qiu, M., Minson, B., and Raeymaekers, B. The effect of texture shape on the friction coefficient and stiffness of gas-lubricated parallel slider bearings. *Tribology International*, 2013, Vol. 67, pp. 278–288.
10. Galda, L., Pawlus, P., and Sep, J. Dimples shape and distribution effect on characteristics of Stribeck curve. *Tribology International*, 2009, Vol. 42(10), pp. 1505–1512.
11. Bompos, D. A., Nikolakopoulos, P. G., and Papadopoulos, C. I. A tribological study of partial-arc bearings with egg-shaped texture for microturbine applications. *Proceedings of ASME Turbo Expo 2012*, Copenhagen, Denmark.

12. Uddin, M. S., and Liu, Y. W. Design and optimization of a new geometric texture shape for the enhancement of hydrodynamic lubrication performance of parallel slider surfaces. *Biosurface and Biotribology*, 2016 Vol. 2(2), pp. 59–69.
13. Fesanghary, M. and Khonsari, M. M. On the optimum groove shapes for load-carrying capacity enhancement in parallel flat surface bearings: theory and experiment. *Tribology International*, 2013, Vol. 67, pp. 254–262.
14. Shen, C. and Khonsari, M. M. Numerical optimization of texture shape for parallel surfaces under unidirectional and bidirectional sliding. Tribology International, 2015, Vol. 82, pp. 1–11.
15. Zhang, H., Dong, G. N., Hua, M., and Chin, K. S. Improvement of tribological behaviors by optimizing concave texture shape under reciprocating sliding motion. Journal of Tribology, 2017, Vol. 139(1), p. 011701.
16. Wang, W., He, Y., Zhao, J., Mao, J., Hu, Y., and Luo, J. Optimization of groove texture profile to improve hydrodynamic lubrication performance: Theory and experiments. *Friction*, 2018, pp. 1–12.
17. Hingawe, N. D., and Bhore, S. P. Multi-objective optimization of the design parameters of texture bottom profiles in a parallel slider. *Friction*, 2019, pp. 1–20.
18. Shinde, A. B., and Pawar, P. M. Multi-objective optimization of surface textured journal bearing by Taguchi based Grey relational analysis. *Tribology International*, 2017, Vol. 114, pp. 349–357.

16 Case Study *Wear Behavior of Different Seal Materials under Dry-Lubricated Conditions*

Corina Birleanu, Marius Pustan, Cosmin Cosma, Mircea Cioaza and Florin Popa

Technical University from Cluj-Napoca, Cluj-Napoca, Romania

CONTENTS

ABBREVIATIONS/ACRONYMS

ASTM	American Society for Testing and Materials
SLM	selective laser melting
CNC	Computerized Numerical Control
AW	antiwear
COF	coefficient of friction
μ	coefficient of friction
ρ	density
RT	room temperature
SEM	scanning electron microscopy
E	Elastic Modulus
HRC	Rockwell hardness
HV	Vickers hardness
EDS	Energy Dispersive X- ray Spectroscopy

DOI: 10.1201/9781003243205-16

16.1 INTRODUCTION

Seals are used on a very large scale with a direct influence on the reliability of mechanical systems (pumps, compressors, turbines, etc.). A sealing is a machine element mounted between two surfaces to prevent or reduce to an acceptable minimum scale leak of liquid or gas from one region to another, also prevents dirt from entering through these surfaces. A dynamic seal is a mechanical device used to prevent liquid leakage when there is rotational movement between the sealing surfaces.

The seal is designed to fit around the shaft or some part connecting to the shaft, dynamic sealing is often referred to as shaft sealing. The operation of many types of front seals, as mechanical systems, from the simplest to the most complex, requires the existence of direct contacts, relative movements or complex interactions.

There are many concrete situations where, in the context of increasing the reliability and performance of technical systems, the use of front seals is a major requirement and simultaneously a unique solution.

Wang et al. [1] developed a C-coating deposited onto a 9Cr18 rotor of the face seal in liquid rocket engine turbopumps, and the tribological performance of each specimen was tested under three fluid conditions (air, water, and liquid nitrogen). Young et al. [2] investigated the macro/micro laser machined characteristics of mechanical face seals, and friction-testing experiments using water for the seal were conducted. Zhao et al. [3] studied the frictional performance of silicon carbide underwater and lubrication-absent conditions by using a Falex1506 tribo-tester and different working parameters. Frölich et al. [4] developed a macroscopic simulation model for the radial shaft seals and the results of simulations using their model showed that the material of the seal ring is intensely influenced by both the temperature and the contact pressure. Cui et al. [5] developed new self-lubricating bronze matrix composites for seals and explored the tribological mechanisms in antiwear (AW) hydraulic oil using a balloon-disk tribo-tester. In summary, previous studies have indicated that the study on the suitability of materials for the seal's pairs and the improvement of materials' surface properties can significantly improve the friction and wear performance of the seal under the extreme conditions, and these results also provided the important ideas and methods for selecting material, treating surface, improving the tribological performance and even monitoring tribological behavior of the mechanical seal [6, 7].

Most dynamic seals, such as mechanical seals, fall into the category where friction surfaces are separated and lubricated with a thin film of lubricant [1]. For the devices executed from thin films is evident that the stress-strain relationship is affected by the relatively high area-to-volume ratio. As a result, typical properties used for description of bulk material strength and deformation do not apply for thin films and micro devices. This size effect relationship is not yet well understood when structural dimensions decrease from millimeters to micro and nanometers [8].

In a frontal seal depending on a number of functional and constructive features, various types of friction can arise. Liquid and mixed friction may occur in the case of hydrodynamic and hydrostatic front seals, dry friction – in the case of overheated seals or used for gas sealing, friction at low temperatures – in the case of liquefied gases, vacuum friction – in the case of seals from jet aircraft or missiles

A schematic presentation of a conventional mechanical face seal is presented in Figure 16.1. Mechanical face seals, principally containing two rings, are adjusted on

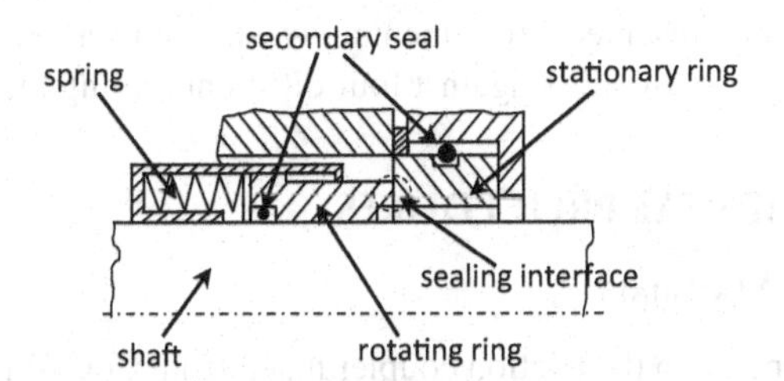

FIGURE 16.1 Schematic presentation of a mechanical face seal as a tribosystem.

the rotating shaft. One part (face) is a rotating ring that is attached to the shaft and rotates with it; the other is stationary and is mounted in the housing. The rotating face is mounted with clearance and may be the "floating face" in the axial direction, usually driven by a spring. So it can move along the shaft axis allowing relative movement and provide flexibility for small misalignment between the parts.

As shown in Figure 16.1, the interface seal is composed of two faces, which interact in relative motion to each other in lubricating conditions; therefore, the seals herein may be considered to be a tribosystem. If the lubricant cannot separate the faces of the seal that come into contact, the asperity contact occurs at the microscopic level (Figure 16.2). The intensity of contact between the asperities depends on the operating conditions, namely the applied load, the slip speed, the temperature, the type of lubricant, the material coupler and the surface roughness. Over time, due to the friction between the friction coupler elements, there is a process of wear that changes the roughness and micro geometry and leads to damage to the tribo-mechanical system.

We specified that the purpose of tribological studies in this work is to establish a scientific approach to predict the friction state of mechanical seals based on higher material performance. The tribological process was also the general idea of developing the progress of the research on mechanical seals, which is also presented in [9–12].

CoCrWMo alloy widely used as a hard material is a potential candidate for mechanical seals, but its tribo-mechanical properties need to be investigated in the aim of the use in high performance mechanical seal working in severe conditions. CoCrWMo alloys offer good mechanical properties (hardness, Young modulus), resistance to corrosion and thermal conductivity that make it appropriate for tribological application in dry and lubricated sliding. Combined with matting face ring from a softer material, the sliding of CoCrWMo alloy can be sustained in severe working conditions [13–17].

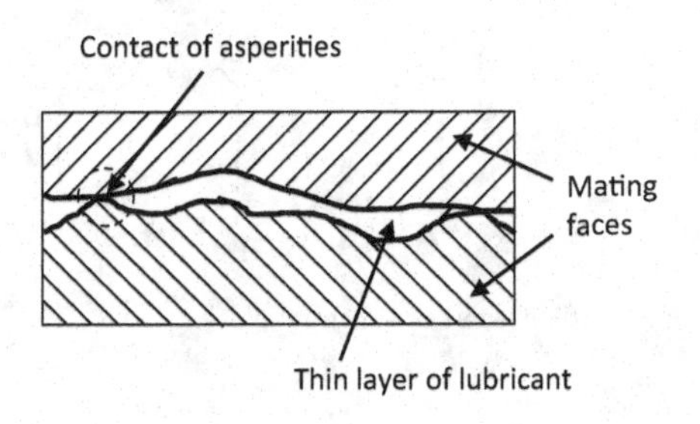

FIGURE 16.2 The contact of asperities in the sealing interface.

In this chapter the presented experimental investigations assesses the tribological behaviors of an CoCrWMo alloy against four different sealing materials.

16.2 EXPERIMENTAL PROCEDURES

16.2.1 SAMPLE MATERIALS

The active semi-ring from the friction coupler made from CoCrWMo alloy was manufactured by the selective laser melting (SLM) process. The SLM process is capable of fusing fine metallic particles together, slice by slice, in a protective high-purity Argon atmosphere. SLM is a complex thermo-physical process that depends largely on material, laser and process parameters (Figure 16.3). The main component is the solid Nd:YAG laser which emits continuous light with a wavelength of 1064 nm in the infra-red spectrum. The system used was Realizer SLM 250 (Germany) and 50 μm the value of spot laser. The following SLM parameters could be set up on Realizer equipment: 20–200 W laser power, 20–100 μm layer thickness, 100–2000 mm/s scanning speed, 0.06–0.20 mm hatch space and various scanning strategies (eq. X/Y, stripe hatch pattern, islands) (Figure 16.3).

The case study in this chapter used two SLM-manufactured rings made from CoCrWMo alloy (commercially named Starbond CoS55) provided by Scheftner, Germany (Figure 16.4). This alloy is highly corrosion-resistant and has a density of approx. 8.8 g/cm³. The solidus-liquidus interval of it is between 1305 and 1400°C.

The chemical characterization/elemental analysis of materials of the samples used in this investigation was done based on the EDS analysis using Oxford EDS – Ultim® Max EDS with AZtecLive software. From the multiple set of test samples, several representatives were chosen for which this analysis was done.

Figure 16.5 shows the SEM and EDS images obtained for SLM-manufactured ring made of CoCrWMo alloy.

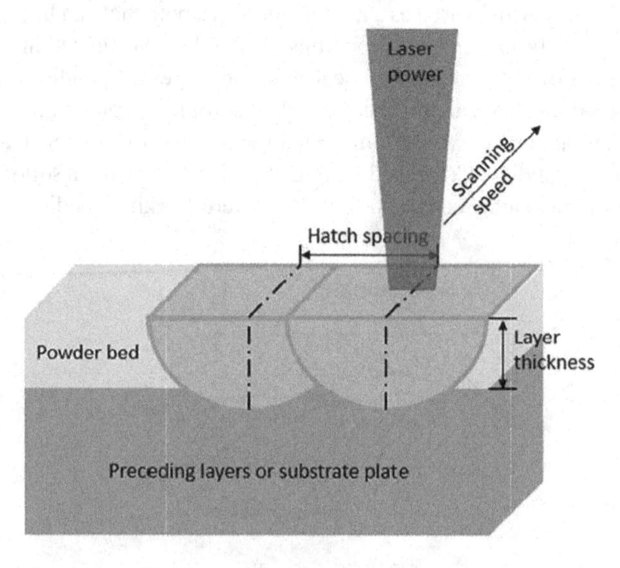

FIGURE 16.3 The main SLM process parameter [9].

FIGURE 16.4 Matting face rings for tribological investigations.

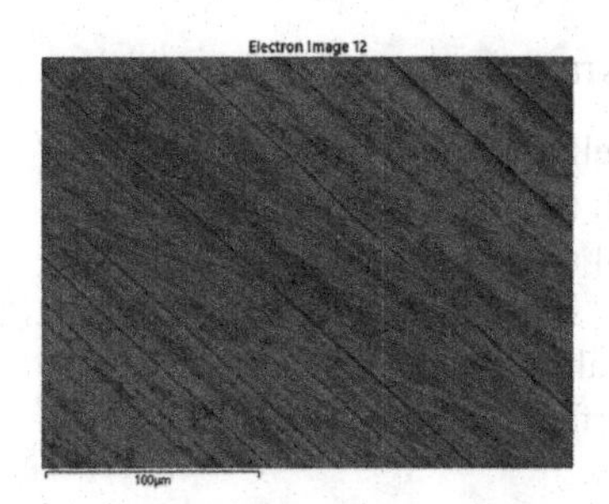

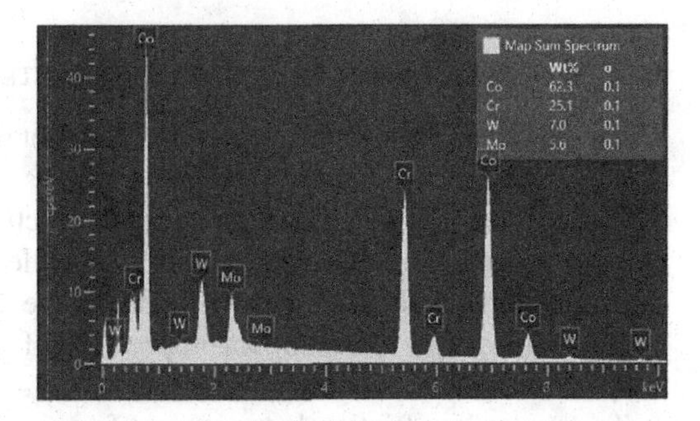

FIGURE 16.5 SEM and EDS analysis of ring surface (magnitude ×500).

The mass percentages of each chemical element are presented in Figure 16.5. The diameter of CoCr grains is between 10 and 55 µm and they are spherical. Using the EDS elemental quantification of elements allows us to deliver a qualitative analysis of the sample surface.

In order to avoid the unfavourable effects of residual stress, after SLM manufacturing the parts were exposed on a thermal treatment. This heat treatment was performed in an electric oven and air atmosphere. The parts were heated up to 860°C with 7°C/min. The parts were heated at this temperature for 1 hour, which was followed by a cooling rate of 12°C/min. After the temperature decreased to 300°C, we opened the door of furnace for natural cooling. Finally, the parts were sandblasted with alumina.

To achieve the required accuracy, the rings were conventional processed (CNC turning – a computerized manufacturing process in which pre-programmed software and code controls the movement of production equipment).

Normally, after conventional manufacturing such as casting, the hardness of this CoCr alloy is between 407 and 601 HV1 [14]. Based on our previous research, the hardness of parts SLM processed is directly influenced by process parameters such as laser power, scanning speed or by the density of energy which is distributed on each powder slice [15].

The dimensions for the primary ring from CoCrWMo are 62 mm for external diameter and 4 mm thickness of hardness 540 HV1 (51,8 HRC). For the second ring (soft sample) diameter is 16 mm and 6 mm thickness (Figure 16.4) and four different materials with a surface hardness presented in Table 16.1. In the chapter were defined as soft samples as follows: sample 1 made from copper-zinc alloy named brass (CuZn40PtSn); sample 2 from tin-copper alloy named bronze (BzA15); sample 3 from graphite bronze; and sample 4 from Teflon – PTFE.

Based on the EDS elemental quantification of elements, we are able to deliver a qualitative analysis of the soft samples surface (named samples 1–4). SEM and EDS analysis of these surface samples are presented in Figures 16.6–16.8 (magnitude ×500).

The hardness values from Table 16.1 are the average of 12–15 determinations per sample. The hardness was evaluated using a Microhardness Tester VMHT with the trace marks shown in Figure 16.9.

16.2.2 Measurement of Surface Characteristics

When the surfaces that come into contact are not well lubricated, they tend to "stick" to each other. Then, a moment is applied to the shaft the surfaces in contact will suddenly separate. This phenomenon causes the actual speed of rotation of the sealing faces to fluctuate in a tribological this is so-called "stick slip" phenomenon [15]. It can cause severe vibration that will damage the sealing faces and lead to excessive lubrication leakage, indicating the importance of surface quality for mechanical sealing performance.

To acquire low friction between seal faces, good surface roughness and flatness should be ensured. The surface of primary ring, SLM made from CoCrWMo alloy, was polished to show an arithmetical mean height R_a less than 0.4 μm.

Four different sealing materials introduced above were selected as matting face rings. In the same way the surface were manufactured and R_a roughness for them is: sample 1–0.12 μm; sample 2–0.35 μm; sample 3–2.2 μm and sample 4–0.6 μm.

16.2.3 Tribometer and Method

Block on Ring test is a widely used technique that evaluates the sliding wear behaviors of materials in different simulated conditions, and allows the reliable ranking of material couples for specific tribological applications. Sliding wear often involves complex wear mechanisms taking place at the contact surface, such as adhesion wear, two-body abrasion, three-body abrasion and fatigue wear. The wear behavior of materials is significantly influenced by the work environment, such as normal

TABLE 16.1
Rings Hardness Values

Rings Material	CoCrWMo Primarily Ring	Material for Soft Samples			
		Sample 1	Sample 2	Sample 3	Sample 4
Hardness HV0.5 (GPa)	5.4	1.906	1.634	0.460	0.033

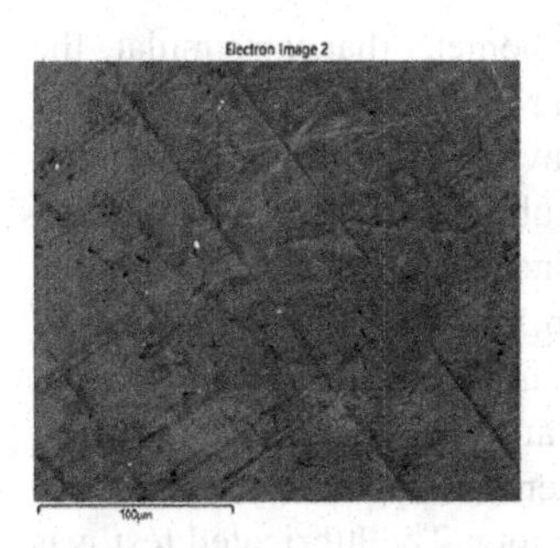

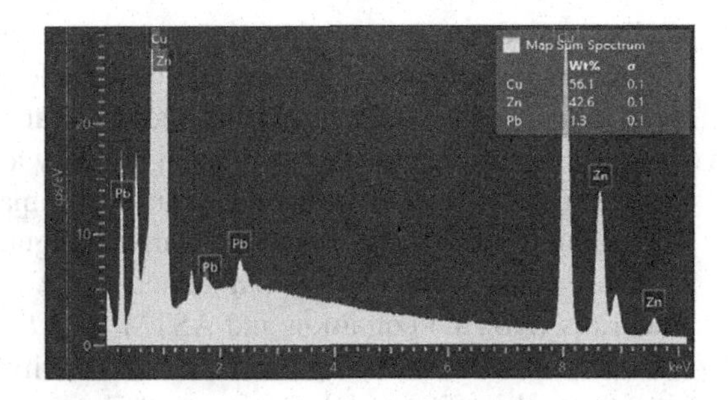

FIGURE 16.6 SEM and EDS analysis of sample 1 surface (magnitude ×500).

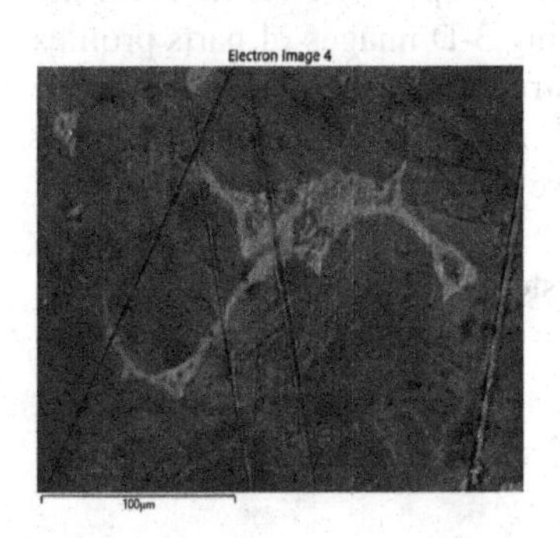

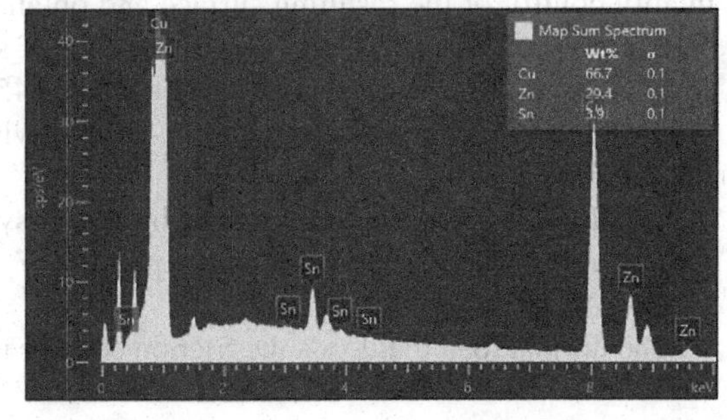

FIGURE 16.7 SEM and EDS analysis of sample 2 surface (magnitude ×500).

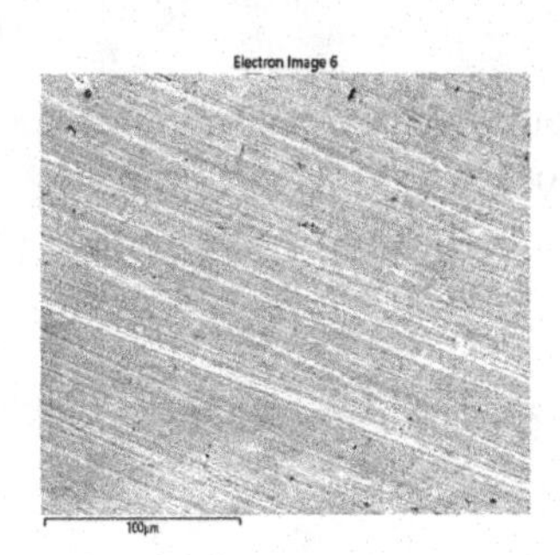

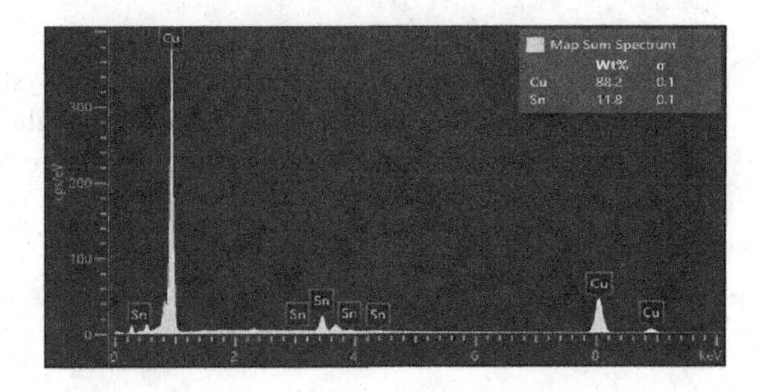

FIGURE 16.8 SEM and EDS analysis of sample 3 surface (magnitude ×500).

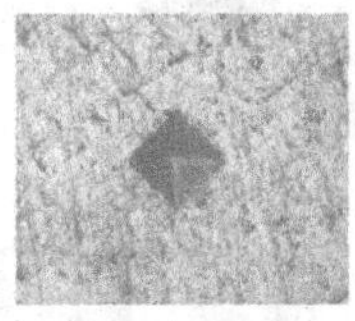

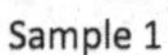

Sample 1 Sample 2 Sample 3 Sample 4

FIGURE 16.9 Indentation marks for the samples (soft samples).

loading, speed, corrosion and lubrication. A versatile tribometer that can simulate the different realistic work conditions will be ideal for wear evaluation.

In this chapter, we evaluated the sliding wear behaviours of a CoCrWMo alloy ring on four different materials rings in both dry and lubricated environments using Falex Block-on-Ring friction and wear testing machine (tribometer) in simulating the sliding wear process of material couples (Figure 16.10).

These tests were carried out under the American Society for Testing and Materials (ASTM) condition and ASTM G77 Standard. The using rings on ring tests were performed in dry and lubricated environments, respectively, in order to investigate the effect of lubrication on the wear behaviour. The lubricated test was performed in the T80W90 mechanical transmission oil. The wear track was examined using a ZEISS Smartzoom 5 smart digital microscope, which is ideal for the quality control of the resulting surface and obtaining 3-D images of parts profiles The microscope's swing arm allows viewing of structures on the sample surface from continuously adjustable angles between ±45° As the head swivels, the pivot point of the swivel axis remains stable, and likewise the focus remains squarely on the sample.

Loading the friction coupler is done by a lever system by mounting the weights on the two arms. The load that sits on the plates is $F_1 = 10\ N$, and the second test with $F_1 = 30\ N$.

The normal load that loads the friction coupler is:

$$F_{n1} = F_1 \cdot \frac{c}{d} \cdot \frac{a}{a+d} = 72.64\ N$$

and for the second test $F_{n1} = 217.91\ N$

Where *a, b, c, d* represents the tribometer lever system which are transmitting the load. The total normal load will consider the weight of the arms as well as plates which, in combination, produce an additional charge $F_{n2} = 10\ N$.

$$F_n = F_{n1} + F_{n2} = 82.64\ N$$

respectively, $F_n = 227.91\ N$.

FIGURE 16.10 Friction test set-up before (sample 4) and after testing (sample 2).

The speed at the vertical shaft is $n = 980\ rpm$ and the peripheral speed is:

$$v = \frac{\pi \cdot D \cdot n}{60 \cdot 1000} = 3.18\ m \cdot s^{-1}.$$

The wear rate, W, was evaluated using the formula:

$$W = \frac{V}{F_n \cdot s}$$

Where: V is the worn volume, F_n is the normal load, and s is the sliding distance.

The wear volume V is determined with both ZEISS Smartzoom 5 smart digital microscope and through the difference between the sample masses before and after wear $(m_0 - m_1)$ [g] using an accurate balance type Kern AEJ with a measurement accuracy of 0.1 mg.

Before and after the test, the specimens are washed with diluent and dried and then weighed.

The surface topography of worn rings was determined using an optical microscope. All measurements were done at room temperature (RT). The test time was $t = 60$ minutes, which means 11,448 m sliding distance per test.

16.3 RESULTS AND DISCUSSIONS

Figures 16.11 and 16.12 show the optical images of the wear scars on the soft samples (1–4) after the dry and lubricated wear tests, respectively. Optical microscopy analyses were performed in order to study the surface of the samples. The tests were carried out on an Olympus GX 51 microscope. All the samples were investigated at a magnification of 50:1.

The wear track volumes and wear rates are listed in Tables 16.2 and 16.3 and were calculated based on the formulae above. For each sample based on mass measurements and exact volume determination, density was determined. The density (ρ)-specific mass values are shown in Tables 16.2 and 16.3.

TABLE 16.2
Result Summary of Dry Wear Tracks Measured Using Different Test Parameters

	Wear Track Height (μm)	Wear Volume (mm³)	Wear Rate (mm³/N•m)
Sample 1 ρ = 8.4515 g/cm³	288.00	13.8082	0.8757×10^{-3}
Sample 2 ρ = 8.2123 g/cm³	202.30	25.0843	1.5908×10^{-3}
Sample 3 ρ = 6.2206 g/cm³	96.60	2.25058	0.1427×10^{-3}
Sample 4 ρ = 2.1310 g/cm³	373.00	44.2984	2.8094×10^{-3}

Note: (F_1 = 10 N, operating time = 60s, RT).

TABLE 16.3
Result Summary of Lubricated Wear Tracks Measured Using Different Test Parameters

	Load (F_1) (N)	Wear Track Height (μm)	Wear Volume (mm^3)	Wear Rate (mm^3/N•m)
Sample 1	10	49.10	0.07099	0.07503×10^{-6}
ρ = 8.4515 g/cm³	30	84.20	0.13015	0.04988×10^{-6}
Sample 2	10	48.70	0.02435	0.02574×10^{-6}
ρ = 8.2123 g/cm³	30	54.00	Inconclusive result	Inconclusive result
Sample 3	10	18.90	Inconclusive result	Inconclusive result
ρ = 6.2206 g/cm³	30	25.70	Inconclusive result	Inconclusive result
Sample 4	10	56.20	0.6912	0.73×10^{-6}
ρ = 2.1310 g/cm³	30	163.50	7.0858	2.7158×10^{-6}

Note: F_1 = 10 N/F_1 = 30N, operating time = 3600s, RT.

The soft samples after the dry wear test exhibits a big wear scar of volume. As expected, the most pronounced wear is at sample 4 (PTFE) in the amount of ~44.3 mm³ and the least is worn sample 3 of bronze graphite in value of ~2.25 mm³. In comparison, the wear test carried out in the mineral oil lubricant creates a substantially smaller wear track with a volume of only ~0.69 mm³ at sample 4 (PTFE) loaded with 10N and ~7.1 mm³ loaded with 30N, respectively.

As shown in the images taken under optical microscope in Figures 16.11–16.13, severe wear takes place during the Block-on-Ring test in the dry atmosphere, compared to mild parallel wear scars on the samples after the much longer lubricated wear test. The high heat (around 80–90°C) and intense vibration generated during the

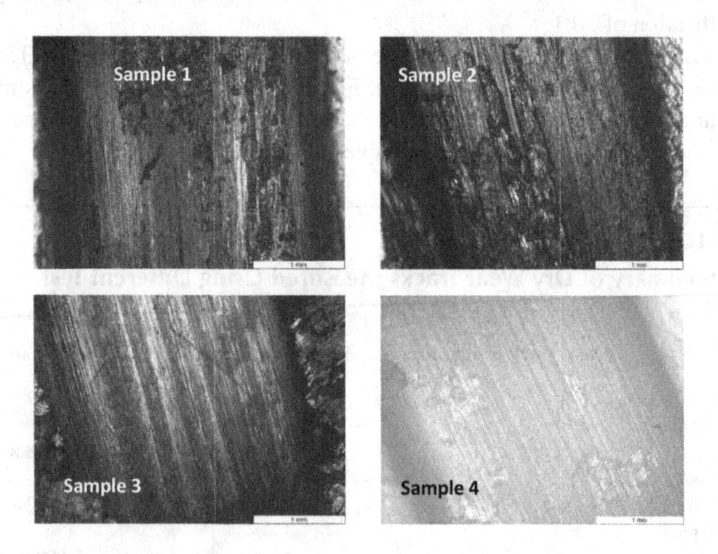

FIGURE 16.11 Wear scars of the soft samples after dry wear tests under 10N applied force and 60s operating time at RT, 50:1 resolution.

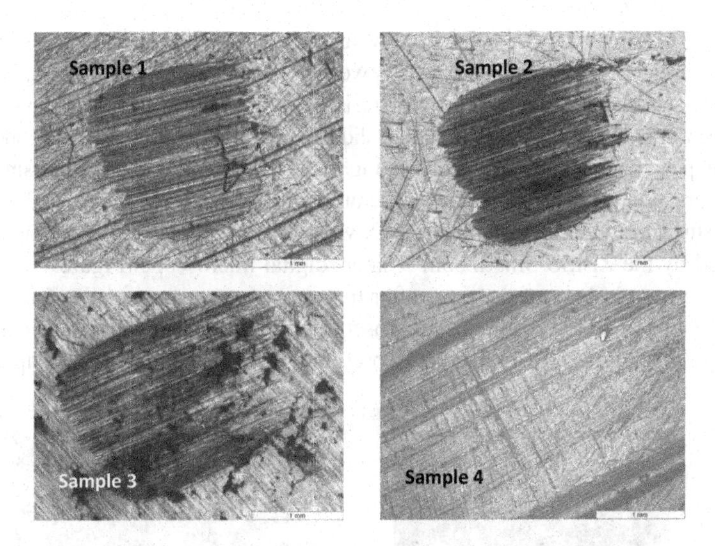

FIGURE 16.12 Wear scars of the soft samples after lubricated wear tests under 10N applied force and 1 hour operating time and RT, 50:1 resolution.

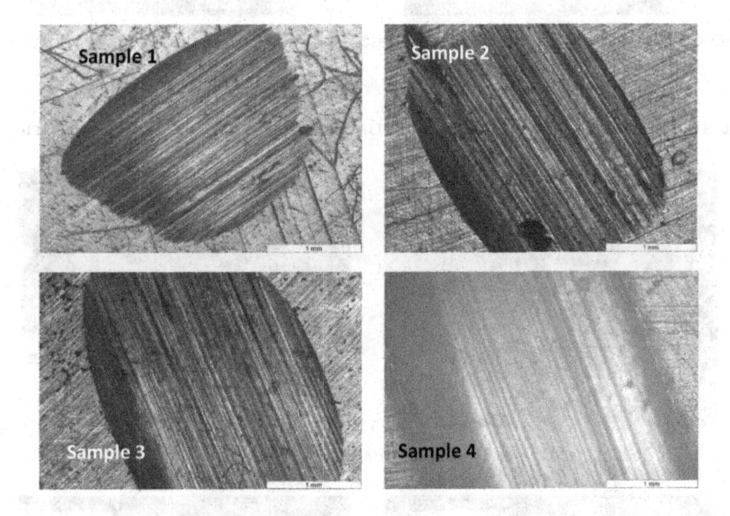

FIGURE 16.13 Wear scars of the soft samples after lubricated wear tests under 30N applied force and 1 hour operating time and RT, 50:1 resolution.

dry wear test promotes the oxidation of the metallic debris and results in severe three-body abrasion. For the dry test it is observed on the surface images that they were affected by the sudden rise of temperature in the contact area between the sample and the rotating ring (sometimes the temperature reached 90–100°C).

In the lubricated test, however, the mineral oil ultimately reduces the friction and cools the contact face, as well as transporting away abrasive debris created during wear, leading to a significant reduction in the wear rate. It was found that for the graphite bronze due to the structure of the material it absorbed oil during the lubricated test and thus the volume of wear material could not be expressed by the mass difference. Therefore, these results were not relevant by this method.

Such a substantial difference in wear resistance measured in different environments shows the importance of proper simulating sliding wear in the realistic service condition.

The experimental results of the dry and lubricated wear tracks measured using different test parameters and RT (ambient temperature) are presented in Tables 16.2 and 16.3.

The surface conditions of the CoCrWMo primarily ring and soft samples were examined by the ZEISS Smartzoom 5 smart digital microscope. Figure 16.13 shows the surface morphology of the rings after the wear tests as an example. The four soft samples were analysed with this microscope under dry and lubricated conditions (Figure 16.14). Figures 16.15 and 16.16 present the depth and profile of samples 1 and

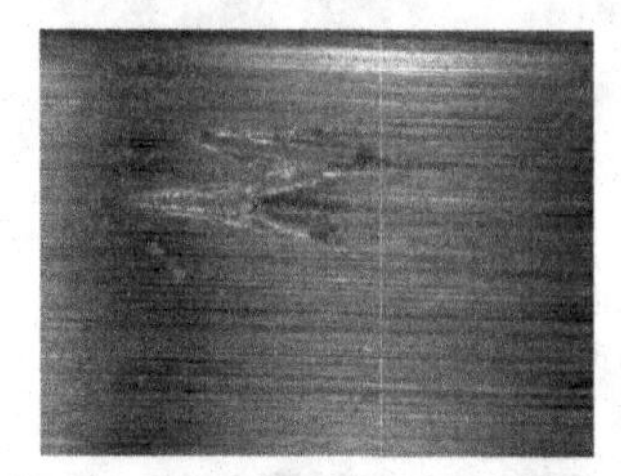

CoCrWMo ring surface after lubricated test (30N)

CoCrWMo ring surface after dry test (10N)

FIGURE 16.14 Wear scar profiles on the primary ring for dry and lubricated wear tests.

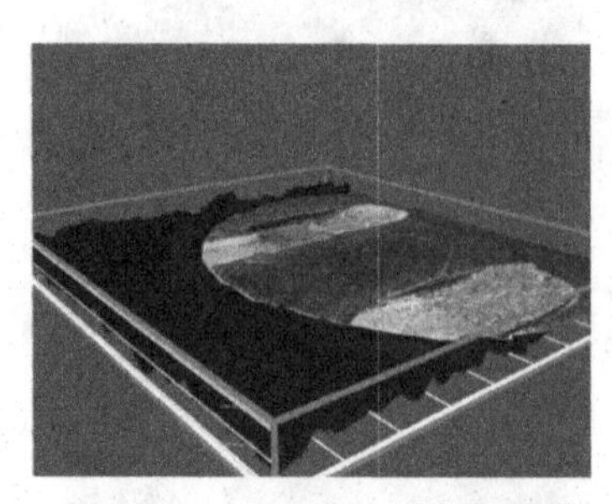

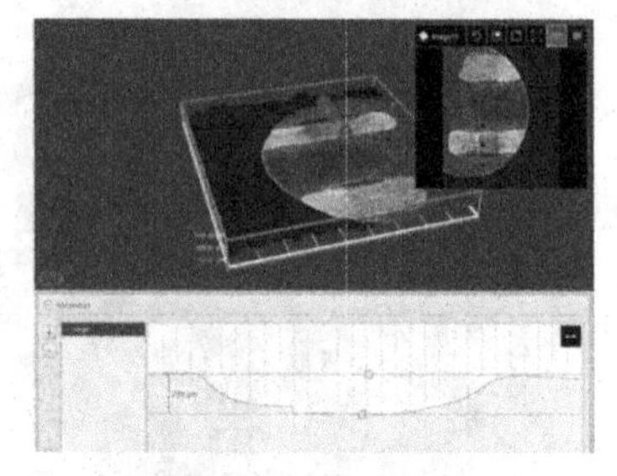

Samples 1 – surface images after dry test (10N)

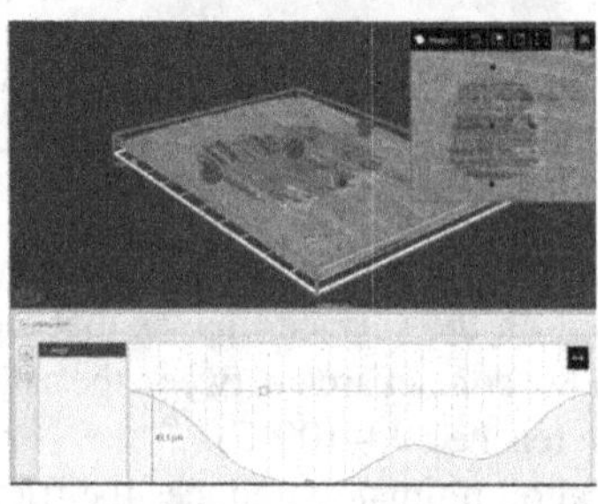

Samples 1 – surface images after lubricated test (10N)

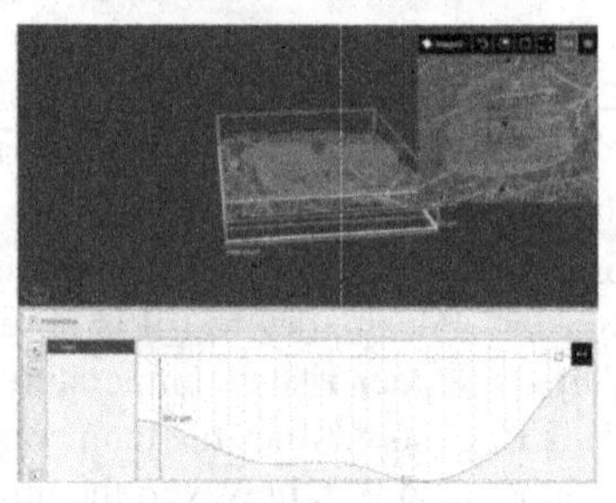

Samples 1 – surface images after lubricated test (30N)

FIGURE 16.15 Optical profilometry of the surface images for sample 1 after dry and lubricated test.

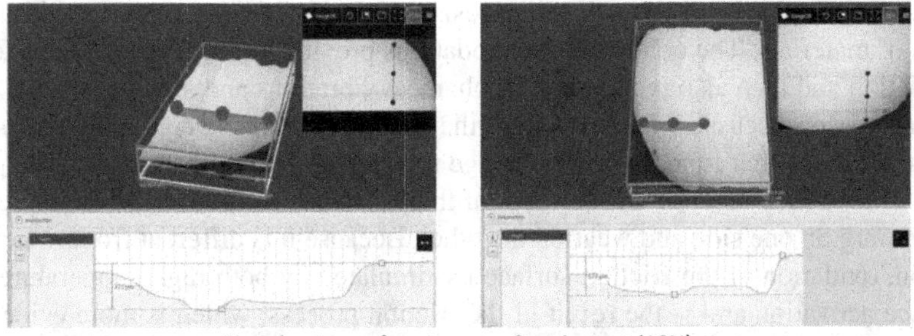

Samples 4 – surface images after dry test (10N)

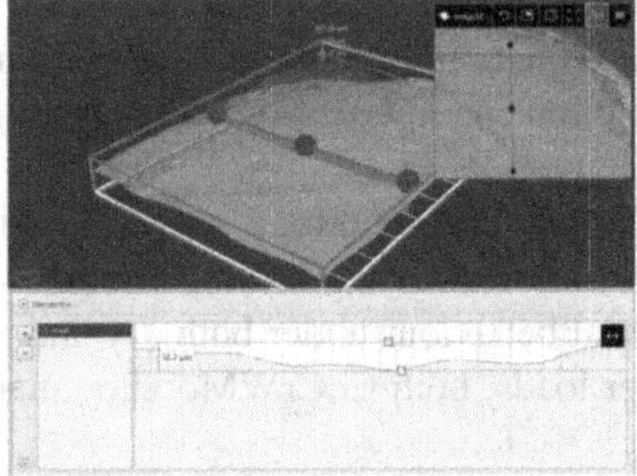

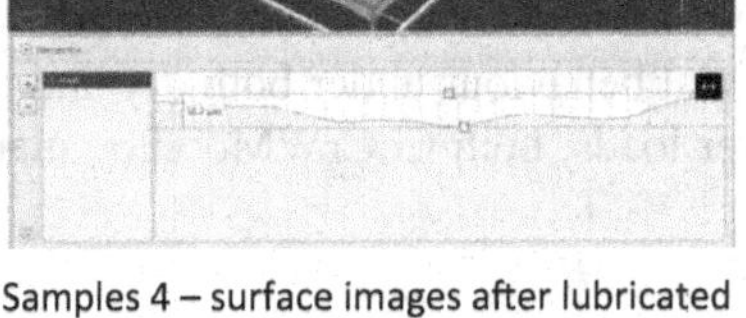

Samples 4 – surface images after lubricated test (10N)

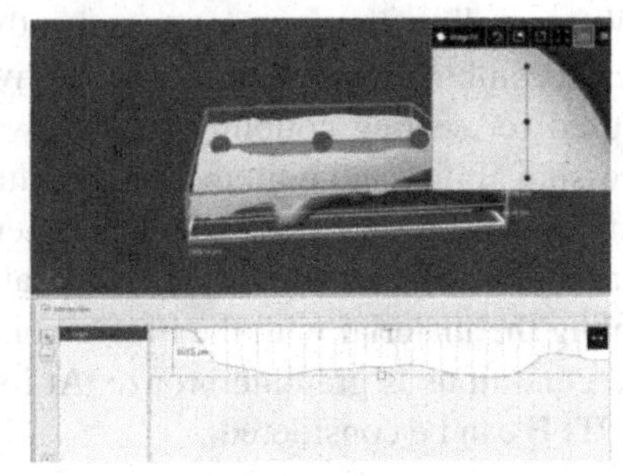

Samples 4 – surface images after lubricated test (30N)

FIGURE 16.16 Optical profilometry of the surface images for sample 4 after dry and lubricated test.

4 under both dry and lubricated wear tests in different experimental conditions. In the case of lubricant tests, two values for the force applied to the friction coupler were used: 10N and 30N. All experimental results obtained on the wear track depth are presented in Tables 16.2 and 16.3. Significant roughening of the surface took place due to the three-body abrasion process during the dry wear test of 980 revolutions.

Severe wear quickly damages the contact surface without lubrication and leads to irreversible deterioration of the surface quality. Wear rate values for both primarily ring and soft samples are calculated from the volume of material lost during a specific friction run. The primarily ring scar depth is measured to calculate the ring scar volume, and soft samples scar volumes are calculated from the samples' weight loss. This simple method facilitates the determination and study of wear behaviour of almost every solid-state material combination, with varying time, contact pressure, velocity, temperature, humidity, lubrication, etc.

16.4 CONCLUSIONS

In the dry slip test for the CoCrWMo alloy/graphite bronze during a running-in period and accommodation, a higher degree of wear loss is observed. A third body is created mainly from the wear of the sealing materials ring. The impregnation of the graphite material plays a predominant role in the tribological performance of the matting face material. This third body acts like a solid lubricant reducing the stick-slip phenomenon inside the contact. Better resistance to the wear is achieved with these pairs of materials. The contact accommodations provide the best behaviour in terms of friction and then also reduces the mechanical vibrations and also the noises.

In all cases discussed, the surfaces of the friction pair are covered with oxides and adsorbed gases that fundamentally change the surface friction properties. The presence of the oxide layer on the surfaces of the pair is the result of a dynamic process of forming on one side and wear on the other. Because it is different from static oxidation, oxidation of the friction surface is stimulated by both high temperature and surface activation and is the result of the friction process, which is more evident in dry friction.

When sealing rings connect each other, the tribological characteristics of the material combination determine the survival or the failure of the seal. The best material used for sealing couplings have low friction, good corrosion resistance, good processing ability and high thermal conductivity.

In conclusion, based on the experimental data performed we appreciate that in contact with the proposed CoCrWMo alloy material as the sealing friction semi-coupling the material with the best tribological behaviour under both dry and lubricated conditions is graphite bronze. At lower loads, both CoCrWMo alloy material and PTFE can be considered.

REFERENCES

1. Wang J L, Jia Q, Yuan X Y, Wang S P. Experimental study on friction and wear behaviour of amorphous carbon coatings for mechanical seals in cryogenic environment. *Appl Surf Sci* 258(24): 9531–9535 (2012).
2. Young L, Benedict J, Davis J. Investigation of a unique macro/micro laser machined feature for mechanical face seals with low leakage, low friction, and low wear. In *Proceedings of the ASME/STLE 2011 International Joint Tribology Conference*, Los Angeles, California, USA, 211–214 (2011).
3. Zhao X Y, Liu Y, Wen Q F, Wang Y M. Frictional performance of silicon carbide under different lubrication conditions. *Friction* 2(1): 58–63 (2014).
4. Frölich D, Magyar B, Sauer B. A comprehensive model of wear, friction and contact temperature in radial shaft seals. *Wear* 311(1–2): 71–80 (2014).
5. Cui G J, Li J X, Wu G X. Friction and wear behavior of bronze matrix composites for seal in antiwear hydraulic oil. *Tribol Trans* 58(1): 51–58 (2015).
6. Towsyfyan H, Gu F, Ball A D, Liang B. Tribological behaviour diagnostic and fault detection of mechanical seals based on acoustic emission measurements. *Friction* 7(6): 572–586 (2019).
7. Zhao W, Zhang G, Dong G. Friction and wear behavior of different seal materials under water-lubricated conditions. *Friction* 9: 697–709 (2021).
8. Towsyfyan H. Investigation of the nonlinear tribological behaviour of mechanical seals for online condition monitoring. Doctoral thesis, University of Huddersfield (2017).
9. Nau B. Mechanical seal face materials. *Proceedings of the Institution of Mechanical Engineers, Part J: Journal of Engineering Tribology* 211(3): 165–183 (1997).
10. Minet C, Brunetière N, Tournerie B. On the lubrication of mechanical seals with rough surfaces: A parametric study. *Proceedings of the Institution of Mechanical Engineers, Part J: Journal of Engineering Tribology* 226 (12): 1109–1126 (2012).

11. Pustan M. Contribuţii privind etanşările frontale cu impulsuri. Doctoral Thesis, Technical University of Cluj-Napoca (2006).
12. Birleanu C, Sucala F. About the tribological behaviour of ceramic materials. *Tribology in Industry* 30: 10–14 (2008).
13. Zaidi H, Paulmier A. Behavior of graphite in friction under various environments: Connection with the surface reactivity. *Surface Science* 251: 778–781 (1991).
14. Young W C, Budynas R G. *Roark's formulas for stress and strain* vol. 7 (New York: McGraw-Hill) (2002).
15. Cosma C, Balc N, Moldovan M, Morovic L, Gogola P, Borzan C. Post-processing of customized implants made by laser beam melting from pure Titanium. *J. Optoelectronics Advanced Materials* 19 (11–12): 738–747 (2017).
16. Yap C Y, Chua C K et al. Review of selective laser melting: Materials and applications. *Applied Physics Reviews* 2 (2015).
17. Birleanu C, Pustan M, Cosma C, Merie V, Dranda O. Tribological behaviour of sealing materials. *The 14th International Conference on Tribology, ROTRIB 2019, IOP Conference Series: Materials Science and Engineering*, Vol. 724, No. 1, 12–17 (2020).

Index